The Menace of Atomic Energy

OTHER BOOKS BY RALPH NADER

Unsafe at Any Speed (*1965, expanded edition 1972*)

Whistle Blowing (*1972, edited with Peter Petkas and Kate Blackwell*)

You and Your Pension (*1973, with Kate Blackwell*)

Corporate Power in America (*1973, edited with Mark Green*)

The Consumer and Corporate Accountability (*1973, edited with Jean Carper*)

Verdicts on Lawyers (*1976, edited with Mark Green*)

Taming the Giant Corporation (*1976, with Mark Green and Joel Seligman*)

Ralph Nader

John Abbotts

The Menace of Atomic Energy

REVISED EDITION

W · W · NORTON & COMPANY · NEW YORK · LONDON

REVISED EDITION

Library of Congress Cataloging in Publication Data

Nader, Ralph.
The menace of atomic energy.

Includes bibliographical references and index.
1. Atomic power. I. Abbotts, John, joint author. II. Title.
TK9153.N28 1979 614.8'39 78-25725
ISBN 0-393-01239-5
ISBN 0-393-00920-3 pbk.

3 4 5 6 7 8 9 0

TO THE GROWING NUMBER OF *dedicated citizens worldwide whose concern for present and future generations leads them to work toward the replacement of atomic fission energy with safer and renewable alternatives such as solar energies. Those who inherit the earth will owe these individuals lasting gratitude when they finally succeed in their humane mission.*

Contents

4. What Can a Citizen Do?

Preface

ONE EVENING IN December 1974, Dr. Alvin Weinberg, former director of the Oak Ridge National Laboratory and the person most identified with nuclear reactor development in this country, was discussing the energy situation. He seemed worried about nuclear fission's risks, though not basically altering his pro-nuclear position. As usual, he emphasized the need for a managerial "priesthood" of experts to contain the perils of the peaceful atom in strictly protected, giant nuclear energy parks. When the alternative of solar energy was suggested, his response was unexpected. If solar electric could be brought down to a cost not exceeding 2.5 times that of nuclear, he said, then he would favor solar over nuclear fission.

Dr. Weinberg continues to boost atomic reactors, as might be the case of one who has devoted more than twenty-five years to their promotion, notwithstanding his oft-quoted admonition that these reactors require society to make a Faustian bargain. But his conditional preference for solar is a telling signal. For he is beginning to suspect that the long-term safety of fission is probably beyond the effective limits of social, if not technical, management in an unstable world. In addition, he acknowledges that a major nuclear plant catastrophe would turn the politics of the country against atomic energy. With this awareness of the yet unresolved safety problems affecting the nuclear fuel cycle, he knows that this "one bite of the apple" risk lends a critical fragility to a nation's atomic energy planning. In December 1976 Weinberg was concluding that, due to lower future energy demand, the economic consequences of a thirty-year nuclear moratorium would be more modest than he once believed.

Not many scientists in the nuclear reactor business voice such doubts in public. In this sense the technology has long been in the hands of a "priesthood" whose private concerns about risks and consequences were rarely permitted to mar public expressions of its faith. The nuclear cultists have been disinclined to engage what Alfred North Whitehead called the "fearlessness of revision." Nonetheless, fissures are appearing at last in the hitherto solid phalanx behind fission. Celebrated resignations from the atomic energy ranks of industry and government have been accompanied by expressions of silent support from their colleagues and a growing unease among the believers. Each of the engineers and scientists who has resigned or defected has revealed a train of serious deficiencies in the civilian reactor industry or its federal regulatory commission.

Other members of the corporate community are having second thoughts. They see severe shortcomings in the economics of atomic power which include a high-cost spiral and new unpredictabilities that undermine stable planning. Banks are reluctant to provide capital for construction because of soaring costs, defective designs, uncertain backfitting, waste disposal unknowns, and lower-than-planned capacity levels in delivering electricity. Utilities are filing increasing numbers of complex litigations against reactor manufacturers. The various sectors of the nuclear industry are blaming one another for systems and parts failures, for costly shutdowns for repairs and for poorly trained personnel. Insiders with whom we have spoken paint a picture of a technology lacking in tight management and inspections and without an arm's length between government regulators and industry in the process of setting and enforcing standards. They have admitted that no one knows enough about what is going on inside those operating reactors. Since there are over sixty operating nuclear plants in the United States—each with one thousand times more radioactive material than the fallout from the Hiroshima weapon—this technical gap is deeply unsettling. The word that recurs in insiders' conversations is "luck" to explain why a more serious reactor accident has not yet occurred. Already in this infant industry, the government records a number of major "near misses," a larger number of "abnormal occurrences" and many smaller-scale spills, leaks and losses of radioactive material into the water, land and air.

Such present and future dangers very often remain invisible unless special detection equipment is used. Radioactive violence is silent and long range, and it transcends the usual physiological alerts that humans possess. The nature of radioactivity, then, makes atomic energy a profound challenge to the civic mind, since its silent violence circumvents the normal human senses of the exposed populace.

It has been less than a decade since a consumer evaluation of atomic energy began to countervail the narrow promotional-producer perspective which shaped that technology. Through a fortunate combination of skeptical scientists, writers, lawyers, and a broadening citizen involvement around the country, this consumer perspective forged a framework of inquiry that pressed nuclear advocates such as Alvin Weinberg to rethink the detriments of atomic energy both within and beyond the confines of the technical arena. The deepening consumer critique applied analytic and normative standards to raise questions in public that had been irresponsibly ignored or covered up with self-serving semantics. As the succeeding pages report, the expanding opposition to atomic energy has drawn significantly from recently disclosed government documents and data as well as information from industry sources. In accordance with the principle that information is the currency of democracy, a deep split has now developed within the scientific community, in large part because of the public availability of studies and experience about the nuclear fuel cycle. It is important to recognize that the conflict between scientists is based on different evaluations of acceptable risk, on different career and economic allegiances, and on different understandings of the practical availability of other energy supplies. The conflict is not based on different schools of scientific thought. Virtually all scientists agree that atomic energy installations are potentially very dangerous and that there is much still to be learned and developed in order to solve longstanding technical problems.

Furthermore, what the courts on other occasions have called an "inherently hazardous instrumentality" affects a wide array of human rights beyond the technical issues, and this is pre-eminently the case with atomic power. How vulnerable is our society to nuclear theft and sabotage, nuclear wastes, atomic proliferation and the secret, garrison-state mentality associated with the diffusion of technologies that present such awesome risks? What are the justifications for continuing taxpayer subsidies and limited liability protection to an operating industry whose electric rates would otherwise suffice to assure stable and large profits? Why are plans to evacuate populations around atomic plants needed but neither publicized nor practiced? What are the full facts about worker exposure to radioactivity from the mines to the reprocessing plants, about low plant capacity, about the costs of decommissioning the plants after they are worn out, and about the sweetheart standards, inspections and compliance routines? Is atomic energy necessary for the nation's electric needs given the massive and costly waste of electricity? Why has government policy since World War II accorded low priority to more tolerant or self-renewing energy alternatives in its research, development and promotional efforts? What are the more vigorous roles required of Congress, state and local governments and citizens to assure the setting of more humane priorities?

The nuclear establishment is not comfortable with these consumer perspectives whether they are encountered in courts, before administrative proceedings, legislative hearings or in public debates. Indeed, a consulting firm in Connecticut has found itself doing a brisk business training utility company executives how to reply to their nuclear critics. It wasn't this way until a few years ago. More than probably any previous civilian industry, the atomic industry was imposed upon the American people through a sequence of actions by government, reactor manufacturers and utility monopolies. The electric atom developed from the wartime secrecy surrounding nuclear weapons—a tradition that insulated it from public scrutiny for years. It did not have to meet a market test, an open information test, an electoral test or, in fact, much of a congressional test. Congress delegated its responsibilities wholesale to the Joint Committee on Atomic Energy until 1977—a committee of great legislative powers which may as well have been composed of the board of directors of General Electric, Westinghouse, and Commonwealth Edison. The one potential restraining force—competing fuels—has diminished sharply in the past decade as cartels and oil industry control over nuclear, coal, and geothermal continues to mount.

Apologists for the atomic industry lay heavy emphasis on the allegedly stringent internal standards which the technology must meet. They do not recognize the past absence or paucity of external standards which could have inspired accountability to the external groups at risk. The tide is turning, however. From a more assertive citizenry to a more inquiring Congress to a more skeptical president, nuclear power will have to respond to external pressures. Coupled with growing doubts about the economics of nuclear energy when compared with alternative sources, many observers believe that even short of a major accident the future of the electric atom is not bright. Their view is fortified by the growing economic sensitivity of many utilities which have cancelled or deferred over a hundred nuclear plants since 1975, in part due to excessive construction and operating costs.

As the public debate on the nation's energy options intensifies, more people are realizing that one form of energy can lead to more concentrated political and economic power in a few hands than another form of energy. The energy source that would most concentrate this power is, without a doubt, atomic power. As a high technology in a big package, it requires highly centralized institutions. As a national security hazard, it invites the heavy exercise of police power. In contrast, solar energy systems have major decentralizing potential, few security risks, and significant opportunities for self-sufficiencies at the energy consumption site. Far from abetting proliferation of weapons, solar knowhow could become a major, humanitarian export by the United States. The future of solar and nuclear is intertwined, with each one likely to advance at the expense of the other. Stopping atomic fission becomes a function of applying solar energy with its abundant supply and superior characteristics. Certainly, the government's research and subsidy programs will have to reflect a harder choice between these two energy sources than has been the case in the past when the lion's share was taken by fission.

In choosing between energy alternatives, citizens in a democratic society should know about the maximum that can go wrong with each choice, both in terms of immediate and long-range impacts. For atomic energy this could mean the devastation of a major metropolitan area, the radioactive contamination of hundreds of square miles of land and water, and a legacy of cancers for future generations. The prevention of any such catastrophe from any cause in a country scheduled to have over five hundred large plants in operation by the year 2000 places an unprecedented strain even on a peaceful world. A society which insists on a technology that is so vulnerable is a society which invites a destructive few to have great sway over the peaceful many. For sabotage and terrorism are not remote phenomena.

Despite the formidable barriers that government and industry have placed between the people and the facts, the truth about atomic energy is breaking through. It is not yet certain, but at this writing it appears that the new administration may reduce the barriers and give fission to a lower priority. The case against nuclear power dictates a much stronger position, however. In the forthcoming chapters, reasons are advanced to explain why atomic fission is unsafe to an unacceptable degree, why it is unnecessary, and why it is also economic folly. These conclusions are all the more compelling, for the discussion takes into account the pro-nuclear arguments.

Thus, the lines between the pro- and anti-nuclear power forces are clearly drawn on the issues. If the critics of nuclear energy prevail but are somehow proven mistaken about the prospect of a major catastrophe (already lower level emissions, accidental spills, and losses of radioactive material occur routinely), the nation will simply have moved more quickly into solar, energy conservation and safer hydrocarbon technologies. In contrast, if the supporters of nuclear power prevail, but are wrong about the "vanishingly small" chances of a reactor disaster, the consequences will produce incalculable casualties whose effects will be felt for generations, and major political and economic turmoil. Here the United States has a profound responsibility. How this country decides the nuclear fission question will greatly determine not only how the rest of the world decides, but also the developmental pace of safer alternative energy sources.

The world's peoples must stand firm against the unforgiving malignant giant that is nuclear fission and the plutonium economy that it is spawning. Fortunately the world's peoples are blessed with the ever-present renewable energies of the sun. Let it be said that our generation moved vigorously to tap this most essential energy source so that our descendants will not curse their ancestors.

RALPH NADER
JOHN ABBOTTS

March 1, 1977
Washington, D.C.

Acknowledgments

A PARTICULAR DEBT is owed to all those persons whose original research provided the information which is this book's foundation. Although the list of persons who have done research on the nuclear issue is too great to number, many of them are mentioned in the footnoted references to each chapter. Additional research for this book was performed by Ralph Kerns, Michael Bancroft, Anita Gunn, and Ronald Lanoue. The contributions of the following persons are also appreciated: Skip Laitner for work on the citizen action section and other chapters; Joan Claybrook, Herbert Epstein, Martin Rogol, and Robert Augustine for work on the Citizen Action section; James Cubie for work on the chapter "Atomic Promotion: The Federal Push"; Susan Campbell for significant editorial assistance; William Gruen for preparing most of the figures; and Margaret McCarthy and Cheryl Koopman for copy editing and typing. The responsibility for the material in the book, however, belongs solely to the authors.

Throughout the book there are several examples of individuals and groups who have been active on the nuclear power issue. It has been impossible to mention all the citizens who have worked for a saner energy future, as hundreds of individuals have made nuclear power activism a full-time commitment.

Glossary of Terms and Acronyms

ACRS Advisory Committee on Reactor Safeguards. A committee established to advise the Nuclear Regulatory Commission on reactor safety matters.

AEC Atomic Energy Commission. From 1946 to January 1975, the federal government agency with the responsibility to promote and regulate nuclear power. In 1975, the AEC was abolished and replaced by the Nuclear Regulatory Commission and the Energy Research and Development Administration.

AIF Atomic Industrial Forum. An industry trade association founded in 1954 to promote nuclear power.

APS American Physical Society. The national association of physicists. They commissioned a review by a panel of twelve scientists of reactor safety issues which included an examination of the Reactor Safety Study.

ASLAP Atomic Safety Licensing Appeal Board. The Board which holds public hearings and rules on any objections to decisions of an ASLB. Members are appointed by the Nuclear Regulatory Commission.

ASLB Atomic Safety and Licensing Board. A three-member panel which holds public hearings and rules on construction permits and operating licenses for individual nuclear plants. Members are appointed by the Nuclear Regulatory Commission.

atomic number The number of protons in an atom's nucleus.

atomic weight The sum of the protons and neutrons in an atom's nucleus.

BEIR Biological Effects of Ionizing Radiation. The title of a committee established in 1970 by the National Academy of Sciences to investigate the effects of radiation.

BTU British Thermal Unit. A measure of heat energy. One BTU is the energy necessary to raise one pound of water one degree Fahrenheit in temperature. It takes about 150 million BTU's to heat the average house for a year.

BWR Boiling Water Reactor. So named because water is allowed to boil and is converted to steam in the reactor itself.

CANDU Canadian-Deuterium-Uranium reactor. A type of reactor manufactured in Canada which uses heavy water (deuterium) as its coolant and natural uranium as its fuel.

Center for Study of Responsive Law A Ralph Nader organization founded in 1969 to perform research and investigation into different areas of public policy. It includes the Freedom of Information Clearinghouse, which gives legal and technical assistance to citizens on the effective use of laws granting access to government-held information.

CEQ Council on Environment Quality. A federal body established to advise the president on environmental policy matters. Unlike EPA, CEQ has no enforcement powers.

CRBR Clinch River Breeder Reactor. The Energy Research and Development Administration's proposed demonstration breeder reactor, to be sited on the Clinch River in Tennessee.

Critical Mass An organization established by Ralph Nader to investigate the nuclear power issue. Critical Mass publishes a monthly newsletter by the same name. Critical Mass is also the title of two national citizens' gatherings to oppose nuclear power, organized by Ralph Nader, in Washington, D.C., in November 1974 (Critical Mass '74) and November 1975 (Critical Mass '75).

ECCS Emergency Core-Cooling System. The major backup safety system for Light-Water Reactor-plants.

EPA Environmental Protection Agency. The federal government agency with the responsibility for establishing and enforcing standards to protect the natural environment.

ERDA Energy Research and Development Administration. Created in January 1975 from portions of the AEC and other agencies, it is the federal agency responsible for conducting energy research into non-nuclear as well as nuclear energy sources.

Eximbank Export-Import Bank. A federal government agency with responsibility for promoting U.S. exports, including nuclear power plants, via loans, loan guarantees, and insurance to recipient countries.

FEA Federal Energy Administration. The federal agency established to administer oil price and allocation regulations and formulate national energy policy.

FFTF Fast-Flux Test Facility. A test reactor which is part of the federal government's breeder reactor program.

fission The process by which an atom (for example, uranium or plutonium) splits and produces heat and nuclear radiation.

fission products Atoms formed from the fissioning of another atom.

GAO U.S. General Accounting Office. A federal agency affiliated with the Congress which investigates various issues upon the request of members of Congress.

GE General Electric. One of the U.S. companies manufacturing reactors.

HTGR High-Temperature Gas Reactor. A reactor which uses helium as its coolant.

IAEA International Atomic Energy Agency. An autonomous, intergovernmental organization, associated with the United Nations, which promotes and regulates nuclear energy on an international basis.

ICRP International Commission on Radiological Protection. An international, nongovernmental organization founded by scientists to make recommendations on radiation exposure standards.

ionizing radiation Refers to the ability of radiation to remove electrons from atoms, thus creating "ions," charged atoms and molecules. The radiation from nuclear reactors is ionizing radiation.

IOU Investor-owned utility. Electrical utilities which are private corporations and whose chief responsibility is to their stockholders.

isotopes Atoms which have the same atomic number but different atomic weights.

JCAE Joint Committee on Atomic Energy. A U.S. Congressional Committee whose membership comes from both House and Senate, with jurisdiction over atomic energy matters.

kg Kilogram. A metric unit of weight, about 2.2 pounds.

kwhr Kilowatt hour. A measure of power output. One kilowatt (1,000 watts) of electricity produced continuously for one hour would produce one kilowatt hour. Utility costs are often measured in mill/kwhr.

LEMUF Limit of Error, Material Unaccounted For. The possible error, due to measuring errors, equipment inaccuracies, etc., which is considered an "acceptable" discrepancy in accounting for Special Nuclear Material.

LMFBR Liquid Metal (refers to sodium coolant) Fast (refers to the fact that fissions are caused by "fast", or energetic neutrons) Breeder Reactor (refers to the theoretical ability of the reactor to produce more plutonium than it uses).

LOCA Loss-of-Coolant Accident. A reactor accident which would occur if the piping carrying cooling water to the uranium fuel were to rupture.

LWR Light-Water Reactor, which uses ordinary water. So named to distinguish it from Heavy-Water Reactors, which have deuterium, a heavier isotope of hydrogen, in the water they use. LWRs in the U.S. are BWRs and PWRs.

metric ton A metric unit of weight, 1,000 kilograms, or about 2,200 pounds.

mill One-tenth of a cent. Utility costs are often measured in mills per kilowatt hour (mill/kwhr).

mrem One-thousandth of a rem.

MUF Material Unaccounted For. This refers to accounting methods for Special Nuclear Material. Any discrepancy between material inventoried and material which is supposed to be on hand, according to records.

MWe Megawatt-electric. This refers to the electrical capacity of a power plant. A Megawatt is 1,000 kilowatts, or a million watts. Typical large, modern coal and nuclear power plants are about 1,000 Megawatts-electric.

NCRP National Council on Radiation Protection. A nongovernmental organization in the U.S. founded by scientists to make recommendations for radiation exposure standards.

NFS Nuclear Fuel Services. A subsidiary of Getty Oil which operated a nuclear reprocessing plant near West Valley, New York, and operates a plant which fabricates uranium fuel rods near Erwin, Tennessee.

NRC Nuclear Regulatory Commission. Created in January 1975 from portions of the AEC, it is the federal agency with responsibility for regulating nuclear power.

NRDC Natural Resources Defense Council. A national environmental organization, which includes as its activities projects on the breeder reactor, plutonium, radioactive waste, and nuclear technology export.

NUMEC Nuclear Materials and Equipment Corporation. A company operating plants to fabricate uranium and plutonium fuel rods in Apollo, Pennsylvania. Now a subsidiary of Babcock & Wilcox, a reactor manufacturer.

OCAW Oil, Chemical, and Atomic Workers International Union. A labor union representing workers in some nuclear fuel cycle facilities.

PHS Public Health Service. A branch of the federal Department of Health, Education, and Welfare. As part of its activities, the PHS has sampled mines and streams in the western U.S. for radioactivity.

PIRG Public Interest Research Group. A Washington, D.C., organization founded by Ralph Nader to perform research, provide testimony, and initiate legal action related to various public policy issues.

plutonium A reactor by-product, not normally found in nature, which could be used as reactor fuel. Pu-239 is the isotope which fissions.

Price-Anderson Act A law, passed first in 1957 for ten years and subsequently extended for ten years each in 1965 and 1975, which limits the accident liability of any company operating a nuclear plant.

Public Citizen Health Research Group An organization based in Washington, D.C., established by Ralph Nader to examine the government's handling of health problems.

PWR Pressurized Water Reactor. The water circulating in the reactor does not boil, but instead transfers its heat via a heat exchanger to another system of water, and that water is converted to steam.

rem Abbreviation for roentgen equivalent man. A standard measure of biological damage done by ionizing radiation. The established exposure limits are an average of 5 rem per year for radiation workers, and 0.5 rem per year for members of the general population.

RSS Reactor Safety Study. A study on the probabilities and consequences of reactor accidents, completed in draft form in August 1974 and in final form in October 1975. The study was conducted for the AEC (draft) and NRC (final).

SC-UCS. Sierra Club–Union of Concerned Scientists. This refers to a joint review of the Reactor Safety Study performed by these two groups.

SNM Special nuclear material. Uranium enriched in the isotopes 233 or 235; or plutonium. This generally refers to weapons-grade material, but can also refer to reactor fuel.

SWU Separative Work Unit. An arbitrary unit to measure uranium enrichment work. A 1,000 MWe nuclear plant requires about 116,000 SWU to meet its annual fuel needs.

transuranics, or transuranium elements Elements with higher atomic numbers than uranium.

TVA Tennessee Valley Authority. A federally-owned electric utility serving customers in the southeastern part of the country.

UCS Union of Concerned Scientists. A Cambridge, Mass., based, non-profit coalition of scientists, engineers, and other professionals concerned about the impact of advanced technology, including nuclear power, on society.

uranium The nuclear fuel for light-water reactors. The isotope which fissions is U-235, or uranium of atomic weight 235. The isotope most found in nature is U-238.

WASH-740 A 1957 study performed for the AEC by the Brookhaven National Laboratory on reactor accident consequences.

WASH-740 update A study performed in 1964–65 by the Brookhaven National Laboratory to refine and update the results of WASH-740.

1. *How It Started and How It Works*

I

The Nuclear Commitment

"Leo Szilard was a Hungarian whose university life was spent in Germany. In 1929 he had published an important and pioneer paper on what would now be called Information Theory, the relation between knowledge, nature and man. But by then Szilard was certain that Hitler would come to power, and that war was inevitable. He kept two bags packed in his room, and by 1933 he had locked them and taken them to England.

"It happened that in September of 1933 Lord Rutherford, at the British Association meeting, made some remark about atomic energy never becoming real. Leo Szilard was the kind of scientist, perhaps just the kind of good-humoured, cranky man, who disliked any statement that contained the word 'never', particularly when made by a distinguished colleague. So he set his mind to think about the problem. He tells the story as all of us who knew him would picture it. He was living at the Strand Palace Hotel—he loved living in hotels. He was walking to work at Bart's Hospital, and as he came to Southhampton Row he was stopped by a red light. (That is the only part of the story I find improbable; I never knew Szilard to stop for a red light.) However, before the light turned green, he had realized that if you hit an atom with one neutron, and it happens to break up and release two, then you would have a chain reaction. He wrote a specification for a patent which contains the word 'chain reaction' which was filed in 1934.

"And now we come to a part of Szilard's personality which was characteristic of scientists at that time, but which he expressed most clearly and loudly. He wanted to keep the patent secret. He wanted to prevent science from being misused. And, in fact, he assigned the patent to the British Admiralty, so that it was not published until after the war.

"But meanwhile war was becoming more and more threatening. The march of Hitler went step by step, pace by pace, in a way that we forget now. Early in 1939 Szilard wrote to Joliot Curie asking him if one could make a

prohibition on publication. He tried to get Fermi not to publish. But finally, in August of 1939, he wrote a letter which Einstein signed and sent to President Roosevelt, saying (roughly), 'Nuclear energy is here. War is inevitable. It is for the President to decide what scientists should do about it'.

"But Szilard did not stop. When in 1945 the European war had been won, and he realized that the bomb was now about to be made and used on the Japanese, Szilard marshalled protest everywhere he could. He wrote memorandum after memorandum. One memorandum to President Roosevelt only failed because Roosevelt died during the very days that Szilard was transmitting it to him. Always Szilard wanted the bomb to be tested openly before the Japanese and an international audience, so that the Japanese should know its power and should surrender before people died.

"As you know, Szilard failed, and with him the community of scientists failed. He did what a man of integrity could do. He gave up physics and turned to biology—that is how he came to the Salk Institute—and persuaded others too. Physics had been the passion of the last fifty years, and their masterpiece. But now we knew that it was high time to bring to the understanding of life, particularly human life, the same singleness of mind that we had given to understanding the physical world."

—Jacob Bronowski, *The Ascent of Man*[1]

THE WORLD was introduced to the Atomic Age with Hiroshima and Nagasaki. Scientists and Congress, stunned by the A-bomb's destructive power, embarked on a program to challenge and tame the atom. They looked to atomic power for a major contribution to the world's peacetime energy needs. The Atomic Energy Commission was created in 1946 to develop humanitarian, peaceful uses of the atom, as well as to continue weapons research.

Under the mandate of the Atomic Energy Act, the Atomic Energy Commission (AEC), a civilian-run agency with substantial insulation from congressional and executive supervision, was given the dual and often conflicting responsibilities of promoting as well as regulating the development of America's atomic energy program. Throughout the early 1950s, the AEC tried to interest private utilities in supporting nuclear development. But despite a vigorous promotional campaign, the commission had not received even one application for a plant construction license by 1955. In that year the AEC launched a "Cooperative Power Demonstration Program" designed to stimulate construction by offering, to any utility willing to construct an atomic power plant, liberal research assistance and the waiver of certain fees.[2]

Still, utilities doubted the investment value of the plants. AEC Commissioner Willard Libby told the congressional Joint Committee on Atomic Energy in 1956 that he was worried about the reluctance of certain companies "to come into this peaceful uses program as quickly and as well as they have come into the weapons thing."[3] The shadow of Hiroshima and Nagasaki continued to linger as concern about plant safety grew. Early in 1957 attention centered on a report commissioned by the AEC to estimate the effects of a reactor accident. The study, *Theoretical Possibilities and Consequences of Major Accidents in Large Nuclear Plants*, called the Brookhaven Report because it was prepared at Brookhaven National Laboratory on Long Island, was intended to allay both public and industry fears.

In fact, it did no such thing. Although the study team concluded that the chances of a major accident were remote, they predicted that if the "worst case" accident were to happen, 3400 people would be killed immediately and another 43,000 would suffer serious injury. In addition, seven billion dollars in property damage would be sustained over an area as great as 150,000 square miles. The AEC insisted that the casualty and damage estimates were meaningless because an accident of that magnitude was unlikely to occur. Commission officials stressed that the Brookhaven Report took the most pessimistic view possible in assessing the situation.

Despite these assurances from the AEC, the utilities were frightened by the prospect of a disaster, however improbable. The industry continued to balk, unwilling to assume financial responsibility in case of a reactor accident.

Insurance companies were similarly unwilling to accept responsibility for an atomic reactor accident. Hubert W. Yount, then vice president of Liberty Mutual Insurance Company, testifying before the Joint Committee on Atomic Energy in 1956, explained the insurance industry's hesitation to enter into the nuclear arena:

> The catastrophe hazard is apparently many times as great as anything previously known in industry and therefore poses a major challenge to insurance companies. . . . We have heard estimates of catastrophe potential under the worst possible circumstances running not merely into millions or tens of millions but into hundreds of millions and billions of dollars. *It is a reasonable question of public policy as to whether a hazard of this magnitude should be*

permitted, if it actually exists. Obviously there is no principle of insurance which can be applied to a single location where the potential loss approached such astronomical proportions. Even if insurance could be found, there is a serious question whether the amount of damage to persons and property would be worth the possible benefit accruing from atomic development. [Emphasis added.][4]

Faced with these obstacles to private investment in the development of civilian atomic power, Congress, under pressure from the AEC, responded with passage in 1957 of the Price-Anderson Act, which served to protect utility companies from full financial liability for nuclear disasters. The act set an absolute ceiling of $560 million on the damages which could be recovered by victims of an accident as a result of losses suffered. Moreover, the federal government itself assumed responsibility for $500 million of the liability, thus limiting the responsibility of private utilities to $60 million* despite the fact that the AEC's own study stated that damage could be as high as $7 billion.

Even today, the Price-Anderson Act does not provide insurance coverage beyond a tiny fraction of the forseeable personal and property damage of a major nuclear accident. It does provide a limitation on the liability of private utilities no matter how extensive the damage. At the time of its passage, the act was effective in its primary purpose. Utility companies, now freed from financial liability in the event of a nuclear disaster, began to make heavy investments in atomic energy.

In addition to the federal insurance program, other incentives were offered. Reactor manufacturers were given access to government research facilities; billions of dollars were spent in federal research and development; matching funds were given for demonstration plants (prior to 1971, all reactors were considered to be "experimental" and thus eligible for such grants). The government offered enrichment services for nuclear fuel at federally owned plants, which charged the utilities only the cost of performing the services (i.e., no profit was made by the government). The utilities were also promised that fission technology would be an inexpensive energy source. Lewis L. Strauss, chairman of the AEC under the Eisenhower administration, claimed that electricity from nuclear

* Since 1957, the applicable figures have changed. See Chapter VI. On March 31, 1977, the federal district court for western North Carolina declared the Price-Anderson Act unconstitutional. The nuclear industry will appeal the ruling.

power would be "too cheap to meter."[5] The government thus artificially contrived an environment in which utilities were made to fear that hesitation to "go nuclear" would put them in an uncompetitive position.

Having made a decision to go nuclear, utilities received further benefits. Capital or construction costs of a nuclear facility are typically 25 percent or more, higher than for a fossil-fuel electrical plant.* Normally this extra cost would present a major deterrent to the development of a technology. However, under the present system of regulation, the profit a utility is allowed to earn is directly proportional to its "rate base," that is, to the amount of capital it has invested.

The rate base is a measure of the value of a utility's power lines, generating plants, buildings, and other equipment required to produce and sell electricity. The "rate of return," usually set by a regulatory commission, is the allowable profit expressed as a percentage of the rate base. If, for example, a commission has set a rate of return at 8 percent, a utility investing in a $500 million nuclear plant will in the first year of its operation be allowed to collect $40 million from customers *as profit*, above the operational costs of producing the electricity. If the utility had built a $400 million coal-fired plant instead, the return allowed would be only $32 million in the first year. In successive years the plants will depreciate in value, making the rate of return proportionately less. A larger rate base gives a utility a greater cash flow, which in turn can attract new investment for future constuction.

Despite the clearly promotional efforts by the AEC, not all policy-makers were enchanted with the nuclear option. During the 1940s and 1950s, excitement had been built up by early successes in the development of solar-heated homes. President Truman's Materials Policy Commission—the "Paley Commission," named for its chairman, William S. Paley, president of CBS, declared in 1952: "Efforts to harness solar energy economically are infinitesimal. It is time for aggressive research in the field of solar energy—an effort in which the United States could make an immense contribution to the welfare of the free world."[6]

The Paley Commission was established to appraise the natural resources available to the U.S. by 1975 and to formulate policies

* A typical large, modern nuclear power plant could cost over $1 billion to build, if construction started today.

which would meet resource demands at the lowest economic cost. The commission's report was published at a time when the American public was increasingly uneasy about the country's postwar dependence on foreign resources, a situation not unlike today's. By 1975, the report predicted, the nation would be importing large quantities of oil. Only two sources of energy supply would be available to alleviate the demand for foreign oil—uranium and solar energies. The Paley Commission opted for the solar alternative, projecting an installation of thirteen million solar heating systems in commercial and residential dwellings by 1975. This would account for 10 percent of the nation's overall energy needs, the report noted. But the anticipated growth of solar technology never occurred. Inherent in the political and technical traditions of the 1950s were a number of prejudices which effectively limited a solar policy.

Technically, solar energy was an ugly duckling. Energy systems in the 1950s depended heavily, as they do today, on coal and oil, both of which are high-temperature, concentrated energy sources. Of the two alternative sources, only uranium would fit the existing engineering systems. For example, with steam-generated electricity, instead of water boiled by the combustion of fossil fuels, the necessary heat could be created by the fissioning of uranium. In using nuclear power, therefore, one had only to change the heat source rather than the entire steam-turbine plant. Solar energy, because of its diffuse, low-temperature applications, could not be easily substituted as a fuel in existing power plants and was summarily dismissed by the engineering community and corporation managers.

In economic terms, the use of solar energy represented a major threat to the energy industry. The delivery system for conventional energy supply was characterized by its increasing centralization. In the case of electrical production, whole communities often depended upon a single utility for electricity. Solar energy, with its ability to provide energy at the site of the individual buildings, meant that consumers would be using less and less of a utility's service, thereby diminishing the need for more generating capacity. If electrical consumption no longer increased, neither would profits.

The utilities and reactor manufacturers were uninterested in solar technologies because they threatened to erode industry's political influence. If more homes were able to have their own power supply where the energy was used rather than depending upon the utility for power, then consumers would be less concerned about electrical brownouts or temporary reductions in service. The interests of the community would no longer be so dependent on those of the industry. The community would be less vulnerable to public relations efforts and less supportive of political decisions favorable to the utility industry.

With these engineering, economic, and political prejudices working against solar development, coupled with the challenge to tame the atom for "peaceful uses," atomic power quickly emerged as the energy option favored by the industry and government. As a result, the United States is now officially committed to a nuclear economy.

President Ford pushed for two hundred reactors to be "on line" by 1985, and his administration called for faster licensing procedures for new nuclear plants. The Ford administration also proposed legislation which might have pre-empted individual state authority over the location of all electrical generating facilities. As of October 1976, sixty nuclear power plants were licensed to operate, representing many billion dollars of investment.

If some utilities are beginning to have doubts, they are, nevertheless, trapped by the huge amounts of capital and years of advanced planning. In testimony before the Joint Committee on Atomic Energy in March 1974, Dr. Bruce Welch, then an associate professor of environmental health at Johns Hopkins University, remarked:

> The nuclear commitment has effectively robbed a large sector of American business of three most basic elements of the free enterprise system: initiative, competition and risk. The pigeons of this planning and economic fiasco will eventually come home to roost. . . . But the damage will have long since been done.
>
> The various solar options would have required far less federal support and would have been amenable to more diversified activities within a more viable and independent free-enterprise economy.[7]

II

Nuclear Power Overview

NUCLEAR POWER raises critical health, safety, and political-economic issues. An examination of these issues must begin with some understanding of the basic raw materials of the nuclear problem: how the electric utility industry works; how nuclear reactors work; and how the fuel cycle which supports these reactors works. From this brief introduction the inquiry can then proceed to an overview of the main issues and then to more detailed discussions of this deeply troublesome technology.

The Electric Utility Industry

The electricity which powers light bulbs and appliances comes ultimately from some type of electrical generating station. At the station, mechanical force is used to turn a turbine, which consists of a shaft with vanes at one end and a magnetic field attached to the other end. A fluid, such as steam, strikes the turbine vanes and causes the shaft to rotate. At the other end of the shaft, the rotating magnetic field induces an electrical current in stationary wires which surround the shaft. This current can be transmitted over wires to individual homes and buildings, where it can be used to power electrical devices and motors. The network of power lines which transports electricity is called a "transmission" system.

Figure 1 is a schematic diagram of an electrical generating plant. In a conventional plant, a fuel such as oil, coal, or natural gas is burned, producing heat which is used to boil water and create steam. The steam flows through pipes to impinge on the vanes of the turbine, causing it to rotate. For the operation of a nuclear plant, to be discussed later, heat from uranium fuel replaces heat from fossil

Figure 1. *Diagram of a Fossil Fuel Power Plant*

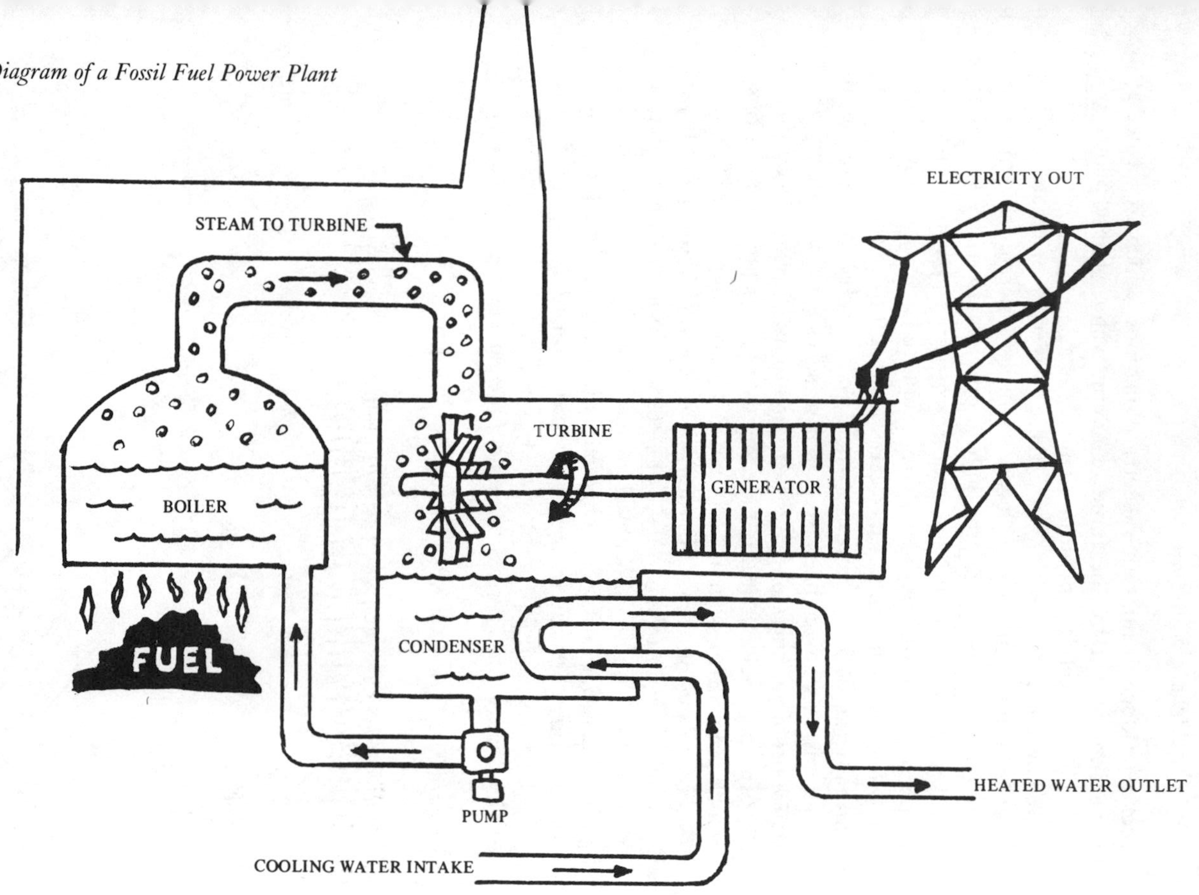

fuel to produce steam. Other electrical turbines, which are located at dams, are turned by the force of falling water. These plants are called hydroelectric generating plants.

The companies which own these power plants make up the electric utility industry. Electric utilities operate as regulated monopolies. Power plants are built to provide electricity to a given utility's service area, although there are points at which different companies' transmission lines are connected. These connections allow one utility to transfer its electrical current to an area it does not normally serve. Such a transfer would be necessary, for example, if Company A's power plant were out of service because of maintenance problems and Company B could supply the extra electricity.

In most states, utilities are regulated by state commissions, which grant permission to build power plants. Utilities are required to serve all customers within their service area. In return, a utility is allowed the right to set "rates," or electricity prices, which will give them the maximum allowable return on their investment, in addition to operating costs. One characteristic of utility regulation is that all operating costs can be passed on to the consumer.

Utilities generally employ a "declining block rate" schedule, of which Figure 2 is a typical example. One can see that as a consumer

Figure 2. *A Typical Residential Rate Schedule*

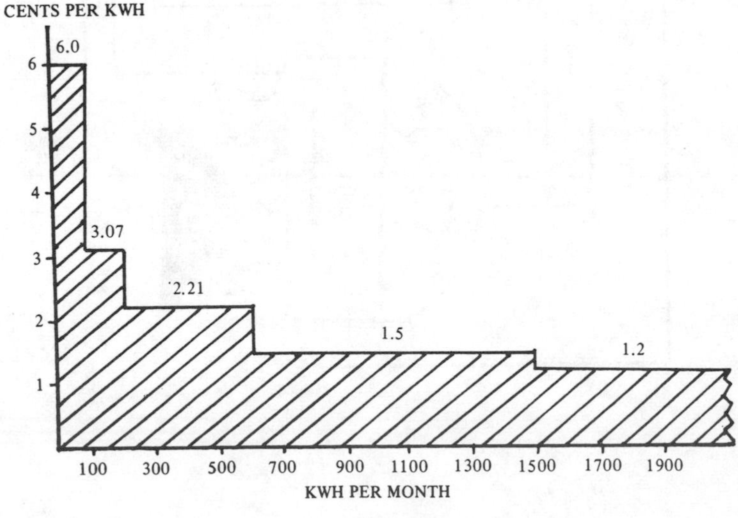

uses more electricity, the average cost per kilowatt-hour drops. Generally, there are different rate schedules for residential (individual homes), commercial (office buildings), and industrial (factories) customers, with residential rates being the highest.

The electric utility industry in the United States is dominated by private corporations, which are referred to as investor-owned utilities—familiarly known as *IOUs*. About 78 percent of the electricity produced in the U.S. comes from about two hundred investor-owned utilities whose chief responsibility is to their stockholders. Examples of IOUs are Consolidated Edison in the New York City area, Commonwealth Edison in the Chicago area, and Pacific Gas & Electric in California.

The remainder of the country's electricity comes from power systems owned by federal and municipal governments (14 percent) or rural electric cooperatives (8 percent).[1] The federal government owns and operates six power systems, including the Tennessee Valley Authority in the Southeast, and the Bonneville Power Administration in the Northwest. These agencies sell directly to IOUs, to municipal utilities, and to rural cooperatives, who then distribute the electricity. Municipal utilities are owned by their customers and operated by the city or regional governments which they serve. There are nearly nineteen hundred municipal utilities throughout the country. Most of these serve small cities, although large cities such as Los Angeles, Seattle, Memphis, and San Antonio also own their own utility systems.[2]

Rural electric cooperatives are customer-owned utilities organized as non-profit private corporations. They were established chiefly to bring electricity to rural and farm customers who are not already served by an existing utility. Some municipal and rural systems own and operate their own generating plants. In most cases, however, they are not able to make the large initial investment necessary to build electrical generating stations. As a result, these systems buy a substantial part of their power from the IOUs and then distribute the power to their own customers.

The growth of the electric utility industry over the past thirty years was very dramatic; but the pronounced slowdown in the industry's growth over the past three years has been perhaps even more impressive. During 1940 to 1973, the annual growth rate in electrical consumption was about 7 percent, which meant that consumption of electricity doubled about every ten years. Since 1973, however, the rise in consumption has slowed. Figure 3 shows the growth in electrical consumption from 1960 to 1975. The lower growth rates in 1974 and 1975 are due to a rapid rise in prices which is not likely to be reversed in the future.

Figure 3. *Historical Growth: Total Electric Demand*

National Energy Outlook (Federal Energy Administration, February 1976), Washington, D.C., p. 223.

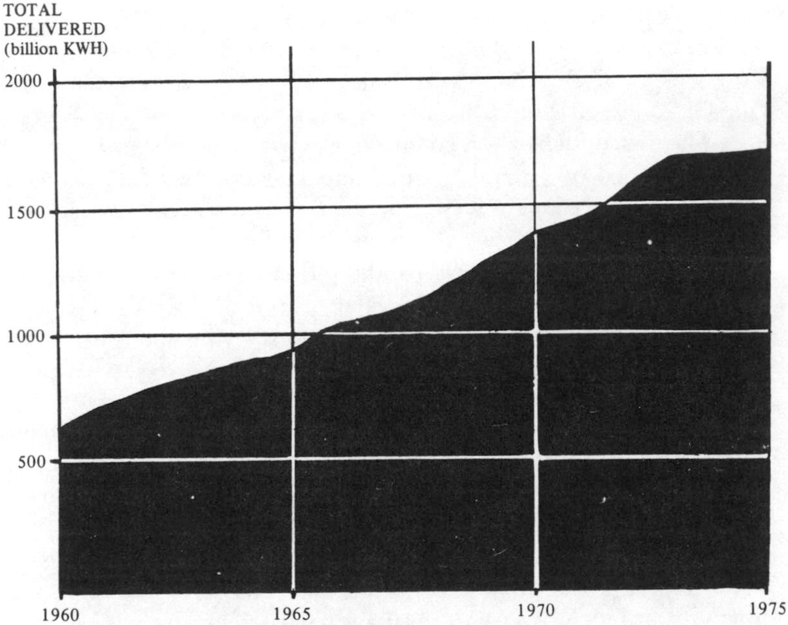

The utility industry has become hooked on the historical 7 percent annual growth rate. In the past, utilities could plan their construction timetables around this growth rate. Since consumption was continually rising, utilities were certain of a cash flow that would allow them to invest in construction of new plants for the future. Since utility rates, as measured by "real" (i.e., adjusted for inflation) dollars, fell during 1940–73, customers were willing to increase their use of electricity.

But, starting in 1972, electric utility prices began to rise rapidly—due to higher fuel costs, construction cost overruns, and higher interest rates. These higher prices made consumers more conscious of their electricity use—and more conservative in their consumption—and the growth rate was curbed. The utilities, which had forecast a continuation of the 7 percent annual growth, were caught by surprise, and were forced to cut back their construction programs for two major reasons. The lower growth rate meant that the utilities would not be receiving the cash sales on which they had planned, and it also meant that plants under construction would not be needed as early as had been predicted. The consequence was that, as of December 1975, the industry had cancelled or deferred 190,000 megawatts of power plant capacity, including 130,000 megawatts of nuclear capacity—the equivalent of 130 large nuclear plants.[3]

The utility industry plans for nuclear power to assume an increasing percentage of electrical production. Nuclear power, with fifty-six operating plants, provided about 9 percent of the nation's electricity (about 3 percent of the country's total energy use) in 1976. The federal government plans to have enough nuclear power to supply up to 50 percent of the country's electricity, from five hundred power plants, by the year 2000. Nuclear power's growth, however, depends on the rate at which electrical consumption rises. If the slowdown in consumption continues, then the optimistic growth projections for nuclear power will fall by the wayside, even if public opposition to atomic power were to vanish.

How Reactors Work

June Allen of Charlottesville, Virginia, began learning about the risks of nuclear power when she read a series of articles written by Environmental Action, a public interest environmental group in Washington, D.C. Her husband was a doctor, and June Allen was an English and music major who worked with the public schools, but she had never thought about nuclear power before. The articles made a deep impression on her. A few days later she saw a notice in the newspaper of a hearing on the nuclear power plant being built by the Virginia Electric Power Company (VEPCO) on the nearby North Anna River. She also noticed a letter to the editor urging citizens to attend the hearing. Ms. Allen called the letter-writer, a woman named Margaret Dietrich, then in her seventies, who lived near the plant site and was a well-informed opponent of nuclear power.

In January 1973, Ms. Allen, Ms. Dietrich, and a few others formed the North Anna Environmental Coalition. They were determined to document and inform the public of the hidden costs and high risks of nuclear power, and to stop, if they could, construction of the North Anna plant.

June Allen went to work full time with the Coalition, working out of her home, without pay, reading stacks of documents and becoming an effective writer and speaker on the issue. Then the Coalition began hearing rumors of a danger nobody had suspected—that the North Anna plant was constructed over a geological fault. In July 1973, Ms. Allen called a lawyer at the Atomic Energy Commission (AEC) in Washington who confirmed the rumors.

In August, the Coalition made the fault public. Their report was widely publicized, to the intense embarrassment of the AEC and VEPCO. Forced by the publicity to act, the AEC called a show-cause hearing ordering VEPCO to prove its plant was safe despite the fault. The Coalition continued its investigations and discovered that the fault had been known at least as early as 1970, when three independent geologists noticed it on visits to the site. Despite efforts by the AEC to lay the issue quietly to rest (its investigation predictably found that the plant was "safe"), June Allen and the Coalition persisted. They appealed the AEC decision that VEPCO could retain its license, and formally charged VEPCO with making "material false statements" in covering up evidence of the fault. The Atomic Safety and Licensing Board found the company guilty of twelve false statements and fined VEPCO $60,000. Appeals panels subsequently reduced the fine to $32,500, but June Allen had definitely established her credibility on nuclear technology.

The geologic fault, June Allen emphasizes, is only one of the issues that she and her colleagues are attempting to expose to their neighbors. The bankruptcy of nuclear technology is, she believes, "the major issue before us now," and for her it has become a "constant and total job."[4]

The technology of nuclear power plants is not simple, but neither is it beyond the ken of ordinary persons. Certainly, June Allen learned enough about nuclear technology to upstage VEPCO and its geological consultants.

Basically, a nuclear power plant is little more than a complicated method of boiling water to generate steam. The steam turns a turbine to produce electricity, just as it does in a fossil-fueled power plant. The difference is that in a nuclear plant, the heat used to generate steam is produced by a nuclear chain reaction, instead of by the combustion of fossil fuel.

A uranium atom of atomic weight 235 (U-235) will split or "fission" if it absorbs a neutron.* As it fissions, a uranium atom produces two new atoms of lower atomic weight, called "fission products." The fission process also produces heat and other neutrons, which in turn can cause fission in other uranium atoms. When as many neutrons are produced by fission as are absorbed by uranium atoms, the nuclear reactor is self-sustaining and generates heat at a constant rate. It is this heat which boils the water to turn the turbine.

Many of the fission products are themselves radioactive † and release heat as a result. During the normal operation of the reactor, the heat emitted by the fission products is much less than the heat generated by fission itself. When the reactor is shut down, however, the fission products still generate heat. As will be shown later, this after-shutdown heat becomes very important in reactor safety.

The reactor's uranium fuel is manufactured in pellet form and packed into long, thin "fuel rods." The uranium fuel is therefore

* Atoms are made up of neutrons, protons, and electrons. The center, or "nucleus," of the atom consists of neutrons, which have no electrical charge, and protons, which have a positive charge. The electrons, which have a negative charge, move in orbits around the nucleus. Each atom contains the same number of protons as electrons, giving the atom a net electrical charge of zero. "Atomic number" is given by the number of protons (or electrons) in an atom.

The neutrons and protons are each about eighteen hundred times as large as an electron, so most of the mass of an atom comes fron the nucleus. "Atomic weight" is a measure of the size of each atom's nucleus. Uranium-235, for example, consists of 92 protons and 143 neutrons (a total of 235). Uranium-238, an "isotope" of uranium, has 92 protons and 146 neutrons. Isotopes are atoms of the same substance which have the same number of protons but different numbers of neutrons (and hence different mass numbers).

Not all atoms undergo fission upon absorbing a neutron. In fact, only a few—such as uranium-235 and plutonium-239—have the ability to fission.

† Atoms which are "radioactive" emit "radiation," which includes alpha particles, beta particles, and gamma rays. Alpha particles consist of two protons and two neutrons. A beta particle is an electron. A gamma ray is similar to an x-ray in its intensity and penetrating power. Most of the heat from fission products, and some of the heat during fission, comes from the emission of gamma rays.

Atoms which emit these particles are said to "decay." Atoms which emit alpha and beta particles become new substances as they do so. For example, Radium-226 changes to Radon-222 when it emits an alpha particle. It is noteworthy that Radon-222 also emits an alpha particle, to become Polonium-218. This demonstrates the fact that many radioactive atoms undergo a "decay chain"; that is, the products of radioactive decay may be atoms which are themselves radioactive. In the radium decay chain, radium and its products undergo several steps of radioactive decay before Lead-206 is formed. Lead 206 is not radioactive and is therefore said to be "stable."

enclosed in metal. About one hundred fuel rods are formed into a fuel "assembly"; a few hundred assemblies are required to fuel a reactor. All the fuel assemblies together are referred to as the fuel pile or the reactor "core." A moderate-sized reactor may require forty thousand fuel rods weighing one hundred tons or more.[5] Figure 4 shows a cutaway view of a typical fuel rod, which is about twelve feet in length.

Figure 4. *A Typical Reactor Fuel Rod*

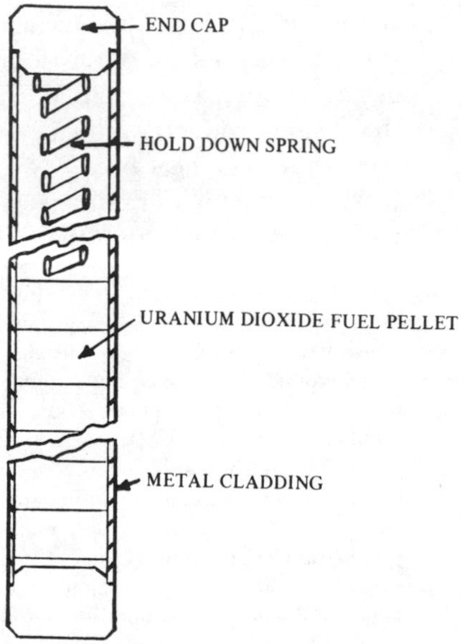

The fission chain reaction is maintained by "control rods," which are interspersed throughout the fuel assemblies. The control rods are made of a substance (such as the element cadmium) which absorbs neutrons. The control rods can be inserted in or withdrawn from the core by means of electrically driven motors. Changing the position of the control rods controls the number of neutrons absorbed by uranium atoms. Proper functioning of the control rods will keep the chain reaction just self-sustaining. In the event of an emergency, the control rods are supposed to jam into the core and shut down the reactor by stopping the self-sustaining fission process. This emergency shutdown is called SCRAM. In an atomic reactor,

Figure 5. *A Fuel Assembly*

WASH-1250, U.S. Atomic Energy Commission

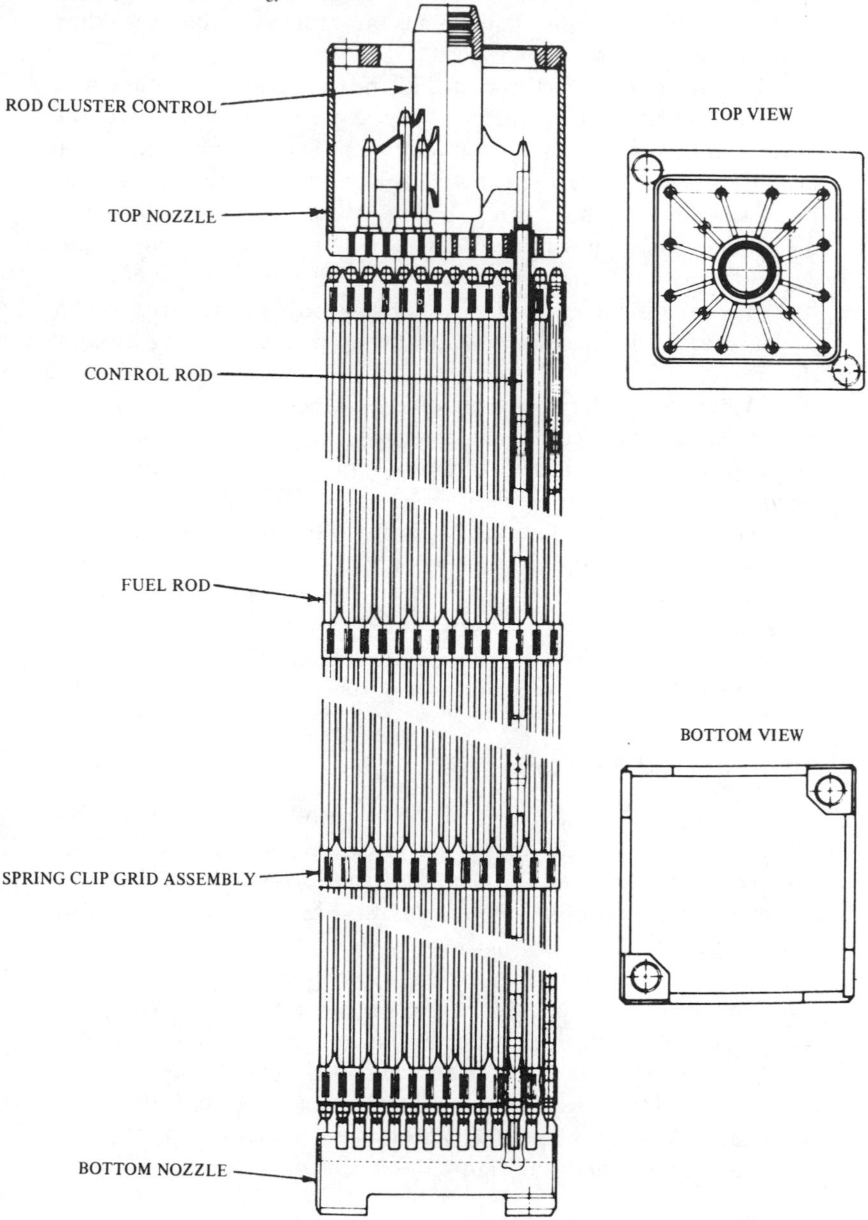

the uranium fuel core and the control rods are all placed in a "pressure vessel." This housing is generally about fifty feet high and twenty feet wide.[6] Figure 5 shows a typical fuel assembly, including control rods.

In reactors manufactured in the United States, water functions as both "coolant" and "moderator." As "coolant," water removes the heat generated by the uranium atom fissions and transfers the heat to produce steam. The heat created by fission is conducted through the metal surrounding the uranium pellets. Water then flows past these metal fuel rods, removing the heat. The temperature of the reactor core during routine operation is about 600° F, with the center of the fuel rods reaching temperatures as high as 4000° F.[7] Massive reactor coolant pumps are necessary to keep the water circulating through the core.

Water also performs a function called moderating. This is necessary because the neutrons created by fission are at a high energy (high speed). To be absorbed by uranium atoms and to cause new fissions, the neutrons must be at lower speeds. The water serves as a "moderator" to slow down the neutrons. Neutrons strike the water molecules and bounce off. Much as a billiard ball loses speed when it strikes the side of the table, neutrons lose speed as they collide with water molecules. As the neutrons slow down, they become more efficient in causing fission reactions. Neutrons at high speeds are referred to as "fast" neutrons, and at low speeds as "slow" neutrons.

Water-Cooled Reactors

In the United States there are two main types of reactors presently operating, the Boiling-Water Reactor (BWR) and the Pressurized Water Reactor (PWR). In a Boiling-Water Reactor, the water passing through the core is allowed to boil after it leaves the core. The steam thus formed drives the turbine. After the steam passes through the turbine, it then passes through a heat exchanger called a condensor. A cold water source—a river, a lake, or the ocean—circulates through tubes in the condensor, cooling the used steam and converting it to water again. This "condensed" water is then returned to the reactor where it becomes steam and the cycle is repeated. BWRs in the United States are manufactured by the General Electric Corporation. Figure 6 is a schematic view of a BWR.

In a Pressurized-Water Reactor, the water circulating through the core is kept at a higher pressure than in a BWR: Thus, the water is prevented from boiling. Heat is exchanged with another water system as the coolant passes through tubes of a steam generator. After passing through the steam generator and giving up its heat, the reactor coolant (or primary coolant) returns to the reactor. The water on the other side of the steam generator tubes, called the secondary coolant, is allowed to boil and in turn drives a turbine, is condensed, and is forced back to the steam generator by pumps. In the United States, PWRs are manufactured by Westinghouse Electric Corporation, Combustion Engineering, and Babcock & Wilcox. Figure 7 is a schematic representation of a PWR.

BWRs and PWRs are often referred to as "Light-Water Reactors" (LWRs). This means that they use ordinary water as coolant, and the term is used to distinguish them from "Heavy-Water Reactors," which use deuterium. Deuterium is an isotope of the hydrogen atom which contains a neutron and a proton in its nucleus. Since the normal hydrogen nucleus contains only a proton, the atomic weight of deuterium is twice that of hydrogen. In a reactor, water with deuterium acts as a more effective moderator than "ordinary" water. This greater efficiency allows "natural uranium" to be used in the Heavy-Water Reactor, in contrast to the "enriched uranium" used in LWR's.

The uranium isotope which fissions is U-235. But the isotope most abundant in nature is U-238, which will not fission. Uranium mined from the ground contains only about 0.7 percent U-235. To be used as fuel in LWRs, uranium must be "enriched," or boosted, to higher concentrations of the isotope U-235. LWR fuel must be about 3.0 percent U-235.

Because deuterium is a more efficient moderator, however, there is no need to enrich the fuel of a Heavy-Water Reactor. The deuterium slows down neutrons so effectively that even with the 0.7 percent U235 in natural uranium, enough fissions will occur to sustain a chain reaction. The Heavy-Water Reactor, which is chiefly used in Canada, does not require enriched uranium, but requires instead the production of heavy water, since most of the water in nature does not have deuterium atoms. Most of the remainder of this book will be a discussion of Light-Water Reactors, since these are the reactors which presently operate in the United States.

Figure 6. *Schematic of Boiling-Water Reactor Power System*

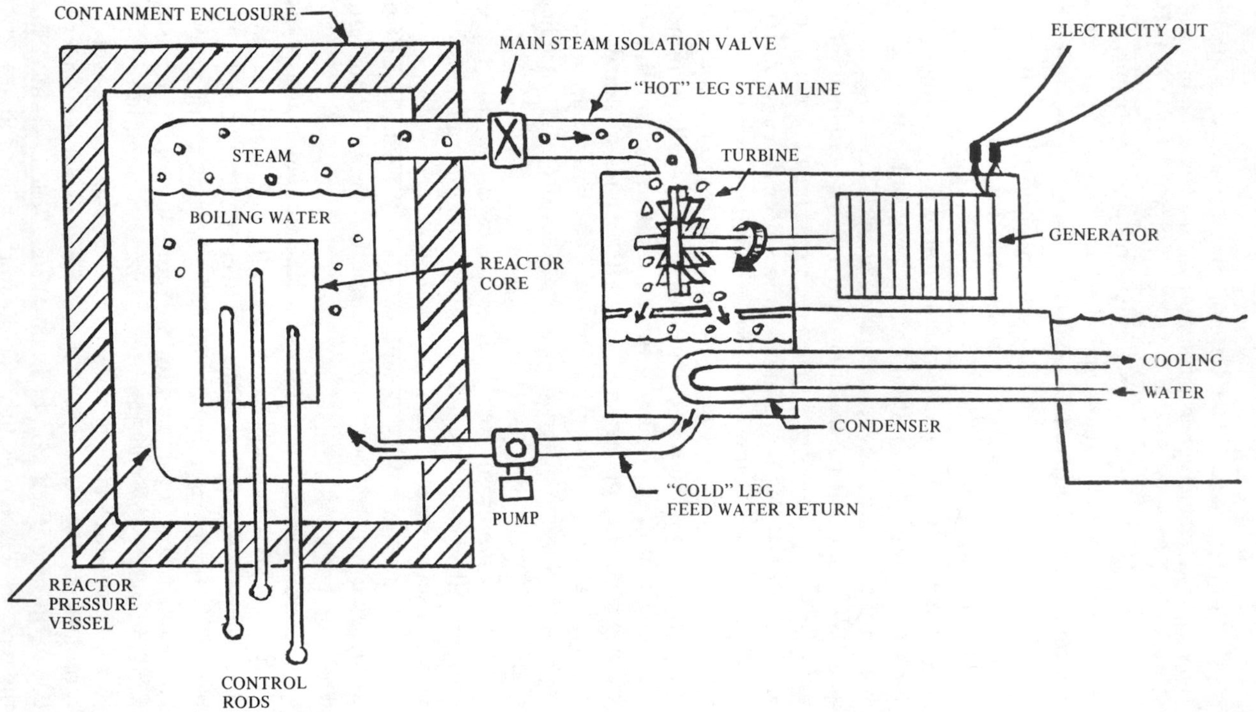

Figure 7. *Schematic of Pressurized-Water Reactor Power System*

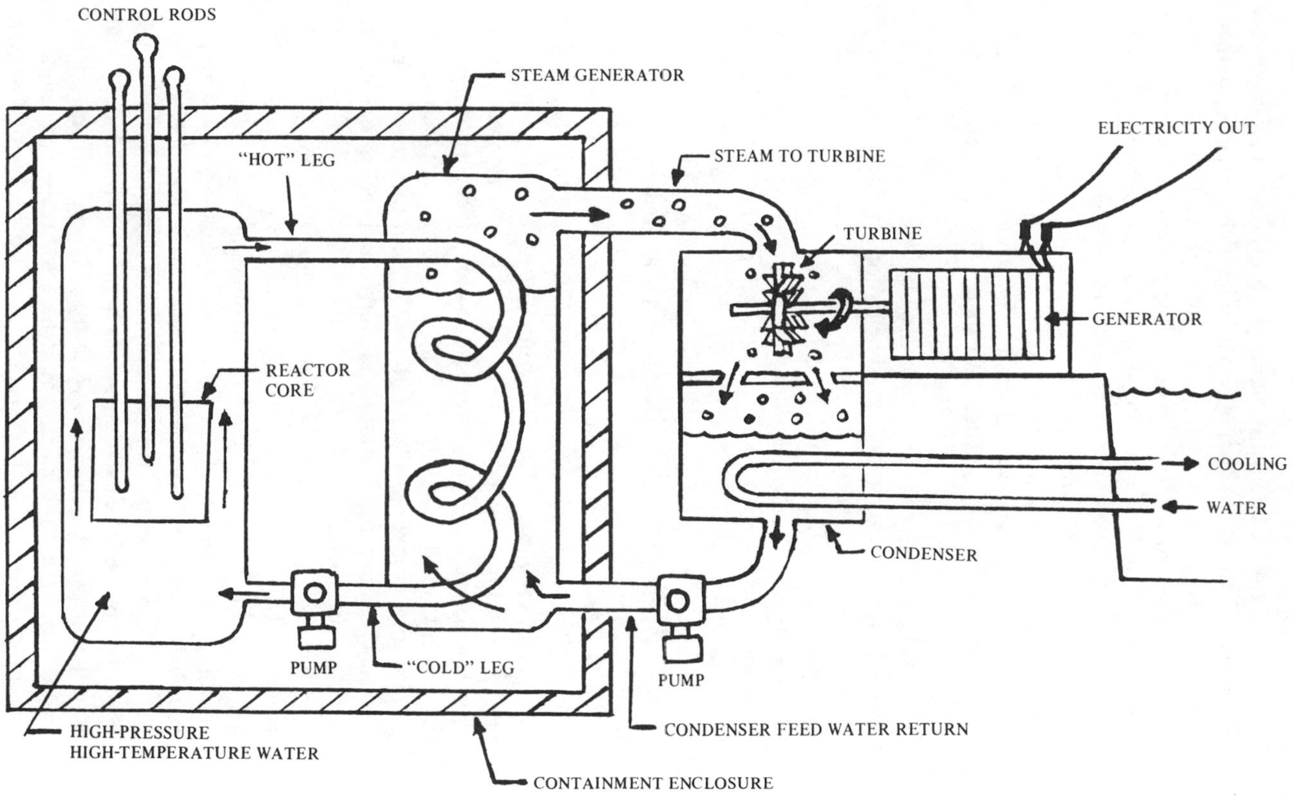

In addition to the equipment which is necessary for the routine operation of the Light-Water Reactor—which includes pumps, piping, and valve systems to direct the flow of steam and water—extra safety equipment is also required. The major backup reactor safety system is called the Emergency Core-Cooling System (ECCS). The ECCS would be required if for any reason there were not enough water to cool the reactor core—if, for example, one of the large pipes carrying primary coolant were to rupture. The coolant, which is under high pressure, would rush out through the break and "uncover" the core. This accident is called a Loss-of-Coolant Accident (LOCA).

Were a LOCA to occur, the reactor would shut down, but the generation of heat would go on. The fission products which build up over the life of the reactor would continue to generate heat. With the core no longer submerged in coolant water, the fuel would continue to heat up until it reached its melting temperature. The core would turn into a white-hot blob, melting through the pressure vessel and then through the reactor building. Such a "meltdown accident" could result in the release of large amounts of radioactivity to surrounding populations, in addition to contaminating land, crops, and water supplies.

The first line of defense if a LOCA were to occur would be ECCS. The system itself consists of different sybsystems which are designed to pump or blow water into the core, thus cooling the core and preventing a runaway meltdown. There is at present a significant scientific controversy over ECCS—many experts doubt its ability to function if called upon. This controversy, which has serious ramifications for the nuclear industry's claim that reactors maintain "defense in depth" against accidents, will be discussed in the chapters on reactor safety.

Another line of defense for an accident is "containment." As mentioned above, the reactor core is surrounded by a steel pressure vessel, and the pressure vessel is surrounded by a steel and concrete building which is intended to contain, or enclose, radioactive material that could be released from the reactor. Were ECCS to fail during a LOCA, however, the containment could be breached. A water reactor cannot explode like a nuclear bomb—its fuel does not contain a sufficient percentage of U-235 to make it weapons material. But a meltdown, as it proceeded, could lead to steam explosions or the generation of gases which, by building up excess pressure, could burst the containment.[8] Even if the containment were not to rupture immediately, the molten fuel would continue to melt through the containment and into the ground.

This meltdown phenomenon is called the "China syndrome." (This name comes from the implication that the fuel would "melt its way to China"—which obviously would not happen, since the molten metal would cool as it came into contact with the ground and would eventually stop melting, albeit several feet into the ground.) About 20 percent of the radioactivity in the core is gaseous. Thus, as the fuel melted into the ground, the gases would diffuse through the soil, allowing radioactive material to contaminate people and land in surrounding areas. Concerns with reactor safety and the consequences of a nuclear accident are also evaluated in the later chapters on reactor safety.

Other Reactors

Two other types of reactors are presently planned for use in the United States. The first is the High-Temperature Gas Reactor (HTGR), which uses a gas such as helium as its coolant. Figure 8 shows that the principle of the HTGR is similar to that of the Pressurized-Water Reactor, except that a gas replaces water as the primary coolant. The fission process heats the circulating gas, and the gas gives up its heat by flowing around tubes in a steam generator. Water passing through the tubes is heated and converted to steam. Helium is used as a coolant because it has good heat-conduction properties, but helium is not an efficient moderator. For this reason, a solid moderator such as graphite is placed throughout the core to slow down neutrons.

At present, only one High-Temperature Gas Reactor is operating in the country—at Fort St. Vrain, Colorado. The future of the HTGR is in considerable doubt. The manufacturer of the HTGR, the Gulf Atomic Company, at present does not have any orders for more plants.[9]

The breeder reactor is planned as the next generation of reactors, and is the federal government's prime energy-research project. No breeder reactors are operating, although the federal government hopes to complete construction on a demonstration unit on the Clinch River, in Tennessee, by 1983. The breeder will use plutonium as its fuel and will operate under different principles from the

Figure 8. Schematic of High-Temperature Gas Reactor Power System

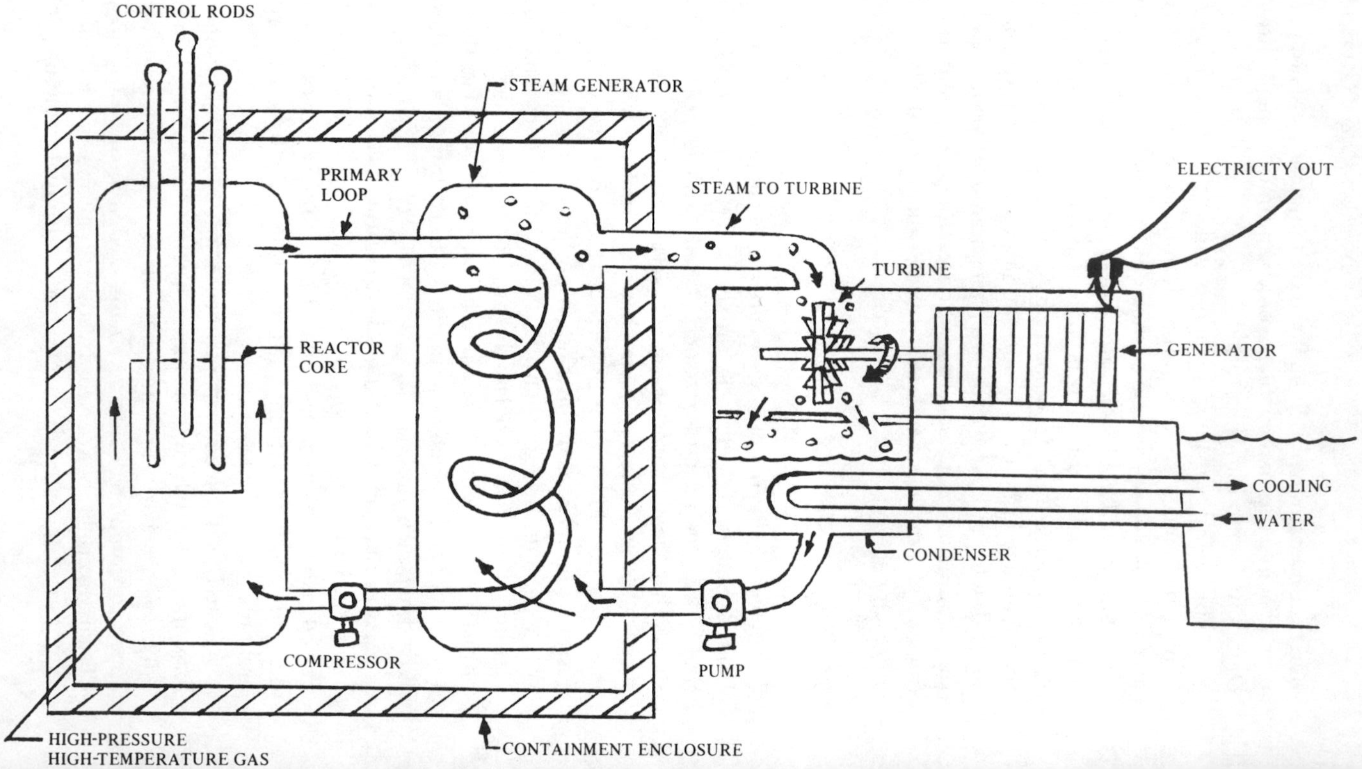

LWR. The breeder gets its name from the fact that it is intended to produce more fuel than it uses. It would do this by the following process: The plutonium fuel atoms would be allowed to fission, just as U-235 fissions in an LWR. The breeder fuel assemblies would consist of plutonium and U-238, the natural uranium isotope of atomic weight 238. Some of the neutrons produced by plutonium fission would be absorbed by U-238 atoms, a process which would change the U-238 to plutonium of atomic weight 239 (Pu-239). The plutonium formed could then be removed from the reactor and eventually serve as new fuel for the breeder. Plutonium is already produced as a by-product in Light-Water Reactors. But the breeder would be designed to produce more plutonium than it used.

With the breeder, the goal is to produce extra neutrons that can be absorbed by U-238 to produce plutonium. This can be performed most effectively by causing plutonium fission with fast neutrons—that is, neutrons which have not been slowed down. "Fast fission" of plutonium releases more neutrons than fission caused by slow neutrons, thus creating the conditions that will allow the reactor to "breed" plutonium from U-238. Although they are more effective at producing new neutrons, fast neutrons are less efficient than slow neutrons in causing fission, a situation which requires plutonium fuel assemblies to be packed closer together than the assemblies in an LWR.

Because neutrons produced by fission must remain fast, the fast breeder cannot have a moderator. The Liquid Metal Fast Breeder Reactor (LMFBR), which is the project the federal government intends to build, uses liquid sodium as its coolant. Sodium was selected because it can transfer heat well but will not inhibit fast fission. Sodium circulates through the core of the LMFBR, and then gives up its heat to non-radioactive sodium through a heat exchanger, as can be seen from Figure 9. This second sodium "loop" then gives up its heat in a steam generator to boil water and drive a turbine.

The need for the second loop stems from the fact that sodium reacts violently with water or air. If radioactive sodium in the primary loop were to contact the water through a leak in the steam generator, a fire or explosion releasing radioactivity to people and land in the surrounding area could result. With the second sodium loop, which is non-radioactive, such an interaction will be less serious. The second loop must be sodium, however, to minimize violent in-

Figure 9. Schematic of Liquid Metal Fast Breeder Reactor

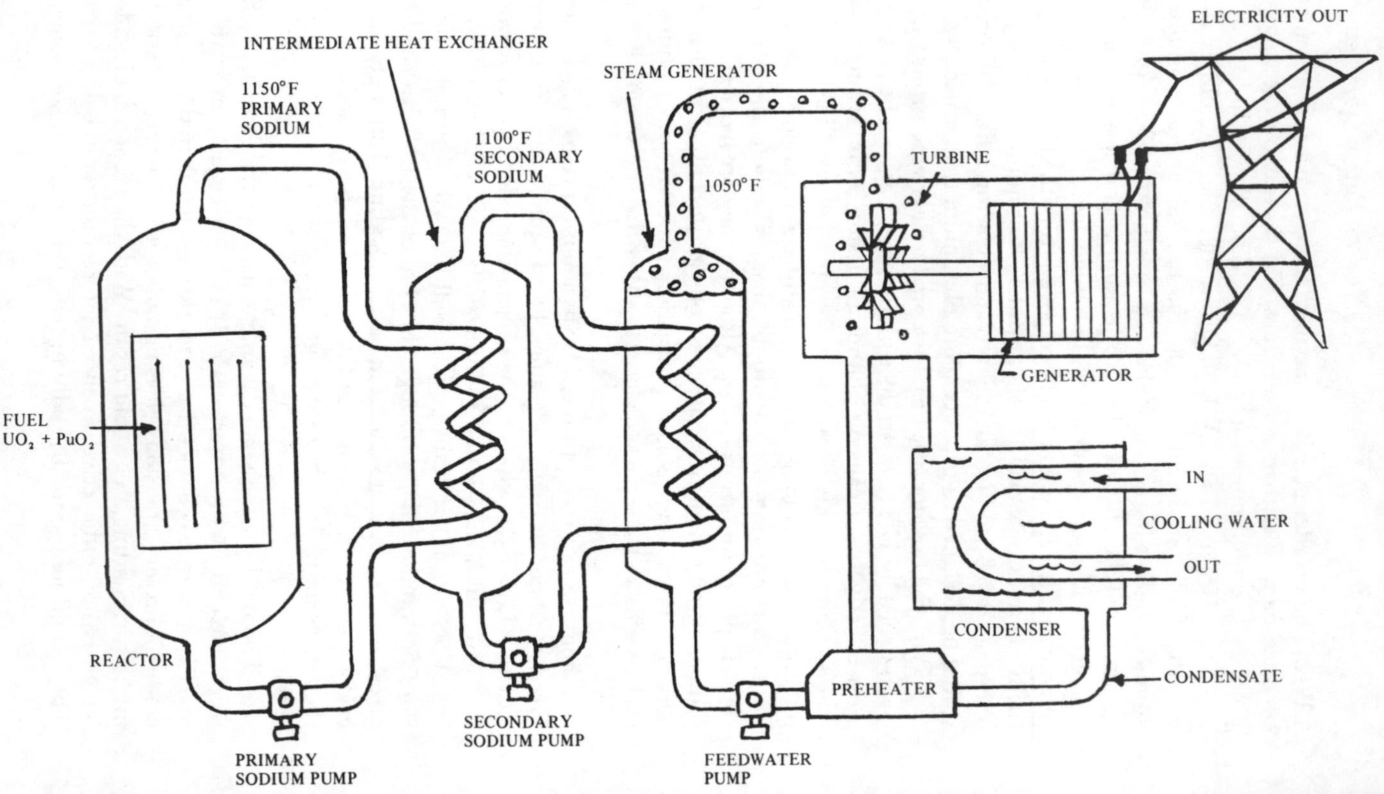

teractions with the first loop, and to transfer heat efficiently. The breeder's technical and economic problems, and the special dangers of plutonium, are discussed at greater length in later chapters.

The Nuclear Fuel Cycle

The nuclear industry extends far beyond the utilities that operate nuclear power plants. A series of operations, called the "nuclear fuel cycle," is necessary to take fresh fuel to the reactor and to remove used fuel. Uranium ore must be extracted, prepared, and fabricated into fuel rods before use in a reactor. Reactor operation depletes the fissionable uranium, requiring the removal of the depleted fuel and insertion of fresh fuel. The depleted fuel must also be processed or stored because it contains radioactive byproducts. Figure 10 is a simplified schematic of the fuel cycle. Its steps are briefly explained below:

1. Mining. Uranium ore is extracted from the ground, via surface mines and deep mines. Most of the nation's uranium—and hence most of the mining industry—is located in the western states, particularly New Mexico, Colorado, Wyoming, and Utah.

2. Milling. The raw uranium ore is crushed and ground, and then treated chemically to extract and concentrate the uranium. The final product, called "yellowcake," is in solid form for shipment. Mills generally are found adjacent to uranium mines and are operated by the same mining companies.

3. Conversion. The yellowcake is converted by a chemical process to the gaseous compound uranium hexaflouride (UF_6). Two companies presently operate conversion plants to meet the commercial nuclear industry's requirements: Allied Chemical in Metropolis, Illinois, and Kerr-McGee, in Sequoyah, Oklahoma. Conversion of the yellowcake to UF_6 is necessary for the next step, which requires a gaseous material.

4. Enrichment. U-235 is the uranium isotope which fissions in a reactor. Natural uranium is 0.7 percent U-235, and the reactor fuel must be 3 to 4 percent. Enrichment is the process which "concentrates" the U-235 in a series of operations, which are carried out by government-owned plants in Tennessee, Kentucky, and Ohio. Each operation forces the UF_6 gas through a porous barrier. Gas with U-235 atoms is slightly lighter than gas with U-238 atoms and hence moves through the barrier faster. As a result, the gas which diffuses

Figure 10. The Light-Water Reactor Nuclear Fuel Cycle

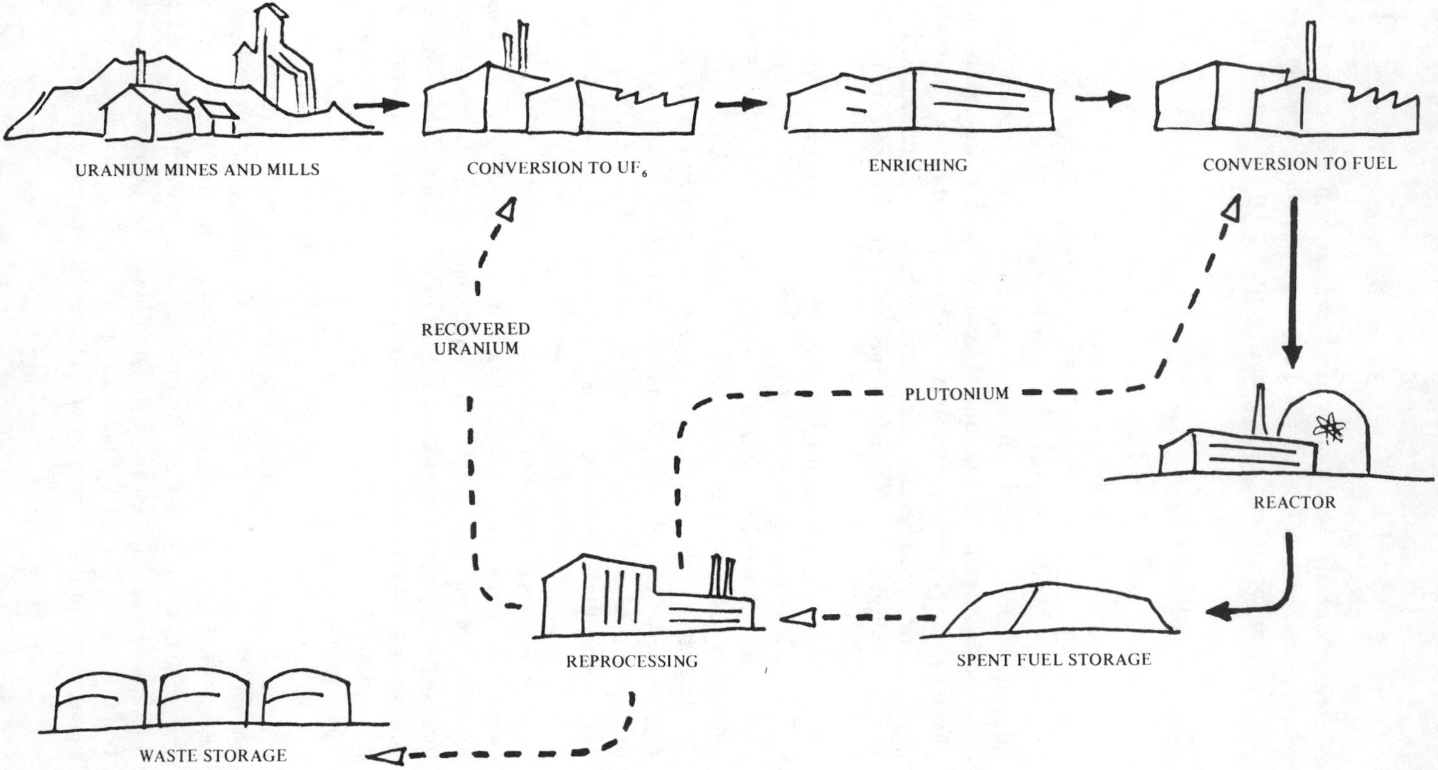

through the barrier has a slightly higher concentration of U-235 than the original mixture. It is necessary to repeat this process more than one thousand times to produce a mixture enriched to 3.0 percent U-235.

5. *Fuel fabrication.* The enriched UF_6 is converted by chemical processes to uranium oxide (UO_2), which is in solid form. The UO_2 is formed into cylindrical pellets and packed into tubular fuel rods. The fuel rods are capped, welded into a fuel assembly, and shipped to a reactor. Fuel fabrication is performed by all the major reactor manufacturers—Westinghouse, General Electric, Babcock & Wilcox, and Combustion Engineering—and by Exxon Nuclear.

6. *Reactor operation* has been discussed in the sections above. Each uranium atom can fission only once, which means that operation of the reactor will deplete the U-235 fuel and require fresh fuel to be inserted. Plutonium and other radioactive materials are produced as byproducts of reactor operation.

7. *Reprocessing.* The depleted, or "spent," reactor fuel must be treated or stored. At present, spent fuel is stored at the sites of the reactors from which it is removed. The next step in the fuel cycle would be to send the spent fuel to a chemical reprocessing plant, but none are operating. One reprocessing plant in West Valley, New York, is shut down—possibly permanently; a second in Morris, Illinois, may never operate, due to technical and financial problems; and a third, near Barnwell, South Carolina, is still under construction. If a reprocessing plant were to operate, it would chemically separate uranium and plutonium from the other radioactive byproducts of reactor operation. (The uranium is present in the spent fuel rods, but in amounts insufficient to sustain a chain reaction. It can be separated and reused with new uranium fuel.) The uranium and plutonium would be returned to the fabrication stage for new fuel to be re-used in reactors. The use of plutonium as reactor fuel, though planned by the industry, must be approved by the Nuclear Regulatory Commission,* which is presently reviewing the question.

* Until January 18, 1975, the Atomic Energy Commission (AEC) was the federal agency which dealt with nuclear power issues. The AEC had the legislative mandate to both promote and regulate nuclear power—a conflicting task which the agency was unable to perform properly. In January 1975, the AEC was abolished and two new agencies formed to take its place. The Nuclear Regulatory Commission (NRC) was given the responsibility of regulating nuclear plants. The Energy Research and Development Administration (ERDA) was given the responsibility of conducting energy research into non-nuclear as well as nuclear energy sources.

8. Waste Management. After the plutonium and uranium have been separated, the radioactive material remaining must be dealt with. Present plans are to store this waste at a government facility—which has yet to be selected—until some "ultimate disposal" method can be devised. Although many speculative "solutions" to the waste problem have been advanced, none have yet proven to be workable in the real world.

9. Transportation. Material must be moved from one plant to the next as it goes through each step of the fuel cycle. Between most steps, this means that the material must be handled by truck, train, or (less frequently) barge or aircraft.

The problems with each of the steps in the fuel cycle are chiefly related to the effects of radiation. Workers can be exposed to radiation, radioactive material can be released to the environment, or major accidents can cause serious exposure to the surrounding population. Some steps in the fuel cycle pose all three of these dangers. The environmental, occupational, and societal problems of the fuel cycle are examined in later chapters.

Industry and Government Whistle-Blowers

Peter Faulkner is a former captain in the Strategic Air Command, a writer, and an electrical engineer with nearly twenty years of experience. In 1974, Faulkner was fired by Nuclear Services Corporation, a consulting firm in California, for his criticism of quality control at nuclear power plants.

Peter Faulkner's particular sin, in the eyes of the atomic industry, was to allow a statement prepared by him to be included as part of the testimony of another individual in March 1974 before the U.S. Senate Subcommittee on Reorganization, Research, and International Organizations. His own experience on a number of nuclear projects had convinced Faulkner of serious problems in three areas: inadequate quality-control during fabrication, poor installation workmanship, and poor utility management. Faulkner recommended that a congressional committee investigate the nuclear industry's quality problems and that the Atomic Energy Commission itself directly supervise the quality of nuclear equipment, rather than delegating responsibility to the utilities.

Faulkner's statement was unsigned when it was inserted into the hearing record, but his name came to the attention of industry representatives.

Three weeks after his statement was released, Nuclear Services Corporation fired him.

As a result of his experiences, Faulkner declared, "I'm convinced that none of my [my recommended] reforms will be implemented, and, as long as reform is unrealistic to hope for, [I'm] adamant that nuclear plant construction should cease now. I didn't hold this position a year ago because I was naive and sincerely believed that the government would force reforms on the industry."[10]

Faulkner also felt that it was important to assemble a group of nuclear critics who had left the industry. Such a group was necessary, he argued, to counter industry propaganda and to tell the public the whole truth about nuclear power. Faulkner himself began working in California with citizen groups trying to curb nuclear power in that state. Within a short time, Faulkner was joined by some other impressive "dropouts" from the atomic establishment.

In September 1974, Carl J. Hocevar resigned his position with Aerojet Nuclear Corporation, which performed reactor safety research for the Atomic Energy Commission. Hocevar, an engineer who was the author of one of the commission's computer codes used to analyze nuclear plant safety, said in his letter of resignation to Dixy Lee Ray, AEC chairman, "In spite of the soothing reassurances that the AEC gives to the uninformed, misled public, unresolved questions about nuclear power safety are so grave that the United States should consider a complete halt to nuclear power plant construction, until we see if these serious questions can, somehow, be resolved."[11]

In January 1976, Robert D. Pollard, a thirty-five-year-old electrical engineer, resigned from the Nuclear Regulatory Commission, the federal agency which took over from the AEC the regulation of nuclear plants. At the time of his resignation, Pollard was project manager in charge of the safety review of Consolidated Edison's Indian Point reactors, twenty-six miles north of New York City. He left the commission because he felt that the NRC was "blind" to unresolved reactor safety issues.[12]

Pollard's resignation was followed in short order by the resignations of three nuclear engineers from the General Electric Company, the world's largest supplier of nuclear equipment. The three engineers said they resigned because they had concluded that nuclear energy represented a "profound threat to man."[13] *The resignation of these engineers could not be taken lightly even by the atomic industry. The three men—Gregory C. Minor, Richard B. Hubbard, and Dale G. Bridenbaugh—had together amassed fifty-four years of experience at General Electric. Like Peter Faulkner, the General Electric engineers chose to work with citizen groups in California to curb the development of nuclear power.*

The pace of unprecedented technical rebellion within the Nuclear Regulatory Commission itself quickened in the latter months of 1976. In December Ronald Fluegge, an NRC engineer who had resigned in October, and five electrical engineers still employed by the NRC came before the Senate Government Operations Committee to testify on problems with nuclear plant safety and with the NRC's safety review process. They were the more visible vanguard of larger numbers of technical NRC employees who feel their findings and evaluations are being ignored or repressed by an agency still bent on promoting an industry. At the time of his resignation, Fluegge wrote that the NRC " "covered up or brushed aside nuclear safety problems of far-reaching significance. We are allowing dozens of large nuclear plants to operate in populated areas, despite known safety deficiencies that could result in very damaging accidents."

The brief explanation of nuclear power in this chapter did not examine institutional and human factors—that fallibility and a misguided profit motive can result in poorly designed and poorly built nuclear plants, as Peter Faulkner and other engineers have discovered. To some extent, the descriptions of reactor operation and the fuel cycle in the previous sections are unrealistic, for they do not touch on the real-world problems of equipment malfunctions, leaks, spills, human error, and incompetence. Later chapters will review both the technical and and non-technical problems of nuclear power, which are significant in number and seriousness. The case will be made that atomic power is unacceptable as an energy source, for this or future societies.

2. *The Case Against Nuclear Power*

III

Overview of the Issues

IT IS IMPORTANT to recognize that there is a growing opposition to nuclear power within the technical community, as well as among the citizenry. But it is also important to recognize that even this technical opposition stems in part from concerns that are non-technical—social, political, and ethical—in nature.

Indicative of the technical controversy is the fact that on August 6, 1975, over twenty-three hundred scientists sent a statement to Congress and the president warning that the dangers of a rapidly expanding nuclear power program were "altogether too great," and urging a "drastic reduction" in the construction of nuclear plants.[1] A similar proclamation in September 1975 addressed the special problems of plutonium, the element to which the atomic industry's future is unavoidably tied, and expressed technical and humanitarian concerns. In this statement, a prestigious panel of distinguished citizens, including fifteen Nobel laureates and twenty-six members of the National Academy of Sciences, concluded that the use of plutonium as reactor fuel is "morally indefensible and technically objectionable."[2]

Nuclear power's problems include the technical controversy over reactor safety; the danger to public health and safety from a major accident releasing radioactivity; and inadequate insurance for the victims of such an accident. The nuclear fuel cycle also poses radioactive dangers to workers and the general environment. Nuclear energy, in addition to generating power, also generates radioactive byproducts dangerous for thousands of years. Then there are the hazards of plutonium, which is weapons-grade material—and as such a target for saboteurs or terrorists—in addition to being extremely toxic and extremely long-lived. The technical and non-technical problems, which have led to the growing opposition to atomic power, are set out below.

Reactor Safety

The technical controversy on nuclear reactor safety began in earnest with revelations about the Emergency Core Cooling System (ECCS), the safety apparatus common to all Light-Water Reactors. During protracted hearings in 1972 and 1973, reactor safety researchers, including the government's own employees and contractors, admitted that they doubted the ability of ECCS to function if called upon. This critical safety system is designed on the basis of computer programs, and those programs have not been verified by experimental tests.

In 1976, the ECCS controversy was reopened by one of the Nuclear Regulatory Commission's (NRC) own consultants. Keith Miller, professor of mathematics at the University of California at Berkeley, is one of ten scientists who review safety computer programs for the NRC. Miller cited a wide divergence between the NRC's public posture of "almost absolute certitude" that ECCS would function and the private doubts that remain with government and industry scientists.[3]

The Reactor Safety Study

The foundation of the atomic industry's safety propaganda is the Reactor Safety Study, released in October 1975 by the Nuclear Regulatory Commission. The report and its conclusions—that the probability of a major nuclear catastrophe was equivalent to the probability of a meteorite striking a major city—were released with much fanfare. But the severe weaknesses of the study have contributed to even greater skepticism over nuclear power.

The study has been reviewed and widely criticized by the Union of Concerned Scientists, the Sierra Club, the American Physical Society, and the Environmental Protection Agency. One of the shaky foundations of the study is its belief that it can predict probabilities of events that have not yet happened. But the study could not have predicted that in March 1975, at the Browns Ferry nuclear plant, a technician with a four-inch candle could accidentally cause a fire that would burn for seven hours, render duplicate safety systems inoperative, and shut down the two largest reactors in the world for nearly eighteen months. The direct and indirect costs of the fire to the Tennessee Valley Authority, the owner of the reactors, were over $200 million.

The Price-Anderson Act

The strongest practical refutation of the Reactor Safety Study, however, is the Price-Anderson Act, which provides for limited liability and government indemnification of the operators of a nuclear plant. Were a nuclear accident to occur, the company operating the plant would pay damages covered by private insurance—at present, $140 million. Every other utility operating a nuclear power plant would be assessed a "retrospective premium"of $5 million, per plant. If private insurance and retrospective premiums totaled less than $560 million, the government would make up the difference.

If this seems to be a great deal of money, it should be remembered that even the Reactor Safety Study concluded that a reactor accident could cause $14 billion in property damage alone, 3,300 deaths, 45,000 injuries, 45,000 cancers in the ten to forty years following the accident, and genetic diseases for five generations. Worse accidents are possible: a 1965 study conducted for the Atomic Energy Commission concluded that the immediate effects of a reactor accident could be 45,000 people killed and 100,000 injured, with radioactive contamination spread over an area as large as the state of Pennsylvania and causing $17 billion in property damage. It is clear that Price-Anderson provides only a pittance of possible damages.

The questions which must be put to the atomic industry, then, as a counter to its public statements on reactor safety, are these: If a reactor accident is really as improbable as the industry claims, why does the industry continue to insist on limited liability? Why will the industry not accept liability for an accident that it contends, in practical terms, will not occur? To these questions the industry has never had a satisfactory answer.

The Fuel Cycle and Waste

In addition to concern about reactor safety, there is concern about the nuclear fuel cycle which supports the reactor's operation. Many of the steps in the fuel cycle pose dangers of environmental degradation through routine releases, catastrophic accidents, or exposure to workers. Some steps pose all three dangers.

But perhaps the most bothersome problem with the fuel cycle is the need to store or dispose of radioactive waste, which can be dangerous for a quarter of a million years or more. There are no demonstrated solutions for dealing with this waste, and the government, which has assumed responsibility for handling the waste, has had difficulties in the past. Over 500,000 gallons of liquid waste have leaked from steel tanks at government storage facilities. Moreover, attempts to put radioactive waste in salt deposits in Kansas were aborted when the Kansas Geological Survey discovered weaknesses in the salt formations that the Atomic Energy Commission had overlooked.

If no technical solution to the waste problem is developed, the federal government will be forced to store and guard the waste for as long as it is hazardous. This means that stable human institutions, as well as stable geological formations, must be found for managing the waste. For many people, this makes the waste problem a moral one: the electricity consumed today leaves radioactive garbage for thousands of generations to come.

Occupational Safety

The atomic industry claims that no one has ever died from civilian nuclear power, but such a statement ignores the silent violence perpetrated on the workers throughout the nuclear fuel cycle. An epidemic of lung cancer has occurred, and will continue to occur, among uranium miners who worked in the 1950s and 1960s. In 1975, it was discovered that workers and their families at uranium mines and mills in New Mexico were drinking radioactive water. Workers at other steps of the fuel cycle—particularly those who came in contact with plutonium, the element intertwined with the atomic industry's future—also face the danger of contracting cancer several years after their occupational exposure.

It should be pointed out that, compared to workers in other energy industries, particularly those in the coal fuel cycle, the numbers of workers injured by the atomic industry are smaller. But it should also be recognized that nuclear power produces much less of the nation's energy than coal power. Moreover, because the occupational dangers of nuclear power include cancers which will not become evident for several years, the full toll of the atomic industry can only be estimated.

Plutonium

Because supplies of uranium are limited, the atomic industry will inevitably have to turn to the use of plutonium, which is formed as a by-product of reactor operation, to extend existing uranium supplies. The industry's reactor of the future is the plutonium breeder reactor, but the industry has also asked the Nuclear Regulatory Commission to allow the use of plutonium as fuel for the present generation of reactors.

Plutonium's major dangers include the fact that it is weapons-grade material, that it is highly toxic, and that it is extremely long-lasting: it will take 24,000 years for half of it to decay. In addition to the possibility that plutonium could contaminate the environment or the population in an accident, there is also the danger that a terrorist group could steal plutonium for the purposes of fashioning an illicit nuclear weapon.

Such a danger was underscored by a 1975 television program in which a twenty-year-old chemistry student designed, on paper, a workable nuclear weapon. The only missing ingredient was stolen plutonium, and the theft of that material is certainly plausible. In fact, in January 1976, the Nuclear Regulatory Commission's own director of safeguards, in an internal memo, expressed his concern that currently operating facilities did not have adequate safeguards against theft of weapons-grade material.[4] Even procedures at government plants are suspect. A July 1976 report by the General Accounting Office concluded that "tens of tons" of weapons-grade material could not be accounted for at thirty-four facilities operated under contract to the federal Energy Research and Development Administration.[5]

International Proliferation

A problem related to concern about the theft of plutonium by a terrorist group is the diversion of plutonium by a country for weapons production. It is widely believed that the plutonium for India's first nuclear bomb came from a Canadian-manufactured reactor located in India. Heavy water exported by the United States, moreover, may very well have been used in the development of India's bomb. The Indian weapon demonstrated that export of nuclear reactors can export the proliferation of nuclear weapons as well.

United States reactor manufacturers are directly responsible for 70 percent of the world's reactors, and controls on both exporting companies and recipient countries are inadequate. In the United States, three federal agencies—the State Department, the Energy Research and Development Administration, and the Export-Import Bank—are active promoters of nuclear power exports. The Nuclear Regulatory Commission, the sole federal regulator of nuclear exports, has not demonstrated that it can counter this promotional bias. The regulations under which the recipient countries must operate all too often stop with the requirements of the International Atomic Energy Agency (IAEA), an intergovernmental organization associated with the United Nations. The weaknesses of IAEA control are so widely recognized that the assurance that nuclear exports come under international controls is an empty defense.

The export problem is so serious that David Lilienthal, the first chairman of the Atomic Energy Commission, has called for an immediate halt to further export of nuclear materials and nuclear technology to stop the "terrifying" spread of atomic weapons. Lilienthal recommended in January 1976 that the United States unilaterally impose a moratorium on nuclear exports in order to put moral and economic pressure on other nations to curb the proliferation of atomic reactors and weapons.[6]

Civil Liberties

Concern over whether adequate safeguards can be devised to prevent plutonium theft has been discussed. A related concern is: if adequate safeguards can be devised, what effects might they have on civil liberties? When the Atomic Energy Commission proposed the use of plutonium as reactor fuel in August 1974, the commission also suggested that a federal plutonium police force be established and wide-scale background security clearances on all workers who might be involved with plutonium be implemented.[7]

Russell Ayres, in an article for the *Harvard Civil Rights–Civil Liberties Law Review*, evaluated what such measures might mean to society. Ayres concluded that the background checks and restrictions on their activities which plutonium workers might have to undergo could distort the workers' view of their own role in society. Workers could come to view themselves more as private soldiers than as civilians, and the industry could develop a garrison-state mentality.

Plutonium safeguards would also have effects on the population at large. The need to prevent plutonium theft could lead to a loosening of standards for surveillance on "suspicious" individuals, which in turn could lead to greater acceptance of covert surveillance. If a plutonium theft ever occurred, Ayres concluded, the measures that would have to be taken by a police force working to recover the plutonium in a short time could "create a situation approaching civil war," with innocent citizens caught in the middle. Since such Draconian measures might necessarily be justified if plutonium were used as reactor fuel, Ayres concluded that "loss or diminution of basic civil liberties" might be a result of the "plutonium economy." He recommended that "all other sources of energy," including conservation, "be proven unworkable or unacceptable" before plutonium is used as an energy source.[8]

Economic Problems

As if the threats to health and safety and civil liberties were not enough, nuclear power also has serious economic drawbacks. Nuclear power plant construction costs are skyrocketing, and the reliability of nuclear plants has left much to be desired. Moreover, nuclear power's advantage is being rapidly eroded in the one area—uranium fuel costs—in which it would supposedly clearly be cheaper than other energy sources. David Snow, financial analyst for the New York investment firm of Mitchell, Hutchins, Inc., concluded in a January 1976 report that uranium prices would continue to rise and could reach $100 per pound—the effective equivalent of oil at $12 per barrel—by 1978.[9] The most powerful statement of nuclear power's economic woes, of course, is that as of December 1975, the industry had canceled or postponed over one hundred atomic plants.[10]

As a result of these economic problems, the atomic industry is looking toward Washington for bail-outs. Nuclear power's subsidies are already significant, though hidden, but the future subsidies of atomic energy will be much larger and more overt. For example, Allied Chemical and Gulf Oil have asked the Energy Research and

Development Administration (ERDA) to give them up to $500 million to complete their Barnwell, South Carolina, nuclear reprocessing facility. Westinghouse has asked the Federal Energy Administration to buy four of its floating nuclear plants, at a cost of at least $1 billion each. The proposed Ford-Rockefeller Energy Independence Authority would give part of its $100 billion subsidies to costly nuclear plants.

Energy Conservation

The alternatives to atomic power are abundant. Over the next twenty-five years, energy conservation—which means energy efficiency—could eliminate any need for atomic power. Denis Hayes of the Worldwatch Institute in Washington, D.C., reported in a study funded by the Federal Energy Administration (FEA) that over 50 percent of this nation's energy consumption is wasted through inefficient use.[11] So the conversation potential dwarfs the supply from atomic power, which presently is only 3 percent of the nation's energy supply.

As just one indication of the potential for energy conservation, a study by the American Institute of Architects estimated that a commitment to develop energy-efficient buildings by 1990 could *alone* save more energy than nuclear power is expected to supply even at historical growth rates.[12]

It should also be recognized that energy efficiency measures are good for the economy. In almost every case, measures to increase energy efficiency can be implemented at less expense, in shorter time periods, and more economically than technologies to supply energy.[13] Energy efficiency will also squeeze expenditures from energy waste and create jobs in addition to reducing pollution and inflation. Outstanding examples of the economic advantages of energy-efficient nations are found in Sweden, Denmark, and Switzerland, which consume about one-half of the per capita energy consumed in the United States, but each of which in 1974 had a higher per capita gross national product than the U.S.[14]

Alternative Supplies

While energy efficiency is the short-term alternative to nuclear power, over the long term the choice must be between the breeder reactor and renewable resources. Renewable resources include solar power in all its forms, including wind and plant material. These and other sources of energy, such as geothermal power, are already providing power on a routine basis, and can be expanded rapidly. Just one indication of solar energy's potential comes from Farno L. Green, a General Motors engineer. Green examined the potential for converting "farm residues"—the non-edible parts of plants—into liquid fuels. This is a form of solar energy because the energy from the plants ultimately comes from the sun. Green concluded that enough liquid supplement for gasoline could be produced from energy crops to eliminate the need to import oil into the U.S.[15]

Solar and other energy sources, however, must compete for research funds with the breeder reactor, which in fiscal year 1977 will consume $655 million dollars, about one-quarter of ERDA's budget for energy research. Thus, one of nuclear power's dangers is that it will be a self-fulfilling prophecy. By gobbling up so much in federal research dollars, the breeder would leave little to be distributed to safer, simpler, and more ethical energy solutions such as solar power and advanced conservation.

This is not to say that solar energy and other energy sources should be viewed as panaceas. There will be technical difficulties in developing non-nuclear energy solutions. But nuclear power's problems, which will be discussed further in later chapters, are institutionally serious as well as technically unresolved. Moreover, even promoters of nuclear power admit that to be a viable energy option, it must be perfectly free from catastrophic accident. One major accident from any plant in any part of the nuclear fuel cycle could create such a public outcry that all nuclear power plants would be shut down—forever. Clearly, such a frail technology cannot be a major energy source, for this or future generations.

Who Decides?

The question for nuclear power and its alternatives becomes one of resource allocation. The nation can continue its present policy of supporting, encouraging, and subsidizing the development of centralized high technology, and costly options for the supply of energy and electricity—such as nuclear power. Or, the nation can turn to a restructuring of its energy system by encouraging energy efficiency and decentralized technologies—such as solar power. The latter option would be better for the economy, would be better for workers because it would create more jobs, and would be better for the taxpayer. The barriers to the development of an energy option based on decentralization and energy efficiency lie with the multi-billion-dollar industrial establishment which profits from energy waste, and its servants in the government.

Thus, a larger question on the nuclear power issue is who should decide whether the country will deepen its dependence on such a questionable energy source as nuclear power. Presently, decisions are made by government bodies which have spent most of their existence promoting nuclear power. As a result, the government has built up a bureaucratic inertia which is reinforced by the urgings of corporations which profit from atomic proliferation.

Citizens who have attempted to influence decisions—for example, by intervening in AEC and NRC licensing hearings on nuclear plants—have found the cards stacked against them. Even cities and states have found the clique of atomic decision-makers tough to join: The state of Minnesota was rebuffed in its attempts to set more stringent standards than the Atomic Energy Commission's, and the city of New York was challenged in court by the Nuclear Regulatory Commission when it attempted to regulate transportation of nuclear material through the city's streets.

The dangers of nuclear power to present and future generations are too great to leave decisions to the so-called experts. There is a growing effort by citizens to remove nuclear power decisions from an estranged bureaucracy and technological elite and place these decisions in the hands of the citizenry. This effort is based on a spreading recognition of the facts about atomic energy, to which our discussion now turns.

IV

Radiation Effects

The Gofman-Tamplin Controversy

By 1969, the Atomic Energy Commission (AEC) program to encourage the development of civilian nuclear power was well under way. The agency kept the public informed about the benefits of the peaceful atom, but little was said about the possible dangers.

Questions on radiation hazards had been fielded by three groups—the National Council on Radiation Protection (NCRP), the International Commission on Radiological Protection (ICRP), and the Federal Radiation Council (FRC). Both the NCRP and the ICRP were non-governmental organizations established by scientists working in the field of radiation protection to make recommendations on radiation exposure standards. For years these groups had been the undisputed authorities on radiation protection issues. In 1959 the FRC had been created to set "official" guidelines on public radiation exposure levels, to offset concern about the fallout from weapons testing. The limits were established, however, with little experience and without accurate, well-developed statistical data.

To respond to doubts which members of the scientific community were beginning to express on the "acceptable" radiation exposure standards, the AEC commissioned a series of studies. John W. Gofman and Arthur R. Tamplin were assigned to investigate the accepted standards. Few lay persons had heard of either man at the time of their appointments, although both were widely regarded in the scientific community. Gofman, a medical physicist and physician, had pioneered research efforts on heart disease in the 1950s and was a co-discoverer of four radioactive isotopes. By 1969, he was an associate director of the Biomedical Research Division at the AEC's Lawrence Radiation Laboratory at Livermore, California. Tamplin, who had a doctoral degree in biophysics, assumed the responsibilities of a group leader in Gofman's division.

In October 1969, the two men, speaking by invitation before a nuclear science symposium of the Institute of Electrical and Electronic Engineers, released their results:

1. Radiation was a far more serious hazard than previously suspected.

2. Twenty times more deaths would occur from radiation-induced cancer and leukemia than had been previously believed.

3. Genetic damage had been underestimated even more seriously.

Far from finding their conclusions welcomed, Gofman and Tamplin experienced a torrent of personal and professional condemnation, not only from the AEC but also from the Joint Committee on Atomic Energy and the nuclear industry. Other pressures mounted, including budget cuts in their research programs, and they reluctantly left their positions with the Lawrence Radiation Laboratory.[1] *In their book* Poisoned Power, *Gofman and Tamplin stated that their experiences had made them aware that the "entire nuclear electricity industry had been developing under a set of totally false illusions of safety and economy. Not only was there a total lack of appreciation of the hazards of radiation for man, but there was a total absence of candor concerning the hazard of serious accidents."* [2]

To better understand the concerns of Gofman and Tamplin, it is necessary to understand what radiation is, how its effects are measured, and how it interacts with the environment. Any substance which naturally emits one or more lighter fragments—producing radiation—is said to be a radioactive isotope (also called "radionuclide"). Each isotope has a different rate of decay, its "half-life." This is the time it takes half of the original amount of material to decay to other atoms. For example, two of the more toxic isotopes in the nuclear fuel cycle, strontium- 90 (Sr-90) and cesium-137 (Cs-137), have half-lives of about thirty years. Thus, two pounds of cesium-137, in thirty years, would turn into one pound of Cs-137 and about one pound of decay products. Sixty years from the starting point there would be a half-pound of Cs-137 left, and so forth.

After ten half-lives of a radioactive substance one would have one-thousandth of the original amount of the substance; after twenty half-lives there would be one-millionth of the original substance. A standard rule of thumb for radioactive materials is that they must decay for ten to twenty half-lives before they can be considered harmless.

Radioactive materials emit what is called "ionizing" radiation. Energy from this radiation disrupts and separates electrons from atoms which absorb the radiation. This produces "ions," atoms or molecules with a net electrical charge. Ionizing radiation produces biological effects, such as damage to cell or genetic material, when the energy from radiation is absorbed by biological tissue.

There are essentially two forms of ionizing radiation. The first, a type of electromagnetic radiation similar to visible light but with a much greater energy level, is found in x-rays and gamma rays. The second form of ionizing radiation is called "particulate" radiation because it includes alpha and beta particles emitted during the spontaneous decay of radioactive elements. Ionizing radiation, regardless of its source, produces some effect on cells or tissue. What is important is how much radiation is absorbed, and at what rate it is absorbed, by a cell over a specific period of time.

In order to correlate the effects of ionizing radiation on tissue, the concept of "exposure" was developed. One unit of exposure, called the "roentgen" (named for Wilhelm Roentgen, who discovered x-rays in 1895), is a measure of the number of electrons torn away from molecules when a beam of radiation passes through air. Although the roentgen is a standard term, it is not a good indicator of the effects of radiation since only *absorbed* energy can have an effect. The unit which measures absorption is the "rad" (which stands for "radiation absorbed dose"). The effect of radiation absorbed in biological systems depends upon the energy level of the radiation, on the type of radiation, on the depth of tissue or cellular penetration, and on a number of other factors. When all these factors are considered, the dose equivalent with its biological consequences is referred to as a "rem" (for "roentgen equivalent man"). For x-rays and gamma rays, 1 roentgen = 1 rad = 1 rem, so the terms *roentgen*, *rad*, and *rem* are often used interchangeably, although the most precise unit from a perspective of biological *damage* is the rem. In this book, ionizing effects will be referenced to "rem" or "millirem" (one-thousandth of a rem).

A single dose of 500 rem will cause death in approximately 50 percent of those persons exposed; 250 rem will cause severe radiation sickness, characterized by nausea, vomiting, infections, and, unless heroic medicine is performed, a large fraction of the exposed population will die. A dose of 100 rem will induce radiation sickness among most members of the population. It is more difficult to measure the effects in the human population of chronic, "low-level" exposures spread out over time, which were the exposures Gofman and Tamplin assessed.

There have always been natural sources of radiation in the world. This "background" radiation had been fairly constant until the arrival of the atomic age. The earth's crust is estimated to contain several tons of uranium and thorium, which are radioactive. Cosmic rays, most of which are filtered by the atmosphere, also contribute to the background radiation. Because the natural radiation similarly affects the air, the water, and plant and animal life, every person has a small but detectable amount of radioactive material inside his or her body as a result of breathing, drinking, and eating. Experts calculate that the average level of background radiation is 100 millirem per year. Gofman and Tamplin estimated that each year in the United States, natural radiation levels cause about 19,000 cancer and leukemia deaths and as many as 588,000 deaths resulting from genetic defects which cause such problems as heart disease and diabetes.[3] Thus, it can be said that people survive in spite of radiation, not because of it.

Nuclear power plants are also sources of radiation because they release radioactive materials routinely. Because of minute cracks in the metal surrounding the uranium fuel pellets, some of the radioactive fission products in the fuel are released to the primary coolant. Other radioactive species are formed by interactions between radiation and water or corrosion particles inside the reactor. These radioactive substances in turn are released to the environment on a routine basis when the plant is operating. Gaseous fission products are vented to the atmosphere through tall stacks at the plant, and liquid and solid radioactive substances are discharged into the body of water that provides the plant's cooling. In 1969, limits on the amounts of radiation that could be routinely emitted from a power plant were established by the Atomic Energy Commission with guidance from the Federal Radiation Council. (Both the AEC and the FRC have since expired. The applicable agencies are now the Nuclear Regulatory Commission and the Environmental Protection Agency, respectively.)

The Federal Radiation Council, in establishing allowable exposure guidelines, had concluded that the average person should receive a radiation dose of no more than 5 rem in thirty years—excluding background radiation and medical and dental x-rays. This made the yearly average allowable exposure 170 millirem (0.17 rem). To ensure that this average dose would not be exceeded, the highest dose allowed any *individual* would be 500 millirem (0.5 rem) per year. Gofman and Tamplin questioned these specific standards, which had been adopted by the AEC, in their October 1969 statement. According to their findings, "If the average exposure of the U.S. population were to reach the allowable .17 rads [the same as 170 millirem] per year average, there would, in time, be an excess of 32,000 cases of fatal cancer plus leukemia per year, and this would occur every year." [4]

The conclusion of the two AEC research scientists was, therefore, that the standards were too high and should be reduced by at least a factor of ten. Such a reduction, however, would mean a drastic modification of the AEC's licensing policies. The proposal was not well received.

The computations of Gofman and Tamplin could not be easily ignored. At that time, the consensus was that the incidence of cancer was directly related to the amount of radiation exposure. Because there is a latency period between initial exposure and the development of cancer, and because the effects of radiation are cumulative, a minimum thirty-years period must be considered in weighing human data. Gofman and Tamplin found in their own studies that absorption of one rem cumulative dose would cause a 2 percent increase in cancer in young adults (twenty-one to thirty years). Thus, if persons thirty years old had accumulated 170 millirem exposure per year, the resultant would be 5 rem multiplied by a 2 percent increase per rem, or a 10 percent increase in cancer plus leukemia expected. Radiation would thus cause about 32,000 extra deaths per year *if* the population received the allowable 170 millirem per year.[5]

The industry and the AEC argued that the exposure assumed by the two scientists was unrealistic since effluents from nuclear plants are limited to only a few percent of permitted radioactive emissions. Gofman and Tamplin readily acknowledged these statements, stating that their conclusion was not that nuclear plants were killing 32,000 people a year but 1) that the effects of low-level radiation had been seriously underestimated; and 2) that if plants *could* be run at lower levels, they ought to be *required* to do so, particularly if more and more plants were to be built. With the country's economic and industrial activities dependent upon nuclear power, it would be difficult, if not impossible, to introduce appropriate standards later.

Moreover, while the atomic industry and the AEC were correct in their claims that routine radioactive releases from the *reactor* were relatively small, Gofman and Tamplin pointed out that radiation standards in 1969 neglected important sources of exposure, which could be more serious than normal emissions from the reactor:

1. accidental releases at the reactor;
2. accidental releases during transport of spent fuel rods from reactors;
3. planned and accidental releases from nuclear fuel reprocessing plants;
4. releases and environmental contamination from low- and intermediate-level waste releases and waste disposal;
5. environmental contamination from storage or disposal of high-level wastes;
6. accidental releases through sabotage.[6]

Following publication of Gofman and Tamplin's findings in 1969, criticism of the allowable radiation standards mounted. Eminent scientists joining the ranks of the critics included Nobel laureates James Watson, co-discoverer of the DNA molecule; Harold Urey, chemist at the University of California; Linus Pauling, chemist at Stanford; and George Wald, biologist at Harvard University.

With this growing criticism, the National Academy of Sciences, at the request of the secretary of health, education, and welfare, in 1970 interceded in the radiation controversy by assembling the Advisory Committee on the Biological Effects of Ionizing Radiation (the BEIR Committee). This committee was to investigate the Gofman-Tamplin data and to review the existing AEC standards. In November 1972, the committee released its report, which concluded that exposure of the population to the 170 millirem limit could cause roughly 3,000 to 15,000 cancer deaths annually, with the "most likely" estimate being 6,000 deaths annually. The BEIR Committee, although it did not come up with the exact numerical estimates of Gofman and Tamplin, agreed that the hazards of radiation had in fact been seriously underestimated previously. They also agreed that the exposure standard of 170 millirem was "unnecessarily high." [7] The BEIR Report had vindicated Gofman and Tamplin.

"ALAP" and the EPA Standards

In 1970, the same year the BEIR Committee began its work, the AEC grudgingly acknowledged the criticisms of Gofman and Tamplin, and initiated steps to tighten the standards for power plant emissions. As part of its early proposal, the AEC announced its intention to keep emissions "as low as practicable" (ALAP). By June 1971 the AEC had essentially proposed a hundredfold reduction in routine emission standards, corresponding to an individual dose of 5 millirem per person per year at the power plant boundary. But the ALAP standard created another controversy, since, by definition, "as low as practicable" implies that the role of the AEC was to determine not the "safest" level of emissions, but the lowest level of emissions that could be attained without causing excessive economic penalties for nuclear power producers.

The Nuclear Regulatory Commission inherited the ALAP question from the AEC and on April 30, 1975, released a final decision on the levels of radioactive emissions which are to be used as "design objectives" for the construction of light water reactors. The NRC's announcement slightly relaxed the proposed standards of the AEC but essentially required that each nuclear power plant be built with a design objective of limiting any individual's exposure to 5 millirem per year.[8]

At about the time the NRC released its "ALAP" guidelines for power plants the U.S. Environmental Protection Agency (EPA) announced that it would convene a rulemaking procedure to set radiation limits for the entire nuclear fuel cycle. Under agreements between federal agencies, EPA has the responsibility for setting general environmental radiation standards. NRC then has the responsibility for taking those general standards and converting them to specific regulations for specific types of nuclear plants.[9] Borrowing heavily from the findings of the BEIR Committee, the EPA issued a draft environmental impact statement in May 1975 in which the agency proposed to limit "the annual dose equivalent to the whole body or any organ, except the thyroid, to 25 millirems, and the annual dose equivalent to the thyroid * to 75 millirens." [10] Emissions from plants in the *entire* nuclear fuel cycle were not to cause those dose limits, for any individual, to be exceeded. These proposed standards became EPA regulations in January 1977.

On the surface the new limits appeared to make major concessions to critics of the nuclear industry: the EPA limits represent a twentyfold reduction in the old standards of 500 millirem to an individual. But the EPA's proposed limits were flawed by some major loopholes. First, in limiting whole-body exposures to twenty-five millirems, the EPA stated that the limit would apply only to "planned releases." A variance from the regulations could be issued to permit temporary operation during "unusual" conditions so as to assure the "orderly delivery of electrical power." [11]

No guidelines were offered, however, as to how releases would be classified as "planned" or "abnormal." Any discharge which proved to be uncomfortably large could be justified by a plant operator as "accidental" or "unplanned." Thus an effective license is given to the atomic industry to release any effluent it cannot otherwise economically or effectively control. In effect, the EPA based the regulation of man-made discharges of radioisotopes on the industry's concept of what constitutes a "planned" release.

The EPA's proposed rule-making was also limited to radioactive substances for which control technology was available. EPA concluded that the proposed standards, which covered materials such as iodine, krypton, and plutonium, would result in no more than 180 "health effects" (cancers, leukemias, and genetic effects) from emissions through the year 2000. These consequences were sufficiently low, according to the EPA, to warrant the exposure limits to be set at 25 millirem for the annual whole-body dose. However, the EPA standards did not include Carbon-14, a reactor byproduct for which control technology is not currently available. According to the EPA, Carbon-14 releases by the year 2000 will eventually cause 12,000 health effects.[12]

In spite of these significant loopholes, the proposed EPA standards were another vindication of the work of Gofman and Tamplin. However, there is still scientific doubt about the adequacy of

* Radioactive iodine, emitted from reactors and reprocessing plants, tends to concentrate in the thyroid gland, which is less sensitive to radiation than other body organs. Thus there is a need to set specific standards for the thyroid, but those standards are less stringent than for the whole body.

radiation standards. The controversy still exists because there are some who believe that the underlying assumptions about radiation damage may be flawed.

The Continuing Controversy

The basis for the EPA proposed standard is the "linear hypothesis theory," which holds that the total number of cancers produced will be directly proportional to the total radiation dose, and this proportionality will be maintained at low doses. Because the consequences of low levels of exposure have not been accurately measured, the effects at low levels are estimated to be proportional to those observed at high doses. For example, if 100 rem of radiation were experimentally observed to produce X number of cancers among a given population of animals (or people), then 1 rem of radiation would be expected to produce X/100 cancers among the same population. This linear hypothesis is generally believed to be "conservative"—i.e., it will overestimate, rather than underestimate, the number of cancers which will actually occur, and thus should be a sound basis for setting standards for exposure of the population.

In fact, there are some scientists who believe that the linear hypothesis may underestimate the cancers and genetic effects that could be caused by low-level radiation. Karl Z. Morgan is Neely Professor of Nuclear Engineering at the Georgia Institute of Technology, editor-in-chief of *Health Physics,* a scientific journal, and a highly esteemed radiation scientist. Dr. Morgan has suggested that existing radiation standards could underestimate the effects of exposure for many different reasons:

1. Extrapolations are made on data with observation periods of no longer than twenty years. Many conclusions are based on studies of animals with life spans of less than ten years. Because many health effects may not be apparent until twenty to thirty years after the initial exposures, or even longer, and because human beings live more than seventy years, on the average, known health-effect rates can only increase as more human data are gathered.

2. The linear model assumes an average exposure. The elderly and the very young may be more susceptible to radiation effects than the middle-aged.

3. Adequate data on the effects of very low exposures have not been developed. Instead, the standards are based on extrapolations

from high or intermediate doses down to zero. But at a higher dose a larger fraction of the exposed cells may be directly killed from radiation, instead of showing signs of genetic damage or cancer. At lower doses fewer cells may be killed and more could be likely to suffer latent radiation damage, such as cancer, as a consequence.[13]

In addition to the general controversy over radiation standards, there is also a more specific controversy over the question of plutonium toxicity. Professor Morgan, for example, believes that the present maximum permissible body doses for plutonium and similar substances should be reduced at least two hundred times.[14]

Plutonium-239 (Pu-239) is a man-made reactor by-product which emits highly energetic alpha particles. Alpha particles are not a highly penetrating type of radiation—they can be stopped by a piece of paper. But they can be very hazardous to tissue if they are taken into the body by ingestion or inhalation. Experiments with dogs show that the inhalation of as little as three millionths of a gram of Pu-239 can cause lung cancer.[15] John Gofman has reported that plutonium and other alpha-emitters, such as curium and americium, when in a form that cannot readily be dissolved by body fluids (called an "insoluble" form), "represent an inhalation hazard in a class some five orders of magnitude [100,000 times] more potent, weight for weight, than potent chemical carcinogens."[16] The fact that plutonium has a very long half-life, 24,000 years, makes it one of the deadliest elements known and one of the most difficult to manage.

Edward Martell, a West Point graduate and a nuclear chemist with the National Center for Atmospheric Research in Boulder, Colorado, is a scientist who has had experience with plutonium and the government's problems in handling it. In 1969, Dr. Martell found that soil samples in the metropolitan Denver area were contaminated by plutonium from the Rocky Flats weapons plant. Some of the plutonium had been released from the plant during a serious fire in May 1969, but most had escaped through leaks, spills, and accidents before the fire. Martell had done his own sampling for plutonium because the Atomic Energy Commission, which owned the plant, and Dow Chemical, which operated the plant under contract, refused to do sampling in the greater Denver area.[17]

Martell's more recent work has some dramatic ramifications for the atomic industry as well as for human health. He hypothesizes that lung cancer from cigarette smoking is actually caused by radioactivity in tobacco. Martell believes the culprit to be polonium, a substance which is found in tobacco and which emits radioactive alpha particles.[18] Because of his theory, Martell also believes that the danger from plutonium and similar alpha-emitters in the nuclear fuel cycle has been underestimated by the government. He recommends that exposure standards for these substances be made 1000 to 10,000 times stricter.[19]

Other scientists propound the "hot particle" theory, which holds that very small particles of plutonium and similar substances, if inhaled and lodged in lung tissue, would release their radiation to a small mass of the lung at a very short distance. The effect of radiation from these "hot particles," giving a concentrated dose to one small area, would be much greater than if the same amount of radioactivity had been uniformly distributed throughout the lung. The Natural Resources Defense Council, an environmental organization for which Arthur Tamplin now works, has argued that existing standards fail to consider the effects of hot particles, and has recommended that the government make its standards for plutonium more stringent by factors of 2,000 to 15,000.

John Gofman has some plutonium findings which should disturb the atomic industry. In a study released in July 1975, Gofman calculated, using estimates of the population's cumulative inhalation of fallout, the number of cancers that have developed or will develop as a result of worldwide fallout from past atmospheric weapons testing. Gofman calculated that 116,000 persons in the United States, and one million persons in the entire Northern Hemisphere, have been committed to plutonium-induced lung cancer from fallout.* Because people were first exposed to weapons-testing fallout several years ago, it is only now and in the next several years that resulting lung cancer deaths will begin to appear in the exposed population. According to Gofman, the cases of cancer will continue to increase until the maximum effect is observed, in the next thirty to forty years. Lung cancers, once induced, do not identify themselves as to cause. This is the reason, Gofman explains, that the "absurd, although common [industry] statement can be made that 'cancers due to plutonium haven't been observed.' "[20]

* About 3 percent of the world's plutonium fallout is attributed to burn-up of a satellite with a plutonium power device, during re-entry in April 1964. Shortly before the satellite burned up, Karl Z. Morgan was told by an AEC official that the chances of such an occurrence were "a million to one" or less.

Gofman expanded his work to estimate the lung cancers that could be expected from plutonium from the nuclear industry. Using the projections of the Atomic Energy Commission, Gofman calculated the amounts of plutonium that the atomic industry would handle through the year 2020. He found that even if the industry could contain plutonium with 99.99 percent efficiency (that is, 0.01 percent of the plutonium in the fuel cycle would reach the biosphere), the industry would still cause 500,000 additional lung cancer deaths per year for about fifty years following the year 2020. Since the current death rate from *all* causes in the United States is about two million per year, Gofman's calculations are quite alarming. He also noted that, considering "the fallibility of men and equipment plus circumstances of accidents," it would be a "miracle" if the atomic industry could actually contain 99.99 percent of the plutonium it handled.[21]

Also of concern are the genetic effects of low-level radiation on the population. Irwin Bross, a biostatistician and cancer researcher at the Roswell Memorial Institute in Buffalo, New York, uses the analogy of a genetic ladder. Bross points out that each member of the population begins life at some point on the ladder and, with exposure to radiation, moves toward the bottom where the lower rungs are missing or broken. Although natural repair processes have been able to handle natural radiation, the body cannot cope with the cumulative effect of radiation from medical x-rays, nuclear weapons, and reactor by-products.

Bross concludes:

> The details of the mechanism are still somewhat speculative. We don't know how many rungs there are in the ladder, how far one must go down the ladder before coming to the broken rungs, or how fast the American population is moving down this ladder. The [evidence] shows that we are moving down the ladder and we cannot wait until we have filled in all the details before we take vigorous action to cut down on the radiation exposure in our environment. Nor can we rely on the AEC or other government agencies to protect us even though they are supposed to do so.[22]

The controversy over low-level radiation standards thus continues. There are many reasons to suspect that current standards may not provide adequate margins of safety. As Irwin Bross states, "We now have solid evidence that low levels of radiation which were considered 'safe' a few years ago are able to produce cumulative genetic degradation which can lead to leukemia and other disease in future generations."[23] It should also be recognized that in 1925, the first standards set by the International Commission on Radiological Protection would have allowed up to 100 rem per year—a standard which might have prevented immediate death, but which did not recognize radiation's ability to cause cancer or genetic effects.[24] Present standards are 0.025 rem per year for individuals, and 5 rem per year for workers. As more has been learned about radiation, the standards have gradually been reduced, so there is reason to believe that even the present standards may be reduced when more information becomes available.

The Front-End of the Fuel Cycle

THE "FRONT-END" of the fuel cycle is a term applied to all the steps involved with the preparation of uranium before it is placed in a reactor: mining, milling, conversion, enrichment, and fabrication (see Figure 1). The major problems with these steps in the fuel cycle include the release of radioactive substances during mining and milling and the economic problems of uranium enrichment. These are discussed in more detail below.

Mining and Milling

Uranium ore is mined and milled chiefly in the western part of the United States. The major part of the mining and milling industry can be found in New Mexico, Wyoming, Colorado, Utah, Texas, and Washington State. Low-grade uranium is also present in Florida's phosphate deposits. Foreign nations with major uranium reserves include Canada, Australia, the Soviet Union, the Union of South Africa, France, and French Africa.[1]

Uranium ore is dug out of rock deposits at both surface mines and deep mines—the total production of uranium from each type of mine is about equal. Mills, which crush and grind the ore and concentrate uranium in the solid "yellowcake" form, are usually located near the mines and operated by the same mining companies. This allows the uranium to be processed near the mine mouth and reduces transportation costs.

Uranium dust represents a respiratory hazard to mine and mill workers, but most of the problems with uranium mining and milling are associated with uranium's "decay products." They present a

Figure 1*. The Light-Water Reactor Nuclear Fuel Cycle*

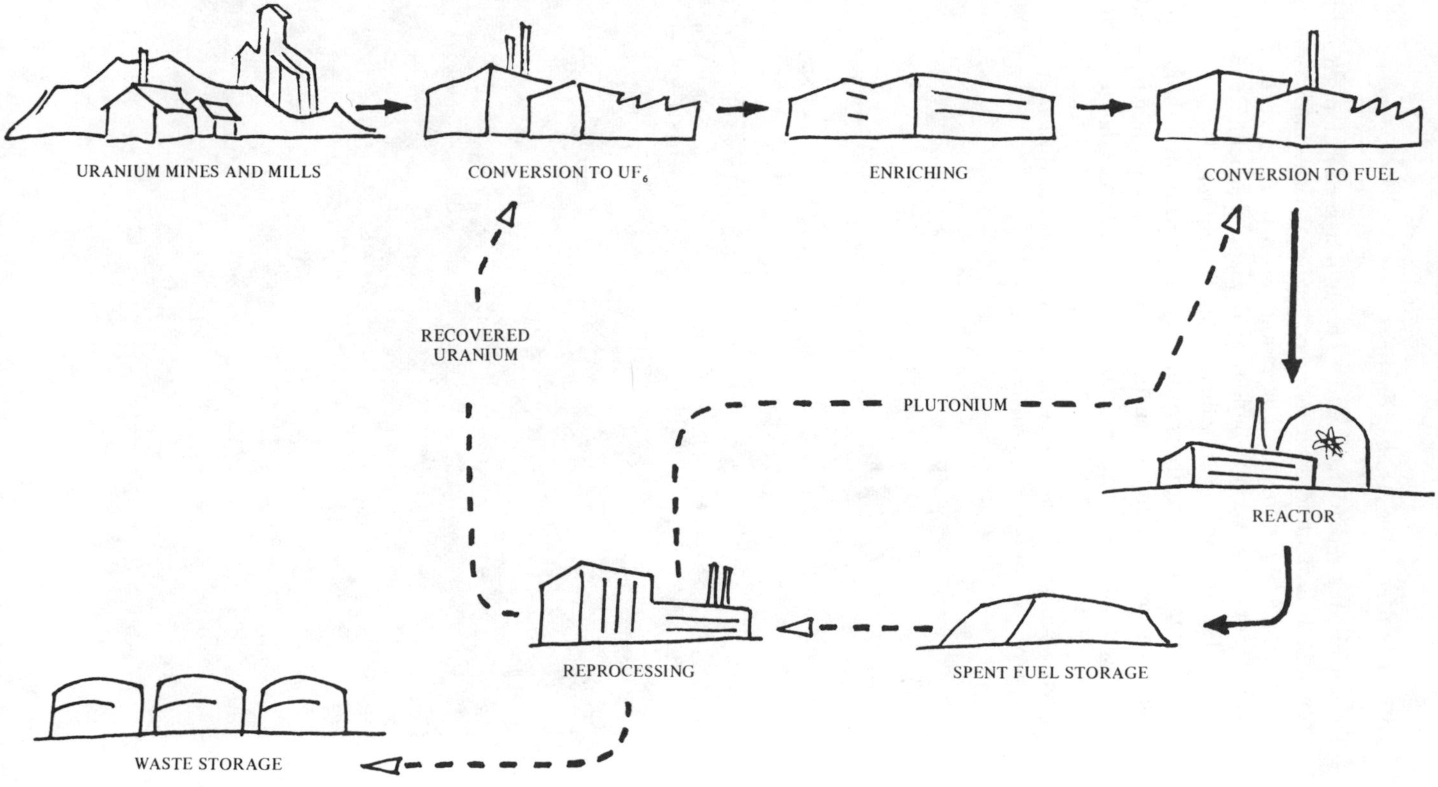

much greater radiation hazard. Through a series of nuclear reactions, uranium undergoes radioactive decay to radium, which in turn decays to radon gas. The radon gas in turn decays to isotopes which can cause serious biological damage, particularly when inhaled. These products are called radon "daughters."

Radon gas in nature is ordinarily trapped in uranium deposits. When the ore is mined or crushed, as it is during milling operations, the radon and its daughters can escape. The chief danger is to the

A sixteen-yard shovel at work at an Exxon uranium mine in Wyoming
U.S. Energy Research and Development Administration

workers, although radon daughters are more hazardous to miners than millers because mills can be more readily ventilated. Poorly ventilated mines have caused miners to inhale the radioactive radon and radon daughters, which has resulted, after a lapse of several years, in lung cancer. The story of lung cancer in uranium miners is one of the more tragic episodes in nuclear power development and is presented in the chapter on occupational hazards.

The radon from uranium operations has also presented dangers to the general public. Normally the waste material, called mill "tailings," left over from grinding and crushing the ore has been left in piles near the mills. The tailing material is a fine, granular substance similar to sand. The piles have been, and still are, free to blow away and contaminate the environment and the population.

In 1958, a U.S. Public Health Service (PHS) biologist was asked to collect fish samples from the Animas River downstream from an AEC-licensed uranium mill at Durango, Colorado.[2] The biologist found damage to river fauna attributable to radium as far as fifty miles downstream.[3] The drinking water in towns along the Animas below Durango frequently contained radioactivity in excess of the limits established by the PHS.[4] Moreover, the radioactivity in the river was concentrated by the food chain: plants in the river had higher levels of radium than the water surrounding them, fish which ate the plants had still higher levels, and so forth. Flora and fauna were found to contain radioactive radium at levels one hundred to ten thousand times the concentrations in the river water.[5]

Similar river contamination was found downstream of other Colorado uranium mills. This state of affairs caused the Public Health Service and the Atomic Energy Commission in 1959 to establish limits on the amounts of radioactive waste material the mills can dump into the rivers.[6] Radium contamination has decreased considerably since the standards were issued, but the problem continues. As late as 1968, the mean annual radium concentration in the San Miguel River near the city of Uraven, Colorado, was above drinking water standards.[7] More recently, 1975 documents from the Environmental Protection Agency indicated that water from mines and mills in New Mexico, contaminated to high levels with radioactivity, was being used as drinking water by workers and their families. This problem is discussed in more detail in the chapter on occupational hazards.

River water contamination is not the only problem with mill

tailings piles. Even if the tailings are not dumped into rivers, the piles are free to blow away or—as happened in Grand Junction, Colorado—to be carried away. Between 1952 and 1966 hundreds of structures were built with radioactive tailings from the Climax Uranium Company mill in Grand Junction.[8] The tailings were used primarily as construction fill underneath or against the buildings, although in at least one instance—at a school—the masonry itself was made with tailings.[9] A 1971 survey in the Grand Junction area indicated that as many as 3,300 buildings may be affected.[10]

In 1972, Congress concluded that the situation in Grand Junction was unsafe and appropriated five million dollars to aid homeowners and schools in removing the tailings. A subsequent survey of 600 representative homes showed that the average level of radiation from the tailings would give occupants a dose of about 2 rem per year to the lungs. Occupants in homes with the highest levels of radiation measured would receive 40 to 60 rem per year to the lungs.[11] In January 1975, after the tailings-removal program had begun, the Energy Research and Development Administration estimated that total costs at Grand Junction would be at least $10.5 million.[12]

The problem of unstable tailings piles exists throughout the western states. A 1976 Environmental Protection Agency (EPA) study found that tailings piles represent a major factor in radiation pollution of the environment and pose risks of lung cancer to large populations. The EPA cited eight states—Arizona, Colorado, New Mexico, Oregon, Texas, Utah, Wyoming, and Idaho—where tailings piles presented hazards to persons living near or downwind of the piles.[13]

A recent study by Robert O. Pohl, professor of physics at Cornell University, evaluated the probable health effects from thorium-230, the isotope in the tailings piles which will decay to radium, radon, and other radioactive products. The decay product radon is a gaseous substance which can escape the piles and disperse throughout the world. The thorium in the tailings piles has a radioactive half-life of 80,000 years, which means that although nuclear power plants will produce power for only about forty years, the effects of mill tailings will remain for thousands of future generations. Pohl's conclusion: the health effects of mill tailings alone from a nuclear power plant are greater than all the effects—including those from air pollution and occupational accidents—associated with the operation of an equivalent coal-fired plant.[14]

All these problems with mill tailings, of course, were not recognized until uranium had been mined for several years. The Natural Resources Defense Council (NRDC), a national environmental group bothered by the "past lax regulatory policies of the Atomic Energy Commission" and concerned that the dangers to public health from uranium mills will be multiplied with an expanding nuclear power industry, petitioned the Nuclear Regulatory Commission (NRC) to address the tailings danger.[15] The NRDC pointed

Aerial view of an Exxon uranium mill near Casper, Wyoming
Norton Pearl Photography

out that there are already tens of millions of tons of uranium mill tailings; that there is "no coherent overall strategy or plan to prevent the critical radionuclides in existing and future mill tailings from entering the biosphere in toxic amounts"; and that the NRC staff itself "admits that current regulations are inadequate to provide for the safe storage of mill tailings." [16]

The NRDC petition recommended that the NRC prepare a detailed programmatic environmental impact statement on the uranium milling industry and on alternatives for regulating that industry. The NRDC emphasized the necessity of completing this impact statement *before*, not after, a large milling industry is built up by an expanding nuclear program. NRDC also recommended that the NRC require operators of mills to post bonds to cover the estimated costs of "interim stabilization" and "ultimate disposal" of the tailings.[17] Such "interim" and "ultimate" measures are required to prevent the tailings from being blown over the countryside and into rivers and potable water supplies, and to prevent radon gas from escaping the piles. Such measures could include covering the piles with several feet of dirt or returning them to unused mine shafts. The NRDC did not recommend which measures should be used, since the responsibility for selecting these measures lies with the NRC. In June 1976, the NRC made a partial response to the petition, announcing that it would prepare a generic impact statement on milling operations. The NRC made no decision, however, on other aspects of the petition, such as the posting of bonds by mill operators.[18]

Conversion, Enrichment, and Fabrication

The conversion step in the fuel cycle is presently performed at two plants in the United States. The Allied Chemical Corporation operates one plant in Metropolis, Illinois, and Kerr-McGee Corporation operates the other in Sequoyah, Oklahoma. At these plants, the yellowcake from the mills is converted to uranium fluoride gas, which is necessary for the enrichment process.[19]

After conversion and enrichment, the fuel is sent to a fabrication plant, where it is reconverted to uranium oxide and packed into tubular fuel rods. Fuel fabrication plants are found in several locations in different parts of the country. Companies with facilities for fabricating uranium fuel rods include all the major reactor makers—

Westinghouse, General Electric, Babcock & Wilcox, and Combustion Engineering—along with Exxon Nuclear Corporation (a subsidiary of the same Exxon which owns oil, natural gas, and coal) and Kerr-McGee (which also produces oil, natural gas, coal, and uranium).

During fabrication, the most significant potential problem is exposure of workers to radioactive material. With uranium fuel rods, the occupational hazards of fabrication are not great. But an expanding nuclear industry would be, to an increasing extent, based on the use of plutonium as reactor fuel. Plutonium's extreme toxicity would make worker exposure at fabrication plants much more serious. The demonstrated inability of commercial plants to adequately manage the small amounts of plutonium already handled and the potential for serious exposures of plutonium fabrication plants is discussed in more detail in the chapter on occupational hazards.

Presently, uranium enrichment is more controversial than the conversion or fabrication steps. Enrichment is the process by which the U-235, which is the isotope that fissions, is concentrated or "enriched" in the uranium fuel. U-235 and U-238 have the same chemical properties, so they cannot be separated by chemical processes. In uranium fluoride gas, however, molecules with U-235 will move slightly faster than the molecules with U-238, which are slightly heavier. If the gas in a container is forced through a hole, the gas emerging from the hole will have slightly more U-235 than the gas on the other side. In actual practice, the UF_6 is forced through porous barriers in several successive containers. After each container, or "cascade," as it is called, the emerging gas has a slightly higher concentration of U-235. Gas that originally contained 0.7 percent U-235 (the percentage in natural uranium) will, after about 1,700 cascades, contain about 4 percent U-235.[20] At this point, the UF_6 is suitable for conversion to reactor fuel. This cascade process of enrichment is also referred to as the "gaseous diffusion" process.

Presently, the federal government owns all three of the uranium enrichment plants in the U.S. and operates them under contract with private industry. Plants in Oak Ridge, Tennessee, and Paducah, Kentucky are operated by the Union Carbide Corporation. The plant at Portsmouth, Ohio, is operated by Goodyear Atomic, a subsidiary of the Goodyear Corporation. The Energy Research and Development Administration (ERDA), the responsible government agency, provides enriched uranium at cost to electric utilities operating nuclear plants. In other words, the government makes no profit on its sale of enriched uranium, thus providing an indirect subsidy to nuclear power. Such a subsidy is not available to other energy sources: An analogous situation would be government-operated refining plants in which the government charged oil companies only enough to recover the costs of refining. The taxpayers' support of enrichment extends to foreign reactors as well as U.S. reactors, because as much as 35 percent of the enriched uranium produced by ERDA is destined for reactors in other countries.[21]

One problem with enrichment plants is that only a fraction of the uranium that goes into the plant is used as fuel. For every pound of uranium that is fed into an enrichment plant, less than one-fifth comes out as fuel.[22] The rest becomes tailings which are depleted in U-235. These tailings are analogous to the piles around mills, but at the enrichment plants they are stored as solid UF_6 in drums. These tailings will be stored until some "future uses" for them can be found, possibly in the breeder reactor.[23]

Perhaps the worst environmental consequences of uranium enrichment come from the electricity that is needed to power the enrichment plants. The plant in Ohio, for example, consumes 10 percent of Ohio's electricity—more than the entire city of Cleveland.[24] Each enrichment plant, for its operation, requires the equivalent of two large power plants to supply electricity.[25] The plants supplying this electricity belong to the Tennessee Valley Authority, the Ohio Valley Electric Corporation, and Electric Energy, Inc. (of Illinois).[26] All the electric plants involved are fossil-fired, and the chief fuel used is coal. One of the ironies of supposedly "clean" nuclear power is that each 1000 Megawatt-electric * (MWe) nuclear plant requires the equivalent of a 45 MWe coal plant—which annually burns 135,000 tons of coal—to supply its enrichment needs alone.[27] In addition to the electricity it consumes, each plant complex also occupies about 1,500 acres of land, which is about 2.3 square miles.[28]

A uranium enrichment problem for the nuclear industry is when and by whom the next enrichment plant will be constructed. ERDA's three plants have a total capacity of 17.2 million kilogram-separative work units, which will be sufficient to supply fuel for

* One megawatt is a thousand kilowatts, which is a million watts of power (this can be compared with an ordinary light bulb, which uses about one hundred watts). The size of typical large, modern nuclear and coal power plants is about 1000 MWe.

enrichment plant at Oak Ridge, Tennessee

F. W. Hoffman, U.S. Energy Research and Development Administration

148,000 MWe—the equivalent of 148 large, modern nuclear power plants.[29] Since mid-1974, the capacity of the three government plants has been fully committed under long-term contracts, including contracts for plants to begin operation in the 1980s. If the nuclear industry is to expand at its own projected rates, new enrichment capacity must be operating by about 1983.[30]

The Ford Administration wanted private industry to construct the next enrichment plant, but industry is reluctant to do so for two major reasons. First, the plant could be very expensive: $2.7 billion for construction costs alone.[31] Second, a gaseous diffusion-type plant could turn out to be obsolete by the time it is constructed, because technologies which may be simpler and cheaper are on the verge of development.

There are at least two other methods of uranium enrichment being developed in the U.S. One, the centrifuge method, would operate in a manner analogous to an eggbeater. The UF_6 gas would be placed in a cylinder with a paddle-type apparatus (the "eggbeater") extending through the center of the cylinder. With the paddle rotating at a very high speed, the heavier U-238 atoms would be forced to the bottom and sides of the cylinder. Gas near the top of the cylinder would be enriched in U-235 and could be drawn off. Like the gaseous diffusion process, the gas centrifuge method would require many stages to obtain the required enrichment.

Laser * excitation is the second method of enrichment under development. Here the energy from a laser would be absorbed by U-235 atoms in such a way that the atoms would become electrically charged. In theory it is possible to select a laser, or combination of lasers, capable of charging only the U-235 atoms, leaving the U-238 atoms unaffected. The U-235 would then be collected on a charged plate.

Centrifuge and laser enrichment both promise to be cheaper than the diffusion method: the centrifuge method because it consumes much less energy, and the laser method because it is simpler and requires a much smaller investment. If government research de-

* Laser stands for "Light Amplification by Stimulated Emission of Radiation." In this case, "radiation" refers to radiating light, not radioactivity. Ordinary light, from the sun or from a light bulb, consists of many frequencies and disperses in all directions. Light from a laser can be produced as a single frequency, can be focused, and can be of high energy. These characteristics give lasers many industrial applications, including, possibly, uranium enrichment.

velops either method, gaseous diffusion could become economically obsolete. For this reason, private industry will not build any new diffusion plants without substantial government guarantees.

The Ford administration's proposal to establish a private enrichment industry generated much controversy. Some critics are concerned about the proliferation of a technology which could be used for nuclear weapons, which require 90 percent or more enrichment in U-235. (For this reason, LWRs cannot explode like nuclear bombs, because they do not have sufficient enrichment.) The same technology that can enrich to 4 percent for reactor fuel can enrich to 90 percent—merely by passing the UF_6 through many more cascades. To minimize this threat, it is argued that the technology of enrichment should remain classified and in the hands of the government, instead of making information on the technology freely available.

Another objection to the Ford administration's proposal was that it was a government giveaway. President Ford sought the authority for eight billion dollars in government guarantees to reimburse any private corporation which builds an enrichment plant, should the plant prove to be uneconomical[32] The General Accounting Office (GAO), when asked in 1975 by Senator John Pastore of Rhode Island to review the administration proposal, was highly critical. The proposal was "not acceptable" to the GAO because it placed most of the financial risk on the government and not enough on private industry.[33] The GAO recommended that the federal government build the next increment of enrichment, if needed at all, instead of providing guarantees to industry.[34]

Opposition from a coalition of fiscal conservatives and nuclear power skeptics narrowly defeated the Ford administration's enrichment proposal in the Senate in September 1976—although the proposal had been passed by the House of Representatives.[35] But nuclear promoters may very well re-introduce legislation to subsidize private enrichment plants during the current session of Congress.

The "front-end" of the fuel cycle, then, has both environmental and economic problems. With the fabrication of enriched uranium into fuel rods, the front-end of the nuclear fuel cycle is completed. The next step is transport of the fuel to the reactor, where the radioactive by-products formed as a result of the fission process represent a major hazard to life.

VI

Reactor Safety: The Technical Controversy

There was no meltdown—this time. But a reactor accident on March 22, 1975, caused by a four-inch candle, shut down the two largest nuclear reactors in the world for over a year and caused one of them to come dangerously close to a catastrophic accident. It happened this way:

Two electrical technicians were performing routine maintenance on electrical cables at the Browns Ferry reactor station, operated by the Tennessee Valley Authority near Decatur, Alabama. (Figure 1 shows the layout of the station.) The technicians were working just below the control room for Reactor Units 1 and 2, trying to seal air leaks in the cable spreading room, where electrical cables that are used to control the reactors are routed into their respective reactor buildings. To do the job the men were installing foam rubber "packing" around the cables. Candles were used to determine whether or not the leaks had been plugged—if the candle flickered when held near the cables, that meant that air was flowing and there was still a leak. One of the technicians carelessly brought the candle too close to a piece of foam rubber packing, and the packing burst into flame. The fire immediately spread to the cables themselves and then into the reactor building, burning for seven hours before it was brought under control. It was an accident that was never supposed to happen, but it did.

For Reactor unit 1, the fire damage to the electrical cables rendered inoperative a number of important subsystems of the Emergency Core Cooling System (ECCS). With both the ECCS and normal cooling out of operation because of the fire, there was no way of removing the heat buildup in the reactor core caused by radioactive decay of fission products. This raised the possibility that the water in the core would boil off, leading to a meltdown as the fuel continued to heat itself up. This "meltdown accident" could in turn lead to the release of radioactive material to the public.

Figure 1. *The Browns Ferry Nuclear Generating Plant*

U.S. Nuclear Regulatory Commission

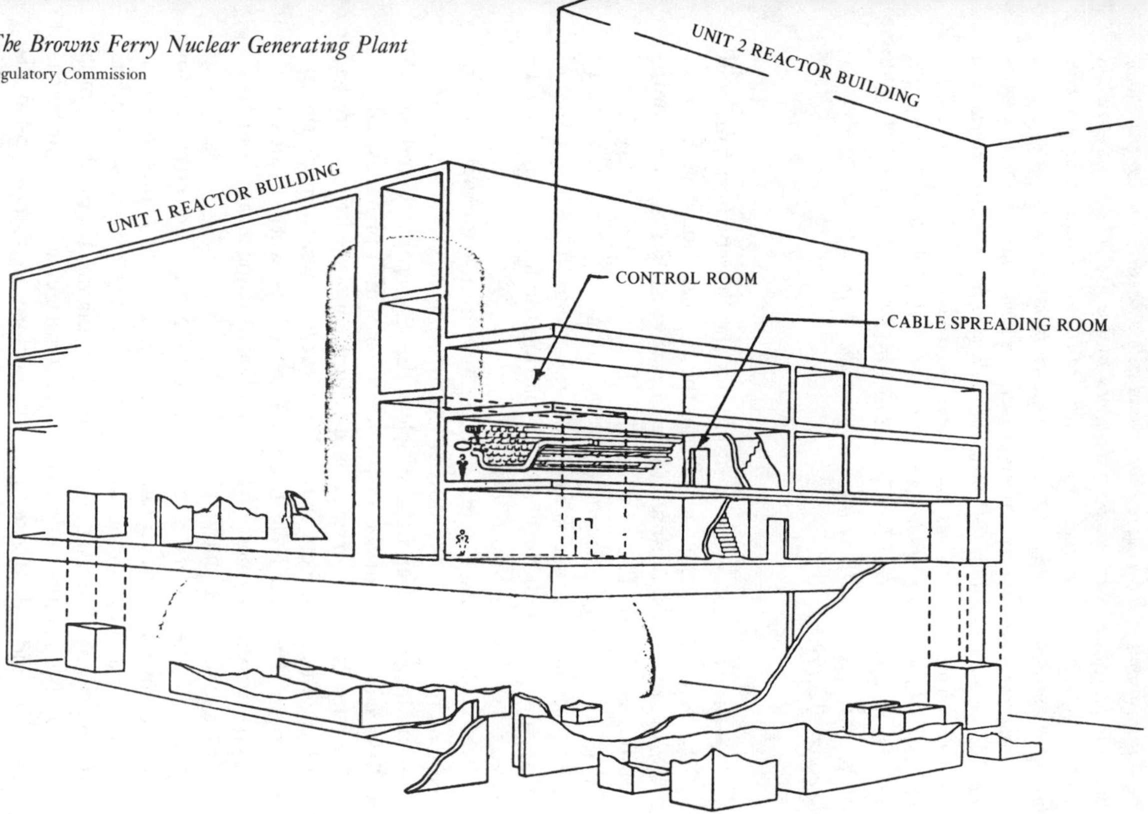

Luckily some relief valves that would allow pressure in the reactor of Unit 1 to be reduced had not been damaged by the fire. Use of these valves lowered the pressure in the reactor and allowed the use of an auxiliary pump, through a makeshift connection, to force water into the reactor. This ad hoc remedy was able to keep the uranium fuel core cool and prevented a meltdown. Despite the success of the makeshift cooling unit, there are experts who believe that Browns Ferry was as close as a few hours away from an accident that would have released radioactivity to the public.[1] *It had thus taken a worker with a four-inch candle to reveal that the nuclear industry had developed for fifteen years without implementing the means necessary to mitigate the effects of a fire.*

AN ACCIDENT WAS AVOIDED, but the Browns Ferry fire will have serious effects on the nuclear industry. For one, Browns Ferry was shut down for repairs for nearly eighteen months after the accident. The cost to the Tennessee Valley Authority (TVA) for buying replacement power alone while Browns Ferry was out of commission was $18 million per month.[2] These costs, of course, along with all the costs of repairing the fire damage, will be passed on to TVA's consumers.

But a bigger black eye for the atomic industry came from inspections by the Factory Mutual Engineering Association, a group of fire underwriters, and the Nuclear Energy Property and Liability Association, one of the insurance pools which provides liability coverage for Browns Ferry. These groups concluded that there had been basic design defects at the TVA reactors which made the fire worse than it should have been.[3] The defects were serious ones, quite apart from the questionable practice of using a candle to check for air leaks near flammable packing.

Some of the defects, apparently defying common sense, are not unique to Browns Ferry. Their correction at other plants may cost the atomic industry millions of dollars.[4] The fire also caused experts to ask questions about sabotage at nuclear plants. If a well-meaning worker with a candle could cause such damage, what might be the effect of an individual attempting to perform deliberate and malicious damage at a nuclear plant?

The Basis for Concern

The Brookhaven National Laboratory completed a study for the Atomic Energy Commission in 1957, indicating that the release of radioactivity from a nuclear accident could kill 3,400 people, injure 43,000, and cause $7 billion in property damage. This study represents the reference point for the debate on reactor safety. The Atomic Energy Commission and its successor, the Nuclear Regulatory Commission, have for twenty years attempted to brush away this Brookhaven study. But its conclusions have never been satisfactorily refuted. The study is often referred to by its AEC catalogue number, WASH-740.[5]

The source of public danger from a nuclear power plant is the amount of radioactive material it contains. As explained earlier, radioactive fission products accumulate in the plant's fuel rods over the life of the plant. In addition to the fission products, isotopes, called "by-products," are also formed. Where the fission products are formed by the splitting of a uranium atom, by-products are formed by "neutron absorption." If a free neutron is absorbed by a U-238 atom, for example, it can be transformed into plutonium (Pu-239) or other isotopes. The by-products are always of higher atomic weight than uranium, while the fission products are always of lower atomic weight. Both types of "products," however, can be highly radioactive, and the amount of radioactivity in an operating reactor core can be prodigious. In fact, the radioactive inventory in a large nuclear power plant can represent 1000 times more radioactivity than was unleashed by the Hiroshima bomb.[6] Here it should be repeated that a light-water reactor cannot explode like a nuclear bomb. The uranium in a reactor is only 3 to 4 percent U-235, while the warhead of a nuclear weapon is 90 percent or more U-235. But radiation can be released to the general population either from the routine emissions of a reactor (referred to as low-level radiation) or from accidents which theoretically could burst the reactor vessel and the containment building, providing a path for catastrophic radioactive release to the public.

The most feared nuclear accident sequence is the Loss of Coolant Accident (LOCA). This would begin with the rupture of one of the pipes carrying water to the reactor. Although the reactor would shut down, heat would continue to be generated by fission products in the core. Without water to cool it, the core would begin to melt until the fuel slumped to the bottom of the containment building, breaching it and escaping into the environment. The Browns Ferry fire, it should be noted, was not a LOCA but could have resulted in a meltdown had the fire not been controlled.

A meltdown sequence could have secondary effects. Interactions of molten uranium and water could cause hydrogen and steam explosions. The molten core could also generate carbon dioxide as it melted through the reactor containment building, with the buildup in pressure from the carbon dioxide gas eventually rupturing the containment. Mechanisms such as these, once breaching the containment building, would release radioactivity to the population within a few hours after the beginning of a LOCA. Even if the reactor building remained intact, the fuel would continue to melt its way through the building and into the ground, eventually coming to rest in the ground underneath the reactor building. Since about 20 percent of the radioactive inventory is gaseous, these isotopes would gradually leak back through the soil to the surface, whence they would be carried by the winds and weather to the surrounding land area and population. This in turn could lead to the consequences predicted by WASH-740, or worse.

Atomic power promoters acknowledge the potential for a catastrophic accident but argue that such an event is very unlikely. First, they insist that the Emergency Core Cooling System (ECCS) and other safety systems would prevent a meltdown should a LOCA occur. Next, they point out that WASH-740 was supposed to examine the worst possible accident and thus made very unrealistic assumptions. Lastly, they suggest that reactor accident studies made since WASH-740 place the consequences of reactor accidents in a more realistic perspective by demonstrating that the probability of an accident is extremely low.

Each of these arguments has a refutation. There is a significant scientific controversy over the ability of the ECCS to function (examined later in this chapter). The latest reactor safety study produced by the nuclear establishment has also caused controversy and has been widely criticized. And then there is the Price-Anderson Act, whose existence indicates that even the nuclear industry does not believe its own reactor safety propaganda.

The Price-Anderson Act establishes a three-tiered system of

limited liability for nuclear plant accidents. Each operating nuclear plant is required to obtain the maximum available private liability insurance, which at present is $140 million. If an accident were to occur, the second tier would be provided by "restrospective premiums" of $5 million, which would be assessed each operating nuclear power plant. (For sixty operating reactors the retrospective payment would be $300 million, and the total of private insurance plus retrospective payments would be $440 million.) If the first two tiers of payment total less than $560 million, then the federal government would assume responsibility for the difference. In any case, the total nuclear accident liability is limited to either $560 million or the sum of the first two tiers, whichever is larger. Either liability limit is clearly inadequate to cover the damages of a WASH-740 type accident. (Even with 1000 reactors operating, the retrospective premium would provide only $5 billion, less than the WASH-740 accident property damage alone.)

If the damages from a nuclear accident were to exceed the liability limit, compensation paid to victims would be on a proportional basis. That is, if the damage estimate were $5.6 billion and the liability limit $560 million, each victim would be likely to receive no more than ten cents for each dollar lost in the accident (due to property damage, medical care for injuries, or deaths.) So it is clear that in the benefit-risk analysis of nuclear power, the industry takes the benefit and the potential victims take the risk. The privileges of limited liability and government indemnity are unique to the atomic power industry; no other industry has such shields. It is also significant that Price-Anderson was passed in 1957 as an inducement to electric utilities to build nuclear power plants, since the utilities were unwilling to use a technology which would lead to large liability payments in the event of an accident.

The Nuclear Regulatory Commission and industry continue to claim that the chances of a reactor accident are very slight: one in a million per year or less. So the question which must be put to the nuclear industry is: If the probability of a reactor accident really is so low, why will the industry not accept full liability for such an "incredible" accident? If the industry statements were, in fact, true, then the industry would be risking very little by accepting liability for an accident that, practically speaking, will not occur. But the fact that the nuclear industry continues to demand the protection of Price-Anderson indicates that it believes a catastrophic accident to be a distinct possibility. Some of the more scientific and technical aspects of nuclear plant safety are discussed in following sections.

The ECCS Controversy

Henry Kandall and Daniel Ford, members of the Union of Concerned Scientists (UCS), in Cambridge, Massachusetts, have been consistently active thorns in the side of the atomic industry. Kendall, professor of nuclear and high-energy physics at the Massachusetts Institute of Technology, has impressive scientific credentials. Ford, not yet thirty years old, is a self-taught expert in nuclear reactor technology. Together they have developed a highly credible challenge to the technical adequacy of reactor safety systems.

The UCS is a coalition of Boston-area scientists, engineers, and other professionals that includes faculty members of MIT and Harvard University. The organization was established in 1969 as an outgrowth of scientific opposition to the Vietnam war, classified university research, and environmental pollution. Kendall was an early member and chairman of the UCS Committee on Environmental Pollution, working on nuclear power and other environmental issues. Ford, an undergraduate at Harvard, was coordinator of environmental research for the Harvard Economic Research Project. In that capacity, he was investigating the financial structure of the nuclear power industry.

In1971 Ford, who was about to graduate from Harvard with a bachelor's degree in economics, went to Kendall with some information which seemed to have a bearing on reactor safety. During his research, he had run across some reports that suggested that Emergency Core-Cooling Systems (ECCS) might not work. That information led to nearly two years of hearings on ECCS adequacy and, in turn, revealed the existence of serious skepticism within the scientific community about nuclear power.

The documents that Ford took to Kendall concerned some tests that had been run at the Atomic Energy Commission's research reactor "reservation" near Idaho Falls, Idaho, to study phenomena related to the functioning of ECCS. In a Pressurized Water Reactor (PWR), one ECCS component would use pressurized nitrogen to force water into the core. Another part of ECCS would use pumps to force water into the core. In a Boiling Water Reactor (BWR), all the water would be force-fed by pumps. Part of the water would go through a spray system to be sprinkled down on the core. Figures 2 and 3 are schematic drawings of PWR and BWR ECCS.

The Idaho tests had been carried out on a semiscale mockup similar to the ECCS in a PWR. The mockup included a simulated reactor core about nine inches high and nine inches in diameter. Heat was provided by electric heaters, to simulate the residual heat in a reactor after it was shut down. The tests initiated a piping break, then operated a simulated ECCS system to flood the core. But only negligible amounts of emergency core-cooling water ever reached the core in any of the tests. In five out of five cases, the mockup ECCS had failed.

Following are the descriptions of the semiscale tests conducted from November 1970 to March 1971, as given by Idaho Nuclear, the company responsible for the experiments:

> Test 845: Early analysis of test data indicates that essentially no emergency core coolant reached the core.[7]
>
> Preliminary analysis for Test 846 indicates little or no core cooling by the emergency coolant.[8]
>
> Tests 847 and 848: Preliminary analysis of the results of these tests indicates little or no core cooling by the emergency coolant.[9]
>
> ECC liquid was ejected from the system in Test 849, as in previous tests with accumulator ECC, and at no time did ECC liquid reach the core.[10]

The reasons that no ECC water reached the core are complex and perhaps not completely understood. In simple terms, it appears that as the pipe break was initiated, the water under pressure in the core suddenly became subject to lower pressures and "flashed" to steam. The steam expanded, preventing water injection into the core. This expansion was greatest in the hottest areas of the core. The ECC water thus appears to have entered the pressure vessel, but was forced right back out the pressure vessel, through the original pipe break.

Information about the Idaho tests did not surface to the general press until about May 1971. By this time the AEC was attempting to downplay the semiscale tests on the grounds that any small-scale mockup was not directly applicable, in size or configuration, to large-scale reactors.[11] While this observation may have been true, it was clearly a case of applied hindsight. If AEC officials had expected the semiscale tests to fail or be inapplicable it is doubtful they would

Figure 2. *Schematic of PWR Emergency Core-Cooling System*

Nuclear Safety, January–February 1974, p. 32

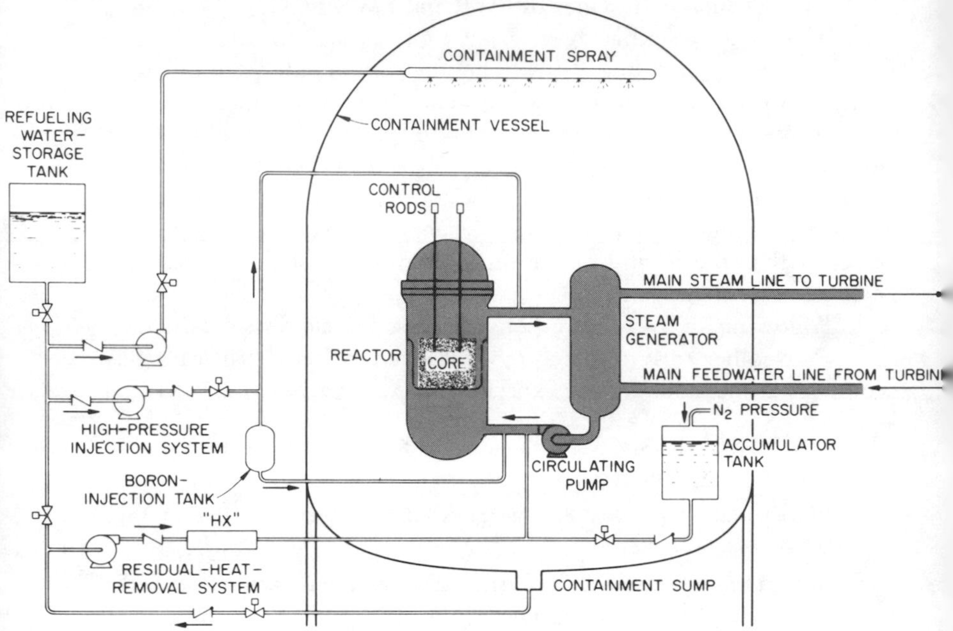

Figure 3. *Schematic of BWR Emergency Core-Cooling System*

Nuclear Safety, January–February 1974, p. 33

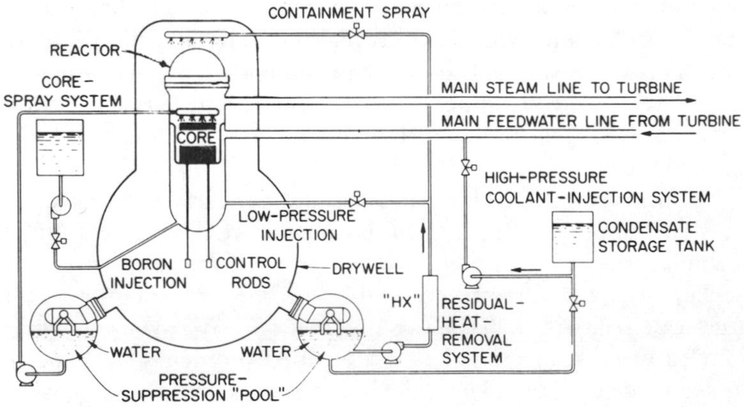

have allowed the tests to proceed. Research engineers could not be expected to keep their jobs very long if they designed experiments they knew would not work. Meanwhile, the AEC was taking some steps that indicated more concern than its public reassurances. In June 1971, the agency issued Interim Acceptance Criteria that would have to be met by ECCS designers, and made the criteria immediately effective, an unusual step for the agency. The AEC had waived the waiting period that normally accompanies proposed federal regulations to allow for public comment.

In July 1971, the Union of Concerned Scientists (UCS) released a report on its evaluation of the Idaho tests. Even under the questionable assumption that the tests were not applicable to large reactors, the UCS wrote, they still were a black eye for the AEC. Furthermore, experimental uncertainties with ECCS had been recognized as early as 1968. Because the semiscale tests were the only experimental tests carried out by the AEC, they were therefore the only indication of how the ECCS would work. UCS called for a total halt to the issuance of operating licenses for nuclear power plants until assurance of ECCS effectiveness could be demonstrated, and until a thorough review of ECCS had been carried out by an independent qualified group.[12]

With the uncertainties about ECCS, the AEC's vaunted "defense in depth" concepts were damaged. The AEC maintained that reactor safety consisted of several lines of defense: that reactor coolant pipes were constructed to high-quality standards to ensure they would not break; that, should the pipes break, the ECCS would prevent a meltdown accident; that, were any radioactivity released from the fuel, the containment building around the reactor would prevent it from reaching the public. But UCS recognized that if ECCS would not function, the same accident that broke a pipe would cause the ECCS to fail, and in turn lead to events that could rupture the reactor building. With the failure of the semiscale tests the "defense in depth" principle was supported primarily by computer models—very little experimental work had been done to back up those models.

The controversy continued to escalate. In October 1971, UCS released another report, which evaluated the AEC's Interim Acceptance Criteria and found them inadequate. In rebuttal to the AEC's claim that the semiscale tests were inapplicable to large, modern reactors, UCS pointed out that the AEC had previously used the semiscale data to develop and check its computer models. The UCS again recommended reactor licensing be suspended until "assured performance" of reactor safety systems could be provided.[13]

Citizen groups intervening in individual nuclear plant licensing cases began raising the question of ECCS effectiveness.[14] By December 1971 one of the AEC's reactor licensing boards, conducting a hearing on Indian Point Unit 2, informed the AEC that the board had serious questions about the technical and legal validity of the ECCS design standards.[15] Under these circumstances, the AEC was forced to convene hearings on ECCS effectiveness as a generic issue.

The ECCS hearings began in February 1972 and were not completed until October 1973. The chief participants in the hearings were the reactor vendors (Westinghouse, General Electric, Babcock & Wilcox, Combustion Engineering); a group of utilities with nuclear plants called the Consolidated Utilities at the hearings; and a coalition of some sixty citizens' groups which had been active in licensing hearings at individual power plants and who joined with the UCS to form the Consolidated National Intervenors (CNI).[16]

The hearings, a tedious process that generated 22,000 pages of transcript, were characterized by considerable legal wrangling. As an example, the Consolidated Utilities in July 1972 objected to Daniel Ford's participation in the hearings as a witness because he was not an engineer but an economist. This objection was raised in spite of the fact that Ford had for months conducted cross-examination of witnesses, demonstrating that he had become an expert on ECCS and winning the respect of other participants.[17] The hearing board subsequently ruled that Ford could appear as a witness, but only in response to questions concerning parts of CNI testimony which he wrote or helped write. The irony of the board's restrictive ruling was that at the time, the chairman of the AEC, James Schlesinger, was also a Harvard-trained economist, not a nuclear scientist or engineer.[18]

As the hearings progressed, it became clear that reactor safety researchers within the AEC had serious doubts about the ability of ECCS to function and about the adequacy of the AEC's research program to evaluate ECCS.[19] Testimony of researchers under cross-examination, and AEC internal memoranda which CNI forced into the public record, supported the Union of Concerned Scientists' fundamental criticism of ECCS.[20] For example:

- *June 1971:* Dr. Morris Rosen, then technical advisor to the director of reactor licensing at the AEC, wrote that the "consummate message" from AEC safety analysis was that reactor safety system performance "cannot be defined with sufficient assurance to provide a clear basis for licensing."[21]

• *August 1971:* George Brockett, a leading ECCS researcher identified by the AEC regulatory staff as one of the country's leading experts in reactor safety technology, had written that present AEC reactor safety analysis was "unverified," "inadequate," "incomplete," and "uncertain."[22]

• *April 1971:* J. Curtis Haire, the man who was in charge of the AEC's primary research effort on reactor emergency cooling systems, testified that the AEC was "censoring" reports from his safety research laboratory to prevent Congress from raising embarrassing questions about nuclear power plant safety.[23]

• *February 1971:* An AEC internal memo from Milton Shaw, at the time head of the civilian nuclear power program, to the general manager of the AEC, stated: ". . . no assurance is yet available that emergency coolant can be delivered at rates intended and in the time period prior to clad and subsequent fuel melting due to decay heat generation."[24]

• *February 1972:* Dr. Alvin M. Weinberg, director of the Oak Ridge National Laboratory, wrote to AEC Chairman Schlesinger: "As an old-timer who grew up in this business before the computing machine dominated it so completely, I have a basic distrust of very elaborate calculations of complex situations, especially where the calculations have not been checked by full-scale experiments. As you know, much of our trust in the ECCS depends on the reliability of complex codes."[25]

As the evidence mounted in the ECCS hearings, other independent scientific groups began to voice strong doubts about ECCS, and in turn about the civilian nuclear power program:[26]

• *May 1972:* The Reactor Safety Committee that advises the German (Federal Republic) government recommended a moratorium on reactor operating licenses until more research on ECCS was completed.[27]

• *February 1973:* The Federation of American Scientists, following a review of the ECCS issue, identified extensive deficiencies in AEC safety research and called for broad restrictions on nuclear power plant operation together with increased emphasis on alternative power generation technologies.[28]

• *September 1972:* The RAND Corporation stated in a report to the California Assembly that ECCS effectiveness was an "unresolved" issue and that "until those questions are resolved, it is unwise to plan for the rapid proliferation of nuclear power plants."[29]

• *September 1972:* The AEC's Advisory Committee on Reactor Safeguards wrote to the AEC with an urgent recommendation on the need for "significant improvements" in ECCS.[30]

• *May 1973:* The California Assembly's Advisory Committee on Science and Technology reported: "Not enough information exists to determine if the [ECCS] systems are adequate."[31]

• *June 1973:* Swedish government scientists investigating ECCS systems on reactors that Sweden purchased from U.S. manufacturers concluded that ECCS effectiveness has not yet been established.[32]

• *September 1973:* The Pugwash Conference consists of a group of distinguished scientists concerned about nuclear weapons and nuclear power. In a report, the Pugwash Conference Working Group on Radioactive Pollution of the Environment concluded: "Owing to the potentially grave and as yet unresolved problems related to waste management, diversion of fissionable material, and major radioactivity releases arising from accidents, natural disasters, sabotage, or acts of war, the wisdom of a commitment to nuclear fission as a principal energy source for mankind must be seriously questioned at the present time."[33]

What was the reaction of the AEC's commissioners to the ECCS controversy? In October 1973, the commissioners heard oral arguments from the hearing participants, and by December 1973 the commission had released its decision on the ECCS hearings.[34] Its Final Acceptance Criteria for ECCS design standards were only slight modifications of the Interim Criteria. The entire question of whether ECCS would actually work, or when experimental verification of ECCS would be available, was left up in the air.

The question of ECCS functional capability, if it is answered at all, will depend on the completion of the Nuclear Regulatory Commission's long-term reactor safety research program. A major component of that program is the Loss of Fluid Test (LOFT) facility. LOFT construction began in 1963, and after schedule slippages totaling nearly ten years, the facility was finally completed in December 1975.[35] Non-nuclear experiments (in which no attempt is made to simulate after-shutdown heat) began in December 1975. If the non-nuclear tests are successful, nuclear fuel will be loaded and nuclear tests will begin about April 1977.[36]

Even persons within the industry recognize, however, that LOFT will leave questions about ECCS unanswered. LOFT is one-sixtieth the size of a large, modern reactor; and some of the components of the facility are not scaled to the same proportions as in a reactor.[37] LOFT will provide more information on emergency cooling systems, but only testing of full-scale systems will determine whether ECCS could prevent a core meltdown. Clearly, the viability of ECCS will remain in question for some time.

The commission decision on ECCS was a disappointment to the Consolidated National Intervenors, but not completely unexpected. In May 1972, a memo entitled "Hints at Being a Witness" had been circulated among AEC personnel who testified at the hearings. Hint number 10 was "Never disagree with established policy."[38] The commission's decision was merely a reflection of that order.

Some atomic energy promoters believe that even the failure of ECCS is not important, that the chances that ECCS would be needed are very remote, and that even if ECCS did fail, the containment building would prevent radioactivity from reaching surrounding areas. It should first be pointed out that this belief is inconsistent with the atomic establishment's own studies: The NRC's Reactor Safety Study* concluded that the chances of a meltdown were much higher than had previously been recognized; the study identified several mechanisms by which a meltdown could breach the containment building and recognized that even if the containment were not breached above ground, melting fuel would pass through the containment floor and into the ground, from which its gaseous components would diffuse through the soil to irradiate land and people. Moreover, ECCS is part of the NRC's "defense in depth": if ECCS failure is conceded, then the "defense in depth" concept, the NRC's basis for reactor safety, is severely damaged.

The Union of Concerned Scientists had in a sense lost the battle but won the war. Their efforts did not stop reactor plant licensing, but they publicly illuminated technical dissent within the atomic establishment. UCS established themselves as technically credible and respected critics of the atomic establishment. Their efforts would continue, drawing increasing support from the scientific commu-

* The Reactor Safety Study is examined in the next chapter.

nity. By August 6, 1975, 2,300 scientists had signed a petition, drafted by UCS, to Congress and the president that called the dangers of nuclear power "altogether too great" and urged a "drastic reduction" in nuclear plant construction, along with greater efforts to develop a non-nuclear energy future for the nation.[39]

In June 1976, the ECCS controversy was re-opened by one of the Nuclear Regulatory Commission's own consultants. Keith Miller, professor of mathematics at the University of California at Berkeley, was one of ten consultants to the NRC on the computer programs that are supposed to predict ECCS performance. Miller cited a wide divergence between the NRC's public posture that there was "almost absolute certitude" that ECCS would function, and the private opinions and doubts of persons within the NRC and the industry. Miller concluded that the ECCS computer simulations were no more reliable than tomorrow's weather prediction. 40

Miller called for a halt in licensing nuclear reactors until there could be experimental verfication of the NRC's computer codes. The NRC, as might be expected, played down Miller's concerns and declined his recommendations. But the commission, try as it might, could not stop the growing technical controversy on nuclear safety.

Reactor Safety: Hidden Documents and Accident Studies

During a debate with Daniel Ford, an atomic promoter accused Ford and others of opposing nuclear power because they were afraid of the unknown. "Quite the contrary," replied Ford. "It is the information that is known that alarms us." He added that some of the most alarming information on nuclear power has come from the federal government's own documents.

Robert Pollard was a thirty-six-year-old engineer working for the Nuclear Regulatory Commission (NRC). With a wife and two children, a house in the Washington, D.C., suburbs, and over six years of experience with the NRC, Bob Pollard might have been just another federal employee on his way up. But he believed that the NRC was ignoring reactor safety problems, following the old Atomic Energy Commission's practice of "doubts in private, reassurance in public." The NRC, he thought, was promoting reactor safety in its public statements, but was shortcutting reactor safety in its day-to-day operation. By January 1976, Robert Pollard could no longer harbor his own doubts in private. He resigned from the Nuclear Regulatory Commission and, taking a pay cut, began working for the Union of Concerned Scientists. He believed he could work more effectively for reactor safety from outside the NRC.

Problems at the NRC

POLLARD'S resignation resulted in the lawful release of more internal government documents on problems with nuclear power. Over the years, much of the information most damaging to the nuclear power program has come from the files of the

Atomic Energy Commission and Nuclear Regulatory Commission. This information has surfaced mainly through leaks or as a result of Freedom of Information requests by citizen groups. During the ECCS hearings, for example, Daniel Ford and Henry Kendall received, from insiders, several documents on reactor safety problems. The implications of the material which has been disclosed have been so serious that one must wonder whether even more damaging information is still hidden.

Before leaving the NRC, Pollard completed a report on safety problems at Consolidated Edison's Indian Point reactor plants 2 and 3, just north of New York City. Pollard had been project manager, with responsibility for the review of reactor safety design at these plants. His resignation report underscored many safety problems which he thought had not been resolved at Indian Point at the time the NRC granted operating licenses to Consolidated Edison. His report was based on sixty-one internal NRC documents subsequently released.

Pollard's criticisms identified four specific issues which in his judgment, had not been resolved when the NRC granted licenses for Indian Point:[1]

1. Containment Isolation. In the event of a reactor accident, it is necessary to isolate piping which penetrates the reactor containment structure, in order to prevent any radioactivity released from the reactor from escaping outside the containment through the piping penetrations. To accomplish this, the piping penetrating containment has valves which may be shut to seal off the piping.

The NRC's own regulations require that containment piping isolation must be accomplished by valves which are locked shut in normal operation, and/or valves which automatically shut in the event of an accident. At Indian Point 3, however, to isolate containment piping, several valves will have to be shut by a reactor operator manipulating switches in the control room. Bob Pollard showed that the NRC had allowed this situation, in violation of its own regulations:

> I questioned those staff personnel with specific expertise in the reactor containment area about their bases for accepting the Indian Point 3 design. Their responses indicated that: a) it was known that the design did not meet the General Design Criteria, b) the design was not different than other licensed nuclear power plants, and c) it was too late to require design changes to the plant.[2]

2. *Submerged Valves.* Many valves located within the reactor containment building are operated by electrical motors, which would be damaged by water. Pollard's report summarizes this problem succinctly:

Basically, this problem is that following an accident, much of the water from the reactor coolant system and from operation of the emergency core cooling systems collects in the containment. Recently, it has been discovered that many valves located inside the containment, including some valves intended to be used to mitigate the consequences of accidents, could become submerged and, thereby, rendered inoperable. Why the vendor, applicant, or staff did not discover this problem over the past years is a question worth explaining for the future, with the aim of preventing similar fundamental oversights.[3]

3. *Pump Flywheel Missiles.* During a loss of coolant accident, water could rapidly pass through a reactor coolant pump causing the coolant pump to overspeed, break apart, and fling missiles throughout the plant, damaging other equipment. Pollard commented on this problem:

. . . the potential for missiles from pump overspeed remains an unresolved safety problem for Indian Point 2 and 3, as well as other plants. Based on the files concerning review of the Westinghouse topical report, WCAP-8163, the status of resolution is that, as of August 13, 1975, the staff is waiting for information. I believe this matter should be reconsidered in connection with continued operation of Indian Point 2 and commencement of operation of Indian Point 3 as well as a similar reconsideration in connection with all PWRs.[4]

4. *Separation of Electrical Equipment.* In general, reactor equipment should be designed so that the failure of one system or component will not affect the performance of its duplicative counterpart. One example of poor separation was the Browns Ferry fire, in which primary and backup safety systems were both disabled by the fire because the cables for the systems were not physically separated from one another. An event such as a fire, which can damage several systems at once, is called a "common mode" failure. According to Robert Pollard, separation of components is also a significant problem at Indian Point:

Based on my knowledge of the Indian Point 2 and 3 designs and the current separation criteria, I conclude that the physical separation provisions at Indian Point 2 and 3 are not adequate for the health and safety of the public. There is no adequate basis for concluding that a common mode failure will not result in a very serious accident other than sheer good luck.[5]. . .

The two reports to the ACRS prepared by the staff and classified as 'Official Use Only' . . . should be reviewed by NRC to determine whether the previous bases for reluctantly accepting design deficiencies are adequate for protecting the health and safety of the public. Based on those reports, it appears that many items were accepted solely because so many other areas of the plant were deficient that it wouldn't do much good to require upgrading only a few. In other cases, it appears that a judgment was made that the cost in time and money needed to provide substantial additional protection for the public health and safety was too great.[6]

Pollard's concern about Indian Point went beyond these four technical items. In an earlier letter to then NRC Chairman William A. Anders, Pollard had stated:

In addition, the Indian Point units should be reviewed to determine: a) the acceptability of the Indian Point site from both a population density and a seismic [earthquake] point of view, b) the feasibility of swiftly evacuating large segments of metropolitan New York, and c) the susceptibility of the plants to sabotage.[7]

Unresolved problems existed at other reactors as well. The NRC published a quartely "Technical Activities Safety Report" listing problems at several reactors. Pollard cited the December 1975 report, which listed nearly 183 specific serious unresolved safety problems as "currently receiving attention, [and] which have an important impact on the licensing review process." Another forty-four equally serious unresolved safety problems are described as "requiring NRR [Office of Nuclear Reactor Regulation] attention, but review has not been initiated because of manpower limitations or information is not available." A third category of eight serious unresolved safety problems involves technical safety activities "planned for the future that would improve the quality of the review to facilitate the review process."[8]

Pollard's conclusion on all these problems was:

These generic unresolved safety problems are so fundamental to the basic evaluation of reactor safety that it is not possible to conclude on a technical basis that operation of any nuclear reactors is safe enough to provide reasonable assurance of adequate protection for the public health and safety.[9]

The technical problems which bothered Robert Pollard were a manifestation of the way in which the NRC did business. Perhaps even more revealing were the complaints Pollard had in his letter to Chairman Anders, on the ways in which dissent within the agency was muffled:

The plain fact is that many of the dedicated government employees in the NRC are deeply troubled about the pervasive attitude in the NRC that our most important job is to get the licenses out as quickly as possible and to keep the plants running as long as possible. . . .

Until you have been a part of the agency at a level where the technical work is performed, it is difficult to appreciate the kind of pressure that inhibits your staff from doing its job. This pressure has become particularly intense in the last year with the increased emphasis from the Office of Nuclear Reactor Regulation to shorten the reviews for the sake of a shorter review and to meet review deadlines at all costs. This has often resulted in accepting what would otherwise be unacceptable solutions to serious safety problems. . . .

For instance, I recently heard of a very senior safety expert who is having great difficulty in releasing a critical safety report because his reviewer wants to tone it down. . . .

Also included with this letter is a Report on the problems of dissent within the agency. . . .

The conclusions of that Report are that by the use of highly effective pressures, middle management totally suppresses most of the dissent within the NRC. In addition to outright threats of adverse consequences for one's job, pressures are applied by the device of requiring numerous rewritings of proposed position papers which conflict with rapid licensing and continued operation of reactors—even though there are no identified technical errors in the papers. Another frequently used device is to claim that when technical positions conflict with the licensing policy, they involve policy issues better left to middle-management to resolve. Those policy issues inevitably involve economic considerations.[10]

Pollard's report has shown that the NRC, which was officially established in January 1975, quickly developed into a poor regulator of the nuclear industry. But that is less surprising when one considers that the NRC was spawned by that monolithic atomic promoter, the Atomic Energy Commission.

The WASH-740 Update

Perhaps the most dramatic example of the AEC's suppression of its internal documents was the WASH-740 reactor accident study update, which the AEC commissioned in 1964. WASH-740 had concluded in 1957 that an accident could kill 3,400 people, could injure 43,000, and could cause $7 billion in property damage,[11] potential damages that overwhelmingly dwarfed the $560 million liability of the Price-Anderson Act.

The AEC apparently hoped that an updated accident study would find the damages from a reactor accident to be much smaller than the WASH-740 figures. If such a conclusion could be reached, extension of the Price-Anderson Act could be greatly facilitated. The act was to have expired in 1967, but by 1965 the AEC wanted to extend it for another 10 years.

By late 1964, the Steering Committee, a group of AEC scientists responsible for preparing the update, realized that the results were not going to favor the nuclear establishment: a reactor accident could cause 45,000 deaths and 100,000 injuries.[12] Property damage could be $17 billion or more.[13]

At a meeting in December 1964 the Steering Committee reached the following decisions:

> The results of the study must be revealed to the Commission and the JCAE without subterfuge although the method of presentation to the public has not been resolved at this time. . . .
>
> The results of the study suggest that the Price-Anderson liability level should not be reduced. Rather, an increase by a factor of 40 is suggested by the calculations (280 billion).[14]

In March 1965, Clifford K. Beck, AEC deputy director of regulation, sent the commission the following information on the update:

> Reactors now being considered for construction are five or six times as large as the 500 MW reactor considered in the first Brookhaven report; the fuel lifetime is three or more times larger than that dealt with in the original Brookhaven report. It is an inescapable calculation, therefore, that given the same hypothetical accidents as those considered in the original BNL study [WASH-740], damages would result possibly ten times as large as those calculated in the previous study.[15]

Beck's memo also reported that the Steering Committee had met with a group from the Atomic Industrial Forum (AIF), the industry trade association founded in 1953 to promote nuclear power. The AIF suggested that the update not be released, but, instead, that the commission simply announce that the update was being expanded and would be ready some time in the future.[16]

The AEC was only too happy to accede to the AIF's request. In fact, the AEC went even further than the AIF had suggested in suppressing the report's details. In June 1965, Glenn Seaborg, then chairman on the AEC, sent to the Joint Committee on Atomic Energy (JCAE) a letter on the possible consequences of a reactor accident, stating: "Reactors today are much larger than those in prospect in 1957, their fuel cycles are longer and their fission product inventories are larger. Therefore, assuming the same kind of hypothetical accidents as those in the 1957 study, the theoretically calculated damages would not be less and under some circumstances would be substantially more than the consequences reported in the earlier study."

But in a strange twist of logic, Seaborg then concluded his letter by recommending that Price-Anderson be extended: "Thus in our opinion, the answers to your questions—that the likelihood of major accidents is still more remote, but the consequences could be greater—do not decrease but rather accentuate the need for Price-Anderson extension.[17]

Seaborg's recommendation that Price-Anderson merely be extended, with no changes, ran contrary to the earlier suggestion of the Steering Committee that the liability level be increased by forty times.

With the support of the AEC and the Joint Committee, Price-Anderson was extended with little congressional debate. Having accomplished its task of further promotion of the atomic industry, the AEC left the documents on the update to gather dust. Apparently the AEC hoped that if the update were ignored, it would go away. But rumors about the update had surfaced, and requests started to come in. The first requests came from David Pesonen, a San Francisco attorney.

August 1965 was not the first time the Atomic Energy Commission had heard from Dave Pesonen. For the previous several years he had been involved in an effort to stop a Pacific Gas & Electric nuclear plant at Bodega Head, about fifty miles north of San Francisco.

In 1958, Pacific Gas & Electric had begun bargaining, quietly, for land on Bodega Head. But one individual, Rose Gaffney, who owned about four hundred acres, refused to sell. The utility went to court and initiated condemnation proceedings on her land, but Rose Gaffney refused to give in. She began a court fight which delayed and publicized PG&E's plans for the plant. Although the utility had tried to keep its plans secret for as long as possible, by 1961 it was forced to announce that the Bodega Head power plant would be nuclear. In the meantime, the Northern California Association to Preserve Bodega Head and Harbor, a citizens' organization whose membership eventually numbered in the thousands, had come into existence, headed by David Pesonen, a young Sierra Club staff member.

The Bodega Head site for PG&E's nuclear plant was extraordinary in at least two ways. First, it had unique scenic, scientific, and recreational values which would be destroyed by the plant. Secondly, it was only 1,000 feet from the San Andreas fault, the most active source of earthquakes in the country. Dave Pesonen was able to marshal public opposition to the plant, as well as an impressive body of expert testimony and independent information on the unacceptable seismic hazards at the Bodega site. The final blow to the utility came in 1963, when, upon excavating the plant site, PG&E discovered a fault which would have run right through the reactor itself. On October 30, 1964, Pacific Gas & Electric threw in the towel. In a brief announcement, the utility stated that it had abandoned plans for a nuclear power plant at Bodega Head.[18]

Although Dave Pesonen had stopped the Bodega Head plant, his activism was only getting started. By 1976, he would be chairman of Californians for Nuclear Safeguards (CNS), a statewide coalition of several groups that campaigned for a stringent state nuclear power initiative. Although CNS lost the vote at the polls, they had brought nuclear power before millions of California voters who had never been faced with the issue before—and 1.8 million persons voted in favor of the initiative. In August 1965, however, these events were still in the future, and Dave Pesonen was only trying to find out about the AEC's new accident study.

Pesonen had attended a speech by AEC Commissioner John Palfry to the Atomic Industrial Forum in San Francisco. In his speech, the commissioner had mentioned the WASH-740 update. In August 1965 David Pesonen requested a copy of the update from Commissioner Palfrey. Palfrey responded by sending Pesonen a copy of Glenn Seaborg's June 1965 letter to the JCAE.

But Pesonen persisted, repeating his request and asking when the update would be made available to the public. In October 1965, Commissioner Palfrey replied that "no report is in existence or contemplated" and that it was the commission's judgment that "no detailed refiguring of the entire report was needed" to provide the Joint Committee with answers on Price-Anderson. Not satisfied with this response, Pesonen wrote an article for *The Nation* in which he concluded that the AEC had abandoned the update for reasons of public relations and a desire to extend Price-Anderson.[19]

Within days, the AEC had drafted a dissembling response to members of Congress and the public who made inquiries as a result of Pesonen's article. The response repeated Commissioner Palfrey's statement that "no new report is in existence or contemplated" and that "no detailed refiguring" of the 1957 report was needed to provide the Joint Committee with information on Price-Anderson. Inquiries were also referred to Chairman Seaborg's June letter to the JCAE.[20]

By 1969, the AEC had elaborated its standard response and had even prepared for members of the commission who might testify on the subject a background paper which advised:

> Attached is a draft of the material you asked me to prepare on WASH-740 for possible use by members of the Commission in answering questions on the 1964 review of the subject by Brookhaven. . . . The background information provided is considered necessary to prevent a Commissioner unfamiliar with the facts from becoming trapped in an untenable position. As you know an important factor in the decision not to produce a complete revision of WASH-740 along the lines proposed by the Brookhaven staff was the public relations considerations. [*sic*][21]

The recommended answer on questions about Pesonen's article was now more elaborate. But the substance of the answer was to refer to the June 1956 letter from Glenn Seaborg and to state that detailed recomputations of accident damages were not justified. The recommended answer was followed by a background paper which stated the reasons for the update's non-publication. "The principal problem," according to the backgrounder, was that if the results of the study were released in summary form there would be no plausible reason for failure to release the full study, including all computations. "At this stage it was decided to abandon any plan to publish a formal report," and only the general language in the Seaborg letter was transmitted to the Joint Committee.[22] The AEC had decided that it was better to suppress the report than to release the potentially damaging accident computations, thus supporting Pesonen's charges.

Requests for information on the update continued to be made, and the AEC continued to rebuff requests from members of Congress as well as the general public. Finally, on April 18, 1973, a Freedom of Information request for the update papers was submitted on behalf of David Comey, a Chicago environmentalist. On the following day, Friends of the Earth made a similar formal request for the update. Faced with the possibility of a suit, the AEC complied by releasing about 2,100 pages of working papers on the update which supposedly had not been "in existence" in 1965. The documents themselves showed a suppression of the 1964 update which was routine and repeated. The AEC's action was so serious that it drew criticism even from Ralph Lapp, physicist and author, and an outspoken promoter of atomic power: "Accordingly, the AEC became more and more vulnerable to high-consequence es-

The Millstone Nuclear Power Station, Unit 1, on Long Island Sound near Waterford, Connecticut.
(Kevin Donovan Films)

timation of damage by its critics and its failure to update the 1957 report, together with its muffling of a 1965 attempt at an update, gave credibility to the charge of 'cover up.' "[23]

The Rasmussen Report

But at least one more AEC nuclear accident study was yet to come. The AEC recognized that the WASH-740 update could not provide support for its plans to promote nuclear power. So in 1970 the AEC began laying the groundwork for another study, partly as a result of the agency's attempts to blunt inquiries on the update. When Senator Mike Gravel of Alaska asked about the 1964 update, the AEC responded in part by referring to a new study that would cover "the subject area included in WASH-740" and which "would be underway in the next year or so."[24]

In August 1974, the AEC released a draft version of its new report. From the fanfare with which the report was released, it is evident that the AEC and the atomic industry hoped the report would lay to rest all concerns about reactor safety. But the hopes of the nuclear establishment were not to be realized. The severe weaknesses of the report have actually contributed to greater skepticism over nuclear power. The AEC's report was the *Reactor Safety Study* (RSS), sometimes called the Rasmussen Report after Norman C. Rasmussen of MIT, the nuclear engineer who directed it.[25]

At the outset, it is important to recognize that the study covered only nuclear reactors themselves. It did not deal with the transportation of radioactive materials by truck, rail, and barge; or the disposal of radioactive wastes; or the risks of sabotage, theft, or terrorism; or fuel reprocessing plants; or uranium mining processes and wastes.

The RSS attempted to determine probabilities and consequences of major reactor accidents. The study predicted that the chances of a meltdown accident, on the average, would be one in 17,000 per reactor per year, but also concluded that most meltdown accidents would result in insignificant radiation exposure to the public. According to the study, the chances that seventy deaths would be caused by a reactor accident were one in a million per reactor per year. The worst accident which the RSS considered was predicted to occur once every billion years per reactor and would cause 2,300 immediate deaths, 5,600 immediate injuries, and $6.2 billion in property damage.

The study's conclusions were widely criticized; the most extensive and severe comments came from a joint review by the Sierra Club (SC) and the Union of Concerned Scientists (UCS).[26] The SC-UCS critique identified significant weaknesses in the study with regard both to the likelihood of an accident and to the seriousness of its consequences. To predict reactor accident probabilities, the RSS used "reliability estimating" techniques developed by the aerospace industry. SC-UCS pointed out that the aerospace industry had abandoned these techniques as a means of providing exact reliability estimates. As many as 35 percent of the failures in the Apollo program, for example, had not been identified as "credible" before they had happened.

SC-UCS tested the validity of the RSS estimating techniques by applying them to a reactor accident which had already occurred at the Dresden plant in Illinois in 1970. The result was an accident-probability prediction of one in a billion-billion. The fact that this accident had already occurred cast doubt on the study's methods, to say the least. It seems that an accurate assessment of reactor accident probability might not be available until after an accident has occurred—which will be too late.

The SC-UCS review found that the RSS underestimated nuclear-accident health consequences by a factor of sixteen or more. This finding was generally supported by the Environmental Protection Agency[27] and the AEC Regulatory Staff, a separate branch of the AEC which performed its own review.[28] With the SC-UCS corrections, the most severe accident considered by the RSS could cause 36,800 immediate deaths and 90,000 immediate injuries. The overall conclusion of the SC-UCS review was that the RSS was deeply flawed.

The American Physical Society, the national association of physicists, commissioned a review of reactor safety issues which included an examination of the RSS by a panel of twelve scientists.[29] This review destroyed whatever credibility the Reactor Safety Study might have retained after the critiques by SC-UCS, EPA, and the AEC Regulatory Staff. To begin with, the APS study, like the SC-UCS review, expressed serious doubts about the study's techniques for probability estimating. The APS group stated that "we do not now have confidence" in the probability estimates generated by the RSS.[30]

The APS panel chose not to review the consequences of the most catastrophic RSS accident, as this had already been covered by the other critiques. Instead, the APS study looked at a "small" accident, which the RSS predicted would cause 62 immediate deaths, 310 genetic defects, and 310 lethal cancers over the long term. The APS study identified serious faults in the RSS analysis of long-term health effects. The RSS had assumed that persons downwind of a reactor accident would receive radiation only through the first day following the accident. In reality, ground-deposited radioactivity would irradiate people over a large area for an extended period. Correction of this factor alone increased lethal cancers and genetic defects by a factor of 25.

The APS study also found that RSS had neglected cancers to the lungs and thyroid that would result from the inhalation of radioactive material. APS corrected for these errors and others and concluded that the "small" accident would cause 10,000 to 15,000 lethal cancers and 3,000 to 20,000 genetic diseases during the five generations following the accident.[31]

Until the RSS, reactor safety researchers had held that the consequences of a meltdown accident would be severe, but that the probability of such an accident was very low—one in a million or less. The RSS concluded that a meltdown was much more likely to occur, but the consequences of an accident might not be severe. Professor Frank von Hippel, affiliated with the Center for Enrivonmental Studies at Princeton University and a member of the APS panel, assessed the RSS in this manner: ". . . two omissions in the calculations made there [with the RSS] resulted in an underestimate of the average number of deaths from the reference accident by one or two orders of magnitude. Thus it appears that the A.P.S. group may have had the dubious honor of restoring the *status quo:* a reactor melt-down accident with containment failure would indeed be very serious."[32]

On October 30, 1975, the Nuclear Regulatory Commission, the AEC's successor, released its final version of the RSS.[33] This was supposedly an improved version which was to have taken into account all the foregoing critiques of the draft. This final version, however, bore a strong resemblance to the August 1974 draft. The final

Reactor Safety Study made grudging but inadequate responses to its critics.

The worst accident considered by the final RSS would cause 3,300 "early" (as opposed to long-term) deaths, 45,000 early injuries, and $14 billion in property damage.[34] The final RSS thus responded to criticisms by increasing early illnesses by a factor of 9 (i.e., multiplying by nine times), property damage by a factor of two, and early deaths hardly at all. These fell short of the revisions recommended by SC-UCS, EPA, and the AEC Regulatory Staff.

The final Reactor Safety Study responded in a similar manner to the APS review. Estimates of latent cancer from the worst accident were revised upward by a factor of 14, to 1,500 per year. These latent cancers would occur during ten to forty years after the reactor accident, meaning that 45,000 total cancers would occur from the accident. The final RSS figures on genetic effects were also increased slightly. The worst RSS accident in the final version would cause 170 genetic defects per year, for a total of 5,100. Alarming though these figures on long-term effects are, they still fell short of the APS recommendations. APS had felt the draft RSS underestimated cancers and genetic defects by factors of 25 to 60.

Although it revised upwards its estimate of reactor accident consequences, the final Reactor Safety Study still concluded—as did the draft—that the dangers from reactor operation are much less than those from other technologies. The final RSS could reach this conclusion because it doggedly maintained the belief that it could predict probabilities of accidents, and that probabilities of reactor accidents are very low. In sticking to this conclusion, the final RSS either rejected or ignored the criticisms of SC-UCS, the APS group, and others.

In rebuttal, the final Reactor Safety Study included letters from the National Aeronautics and Space Administration (NASA) and the U.S. General Accounting Office (GAO). Norman Rasmussen, director of the study, claimed that these letters support the RSS as a valid use of reliability estimating techniques. At least one observer responded immediately that Rasmussen "may be overstating the case."[35] The NASA letter was at best a lukewarm approval. It called reliability estimating "an effective technique," but also concluded that NASA "is not in a position to validate the numerical assessments in the Rasmussen Study because of the extensive efforts such a validation process would require."[36]

Nor can the GAO letter be considered as strong support of the RSS. In fact, the GAO made many comments which could be considered critical of the RSS methods:

> Many problems are due to design "unknowns" not predictable or quantifiable during development. For example, one NASA official told us that six redundant components had failed on one system.[37]
>
> NASA experts believe that "absolute" reliability numbers are misleading. . . .[38]
>
> As far as we could learn during this brief review, DOD and NASA officials can offer little guidance as to how very rare failures or catastrophic accidents to systems can be anticipated, avoided, or predicted.[39]
>
> NASA goes to extraodinary lengths—reliability cost is hardly an object—to prevent disasters in manned space vehicles and has the singular advantage of vehicle occupants prepared to make onboard repairs. Still, three astronauts were lost in one vehicle. The Soviets suffered similar losses in other attempts. No one can tell if and when such catastrophic failures will be repeated.[40]

Another interesting aspect of the final RSS is that its accident figures of 3,300 early deaths, 45,000 early injuries, and $14 billion in property damage are practically identical to the numbers in WASH-740, the 1957 Brookhaven Lab report. It seems that the NRC and AEC went through a lot of effort just to come up with the same numbers they had produced eighteen years before.

The RSS also made serious omissions when it graphically compared reactor accidents with other risks to society, as can be seen from Figure 1. The RSS defined overall risk as a combination of probability and consequences of accidents. Curve A in Figure 1 comes directly from the final RSS, and is supposed to support the study's conclusion that reactors present a much lower risk than hazards such as fires, air crashes, explosions, dam failures, and chlorine gas releases. Curve A represents only the short-term deaths from reactor accidents—it does not reflect the study's own figures on long-term cancers, which have been drawn in as curve B. When latent cancers are added, the 3,300-death reactor accident becomes a 48,000-death accident, and other reactor accidents also become more serious. When curve B is considered, under the safety study's own definition of risk, nuclear power is similar to other potentially hazardous activities—even without considering the criticism of the RSS probability and consequences calculations.

To be sure, the criticism remains. The Environmental Protec-

Figure 1. *Comparison of Nuclear vs. Non-nuclear Risks*

WASH-1400, *Reactor Safety Study,* U.S. Nuclear Regulatory Commission, October 1975, Main Report, Figure 6-1, p. 119; source for curve B: WASH-1400, Appendix VI, figure VI 13-33, p. 13-45

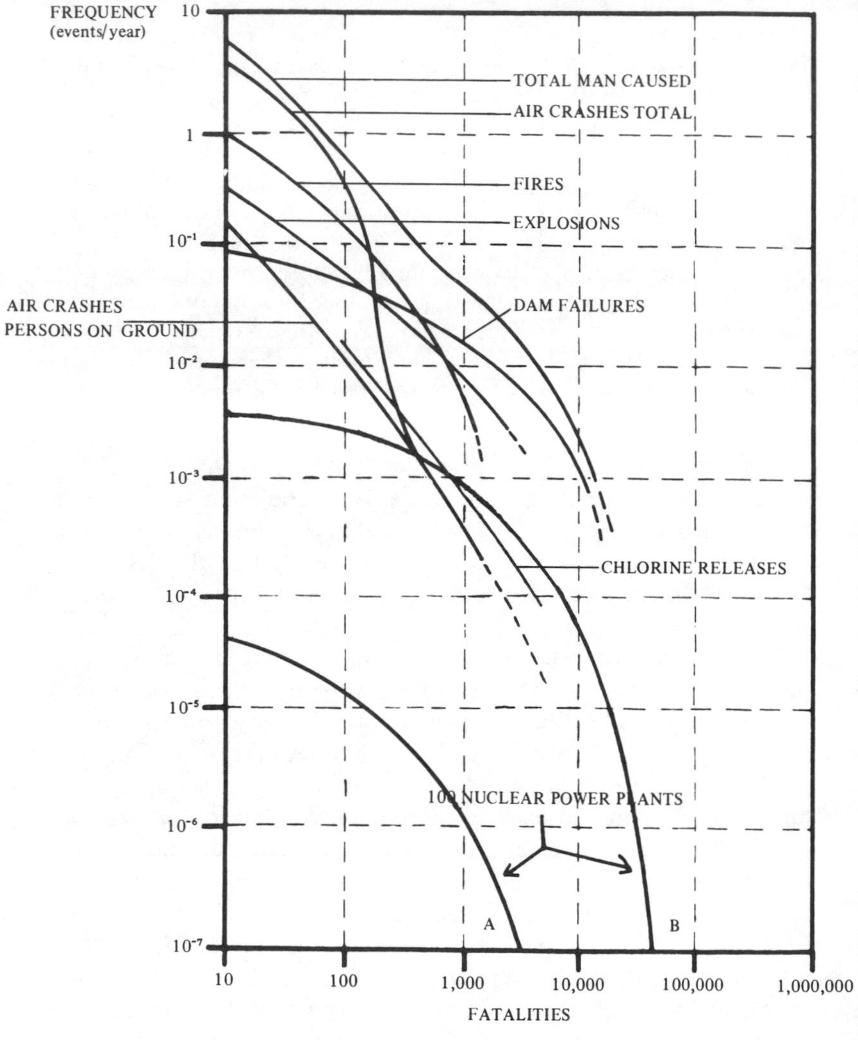

tion Agency completed its review of the *final* RSS in June 1976. EPA concluded that the final RSS may still underestimate the latent cancers from reactor accidents by a factor of 2 to 10—which would mean that the worst RSS accident could easily cause 100,000 or more long-term cancers.[41] Frank von Hippel, a Princeton professor and member of the APS review group, has written that the misleading manner in which the final Safety Study presented its conclusions had "severely damaged" the Nuclear Regulatory Commission's credibility.[42]

The Rasmussen study has become a major part of the atomic industry's propaganda machine. But the study, because of its severe weaknesses is at best a very shaky foundation for the industry. Nor can the study improve the industry's inability to answer the central reactor safety question: If reactors are so safe, why does the atomic industry continue to insist on limited accident liability that is a pittance compared to the potential damages?

Another aspect of reactor safety ignored by the RSS is the interrelated nature of nuclear technology. Even some proponents of nuclear power such as Alvin Weinberg, former director of the AEC's Oak Ridge National Laboratory, acknowledge that for nuclear power to be a viable energy alternative it must be a technology free of catastrophic accident. Were a public catastrophe to occur, at any plant in any step of the nuclear fuel cycle, what would be the reaction of a citizenry who had been told for years that "It will never happen" by government officials and an industry grown smug from its own propaganda? There might be such an outcry that nuclear power plants all over the country would be shut down forever. Even John O. Pastore (D., R.I.), the U.S. Senate's most ardent promoter of nuclear power, has said: "Now, Senators say to me, 'But what if you have an accident where the damages are over $560 million?' I say, 'God forbid. Do you know what would happen? We would close down the whole caboodle.' "[43] Clearly, such a frail technology cannot become a major or reliable energy source, now or in the future.

VIII

The Environmental Effects of Reactor Operation

EVEN WITHOUT a catastrophic accident, the nuclear industry will have significant effects on the environment and future generations. Nuclear power plants and supporting plants in the fuel cycle discharge radioactive substances—through routine releases and unplanned spills, leaks, and malfunctions—to the air and water. The effects of routine radiation have already been reviewed in chapter 4. Nuclear power plants also consume prodigious amounts of water and release enormous amounts of waste heat—even more than fossil-fired power plants.

It should be recognized here that coal- and oil-burning utility plants also have serious environmental effects, present occupational hazards, and are the cause of respiratory diseases for members of the general population. These effects should certainly not be played down, but there are major distinctions which should serve as the basis for comparing nuclear and non-nuclear fuels. Nuclear power development could be stopped almost overnight by one catastrophic accident—a problem which is not a consideration with oil or coal. Fossil-fuel problems moreover, are more amenable to technical solutions—safer mines, land reclamation, cleaner fuels. Nuclear power is plagued by a multitude of serious and unresolved problems, both institutional and technical.

It is also important that energy efficiency, the option which could allow nuclear power to be bypassed for at least the next quarter-century, has positive environmental effects, and is beneficial to the economy.* The routine environmental effects of nuclear power are important in and of themselves, and should be viewed as

* The alternatives to nuclear power are discussed in a later chapter.

part of the entire framework of problems which makes nuclear power unacceptable.

This chapter reviews three aspects of the environmental effects of nuclear power. First, the power plants' consumption of water and their release of waste heat are examined. Next, the effects of "nuclear parks"—collections of several power plants—on the natural and social environment are discussed. The last section of the chapter reviews the question of reactor decommissioning—what to do with radioactive nuclear plants at the end of their useful lives?

Thermal Pollution

By 1974, the Oyster Creek nuclear plant, near Toms River, New Jersey, was threatening the livelihood of Henry Kurtz. Kurtz was the owner of a marina on Oyster Creek, from which the nuclear plant took cooling water and its name. The cedar pilings which held up the piers of the Kurtz marina were being eaten away by shipworms. The larvae of the worms would bore into the wooden pilings, eating their way through as they grew to full size—up to two feet in length and a half-inch in diameter. When the shipworms finished their work, there was little left of the cedar pilings but a hollow shell that could snap easily.

Henry Kurtz had not had problems with shipworms before Jersey Central Power & Light began construction of the Oyster Creek plant. Dredging of the creek, which was necessary for plant construction, had changed Oyster Creek from a freshwater to a saltwater body by allowing water from Barnegat Bay to flow into the creek. Once the plant began operation, it dumped hot water—up to 104° F—into Oyster Creek. Although the hot water is deleterious to many forms of marine life, it was perfect for the shipworms. The hot, saline water gave the worms the conditions they needed to breed, much to the dismay of the marine owners in the area.

One marina operator, two miles south of the plant, had not had to replace any of his pilings in twenty-two years, until 1973, but he had to replace fifteen between July 1973 and July 1974. The shipworms did even more damage to Kurtz's marina, as he had to sink seventy-six new pilings in the eighteen months before July 1974, at a cost of more than $8,600. Until 1973, Kurtz had operated his marina for thirteen years without the need to replace any pilings. By September 1974, Henry Kurtz was ready to leave the Oyster Creek area: "I'd like to live about two thousand miles from the nearest nuclear reactor, it's simple as that. There're just too many unknowns."[1]

Jersey Central Power & Light's "solution" to the shipworm problem *was to buy up four marinas—at a cost of $2 million—and begin tearing them down. In December 1974, the Atomic Energy Commission had ruled that the nuclear plant operation had changed the ecology of the creek, allowing the shipworms to breed. The AEC gave Jersey Central nine months to clear its discharge canal of wood and to replace pilings and piers with material that would not harbor shipworms. After Henry Kurtz was paid for his marina, he moved to Florida.*

Although Jersey Central had satisfied some of the claims of the marina operators, the shipworms continued to spread. By May 1975, twenty homeowners on Oyster Creek were complaining that their piers and boathouses were being ruined by the pests. Moreover, marine biologists were warning that the shipworm plague could spread throughout Barnegat Bay, all along the New Jersey coast.

In 1972, the Oyster Creek plant had also proven hazardous to fish—by killing thousands of menhaden, a non-edible species used for fertilizer and livestock feed. The routine operation of the plant during the winter raised the water temperature at least twenty degrees and attracted the menhaden to the area. Menhaden cannot survive in water colder than 39° F, and under normal circumstances they migrate south in the winter. The warm water kept the fish near the plant, but when Oyster Creek shut down in January 1972, the water temperature dropped from 60° F to 34° F in four days. The result was that 250,000 of the menhaden were killed, circumstances which caused the New Jersey State Department of Environmental Resources to sue the utility for damages.

The case eventually reached the New Jersey Superior Court, which ruled in favor of the utility. Although New Jersey Chief Justice Richard J. Hughes noted that "Jersey Central had convinced the fish they were in Florida," the court found in January 1976 that the utility could not be held liable for the fish kill because they had been acting in accordance with Atomic Energy Commission regulations in shutting down.[2]

The Oyster Creek plant thus had a dubious distinction—it caused environmental problems when it operated, and it caused environmental problems when it shut down. Both the shipworm and menhaden "incidents" are examples of the unexpected changes that can result when man interferes with the natural scheme of things.

Early criticisms of commercial atomic reactors focused on their threat to acquatic life and ecology. Public attention was sharply drawn to this problem in 1963 when the early operation of the Indian Point nuclear power plant in New York was accompanied by massive fish kills. Tens of thousands of striped bass, attracted to the flow of the hot water being discharged from the plant, became trapped by the water intake structures and subsequently were killed, probably by a combination of heat and the mechanical action of the intake system.[3]

Jersey Central Power and Light Company's Oyster Creek nuclear power plant, near Toms River, New Jersey.
(The Scheller Company, Hackettstown, New Jersey)

All steam-generated electrical power plants require massive amounts of cooling water to dissipate their waste heat, but nuclear plants require much more than fossil plants. A nuclear plant typically operates at a thermal efficiency of 33 percent, which means that one-third of the heat produced in the reactor actually produces useful work—in the form of a spinning turbine which produces electricity. The remaining two-thirds is discharged as waste heat. As a staff report from *Environment* magazine has stated, "It is as if for each plant built for electric power, two more plants of equal size were built at its side for the sole purpose of heating water."[4] A fossil fuel plant, in comparison, operates at a thermal efficiency of about 40 percent.

The reason for the different efficiencies lies in the way the plants are operated. With a greater heat ratio between the incoming steam flowing to the turbine and the outgoing steam leaving it, more useful work is done—and more electricity can be produced—for a given steam flow. A condenser transfers the heat from the outlet steam, thereby reducing the pressure. This allows the incoming steam, at higher temperatures and pressures, to move through the turbine more efficiently, increasing the overall plant efficiency. In nuclear plants, reactor temperatures—and therefore the steam pressure to a turbine—must be kept lower than in a fossil fuel unit as a safety measure. Because the pressures in a nuclear plant are lower, the steam is unable to give up as much energy to drive the turbine, thus requiring that more heat be dissipated through the cooling water passing through the condenser. A fossil plant, in addition to its higher efficiency, also discharges some of its waste heat to the atmosphere. These factors mean that a nuclear plant requires about 50 percent more cooling water than a comparable coal-burning plant.

The volume of cooling water needed for a given power plant depends upon a number of variables: the thermal efficiency, the type of plant, the design of the cooling system, and the desired impact upon the environment. Baltimore Gas and Electric Company's Calvert Cliffs reactors, two 845 MWe units in Maryland on the Chesapeake Bay, use the "once-through" cooling method. This method, by which water is removed from a body of water, circulated directly through the plant, and then returned with waste heat to the body of water, is the most economical and most consumptive. Each of the Calvert Cliffs plants requires a water flow of 1.2 million gallons per minute.[5] Four such plants would require nearly 5 million gallons per minute. Few rivers in the U.S. can provide this much water, especially during periods of low flow.[6]

Temperature has an important biological impact along a waterway. As the temperature of the water rises, the metabolic rates—and therefore the demand for oxygen—of fish speed up. Ironically, the ability of water to hold usable oxygen diminishes as the temperature rises. These two factors may result in fish kills, or in increased growth of algae, which compete with fish for oxygen. There may also be a gradual takeover by less desirable species, such as shad or carp, which can more readily adapt to the higher temperatures than, for example, bass or blue-gills. Bacteria growth may also increase, creating problems for communities which depend upon a body of water for their drinking supply, or for commercial fishing industries, since diseased fish cannot be marketed. Even small temperature differences, in combination with other environmental problems, may cause a serious imbalance in the ecology of the waterlife.[7]

To avoid the problems created by heat discharges, especially when adequate water supplies are not available, cooling methods other than the "once-through" system are being implemented. Each has its economic costs. These systems, with their variations, include:[8]

1. Cooling ponds are similar to the once-through system except that the heated water is returned to an artificially created reservoir rather than to a lake, stream, or ocean. The dissipation of heat depends upon a large surface contact between the air and water. The biggest drawback for very large facilities is the amount of land needed for the pond, usually one to two acres per megawatt of generating capacity—or up to 2,000 acres for a typical 1,000-megawatt electric plant.

2. Spray ponds and canals can reduce the amount of land needed for the reservoir by spraying the heated water into the air, thereby increasing the heat-transfer efficiency between the water and air since the surface-to-surface contact ratios are increased. A well-designed system can reduce the land area by as much as 95 percent. Energy costs are higher than for other methods.

3. Wet cooling towers are more complex than cooling ponds or spray canals. Here the heated water is pumped to the top of a tower, where it is then mixed with air, either by forced ventilation or by natural draft. The cooled water is then collected at the bottom of the tower and returned to its original source or to the condenser. Large quantities of water—as much as 1 percent of the original intake—are evaporated to the atmosphere. This requires a continual withdrawal from a lake or stream. The evaporation process may also contribute to fogging and icing conditions in the area of the plant. If the humidity is particularly high, the effectiveness of the cooling process is reduced. The size of natural draft towers may range from 250 to 400 feet in diameter at the base, and from 300 to 500 feet in height, although forced ventilation towers are smaller.

4. Dry towers work much like the familiar automobile radiator, transferring heat by convection rather than evaporation. Without direct contact between air and water, greater air movement is required to attain the proper cooling levels. Because of the nature of its operation, a dry tower can reduce water demand, evaporative losses, and thermal discharges to the river. Capital costs, however, are highest for the dry tower.

Because of their size and cost, dry cooling towers are not likely to be extensively used. This means that the demand for cooling water will continue to grow. In fact, it has been estimated that electric power plants may require one-sixth of the nation's total available freshwater runoff by 1980.[9]

The serious thermal effects of individual nuclear power plants obviously will be magnified if there are several plants in one area. This idea has been labeled an "energy park" by a utility industry searching for a public relations phrase to defuse public opposition to such a scheme. Although the energy park, described in the next section, would have serious effects on natural and social environments, its development will be inevitable if this country deepens its dependence on nuclear power.

Energy Parks

Judy Johnsrud was appalled by utility company plans for "energy parks" in Pennsylvania. She knew that such a complex could devastate the rural Pennsylvania countryside. The "park" could actually be a large island of twenty nuclear power plants, surrounded by a sixty-square-mile lake. The lake would be shimmering with the waste heat put out by the twenty plants, which could be equivalent to the heat output from a large city. The heat release from the complex could be sufficient to cause weather changes or even start an occasional tornado. A few miles from the "park," a city of 40,000 to 60,000 people would have to spring up in an otherwise pristine area to support the power plant complex. Although an energy park might not be completed until the year 2000, four private Pennsylvania utilities—Philadelphia Electric, Pennsylvania Power & Light, Metrpolitan Edison, and Pennsylvania Electric—wanted to lay the groundwork for parks in 1975.

Judy Johnsrud had done battle with the nuclear industry before and was ready to fight them again. She is co-director of the Environmental Coalition on Nuclear Power, a federation of about forty citizens' groups representing about 10,000 members. The Coalition operates chiefly in Pennsylvania but has member groups in New York, New Jersey, Delaware, Virginia, and Maryland as well. Since the Coalition was established in 1970, it has in-

volved itself with nearly a dozen nuclear projects, and has actually stopped some of them. One of the keys to the Coalition's success has been the tenacity of members such as Judy Johnsrud.

In the spring of 1975, Judy, Chauncey Kepford (a former chemistry instructor with the Pennsylvania State University extension system), and other members of the Coalition organized around the energy park issue. As they traveled through the state, Coalition members found that in the rural areas which had been selected as prospective energy park sites citizens were bothered by the thought that their counties had been chosen as "energy sacrifice areas" to provide power to the big cities. One of the proposed park sites was near Pine Glen, in Centre County, in the middle of Pennsylvania. During a town meeting at the Pine Glen Community Center, Herbert E. Probst, a local employee of the State Fish Commission, expressed the concerns of rural Pennsylvanians:

> It's what we have here, which is few and far between, the natural beauty we've tried to conserve all these years, our forefathers and everyone else have fought to keep, what little bit there is for people. . . . And this power plant, as far as I'm concerned, is just the beginning. It's just the beginning of the end of the community we have. It isn't just the power plant itself. They say they will entice other industries. But we have fought to keep what we have here. And if we let it go, we'll never get it back.[10]

Faced with the opposition of Pennsylvania citizens, the Pennsylvania utilities have withdrawn their tentative energy park proposals and retreated to their boardrooms to rethink the idea. Although energy parks are not yet reality, the concept is under active study by the atomic establishment. Studies have been carried out by General Electric, the AEC, and the NRC, as well as by the Pennsylvania utilities.

An energy park would be a grouping of ten to forty power plants in one large area. Some of the plants might be coal, some nuclear, or they might all be of one type. An all-nuclear plant complex has been called a "nuclear energy center." The plants would have to be constructed on a gradual but steady basis—one or two would be completed each year. A variation of the energy park would be the "integrated fuel cycle facility" (IFCF), a complex of nuclear reprocessing and fabrication plants in one area. One rationale for the IFCF is that it would eliminate transportation from the reprocessing plant to the fabrication plant. This could make nuclear materials—particularly plutonium—less susceptible to sabotage or theft.

The utilities believe, among other things, that energy parks will save them money. General Electric, which will benefit by increased reactor sales, concluded that a twenty-plant nuclear park would be 10 percent cheaper than twenty nuclear plants at ten dispersed sites (two plants to a site).[11] These cost advantages, however, are largely uncertain because they depend on a high degree of plant standardization, and industry has a poor record of implementing standardized plants and components. There are also practical uncertainties, as much of the cost saving is to come from a steady construction schedule, allowing a constant labor force—though smaller than would be required for dispersed sites—to be maintained at the complex. If the construction schedule should falter—because of either changes in electrical demand or an inability to raise the enormous capital investment required—then the labor force would fluctuate and the alleged savings would be lost.

While the industry expresses a great deal of interest in the benefits of nuclear parks, there does not seem to be appropriate concern for the citizens who will live near the parks. A favorite tactic of the nuclear industry has been to site plants in rural areas, where opposition is less likely to mobilize effectively than in urban communities, where well-organized citizen groups may be able to muster formidable resources.[12] Often, opposition in rural areas can more easily be blunted by promises of the new facility's substantial contribution to local tax bases. As a practical matter, the land area and water supply necessary for an energy park are more likely to be available in unspoiled areas. For these reasons the effects of nuclear energy centers, and energy parks in general, are likely to fall most heavily upon small towns.

One impact is the need for a construction force of up to 9,500 workers for a twenty-plant park.[13] The workers will also bring families and a need for services such as schools, housing, and sewers. The General Electric study concluded that a community of 40,000 to 60,000 might be necessary to support the working population.[14] Further, the large amounts of electricity available from an energy park might induce industries with heavy electrical demands to build their facilities near the park. One report has stated, "A town in a neighborhood of a park faces possible explosion into cityhood as industries co-locate with the park."[15]

The potential negative impact of an energy park is so great that even the industry publication *Electrical World* has expressed concern over General Electric's twenty-unit park model:

> First, the social impact would be immense. Serious land-use questions are raised by the up to 42,000 acres required, by the dozen or more 765-kv lines that would radiate from the park, and by the inevitable accretion of satellite housing and industry.
>
> Potential environmental effects, too, are awesome. Heat and moisture releases are comparable to those of such natural energy-related phenomena as thunderstorms. Therefore, the meterological effect of a park might be felt over an area similar in extent to those affected by storms. There is also the possibility that such a concentrated energy source could modify certain types of weather systems in such a way as to create a single strong vortex, such as a tornado. Further research into these areas is essential.[16]

Heat release would, in fact, be a major problem. Pumping water from a river through the plant (the once-through cooling method) would not be sufficient because of the large number of plants. Each plant would have to have its own 400-foot cooling tower, or a cooling pond, which for a twenty-plant park could surround the park and cover sixty square miles.[17] The park itself would be a three-square-mile island in the middle of the "pond." A forty-plant park, with a cooling lake, would cover about the same land area as the city of Washington, D.C., but would release ten times the waste heat of that city. A park without a cooling lake (i.e., with cooling towers) would take up one-fifth the area of Washington, D.C., and would have a much more concentrated release of heat.[18]

Other nuclear park environmental problems are the effects of routine radioactive releases, spills, and other unplanned releases that could occur from several plants but whose effects will be concentrated in one area. An accident at one plant could affect the operation of other plants—by either causing another accident or requiring shutdown and evacuation of the remaining plants. In addition, disasters such as earthquakes, sabotage, or military attack would cause the release of more radioactivity from a nuclear energy center than from a similar occurrence at a single plant.

The threatened effects of energy parks deeply disturbed the citizens of Pennsylvania. Many of the residents who have made a conscious choice to live in small towns viewed the energy park as a threat to their way of life, as Herbert Probst explained.

Probst's concerns certainly are not unfounded. Fairbanks,

Alaska, and Colstrip, Montana, are two dramatic examples of the way an energy facility can affect small towns. The influx of workers from the Alaska pipeline caused a rise in crime rates, skyrocketing rents, and strains on services in Fairbanks.[19] Besides increasing the oil industry's power, the Alaska pipeline has resulted in a profound shift of power within the state—by dumping large sums of money into the coffers of the Teamsters Union. Alaska is faced with the threat of becoming a lawless company state under the control of large energy companies and the Teamsters.[20]

Colstrip, Montana, a town of 200 people in 1972, now has a population of 3,000, chiefly because of the operation of two coal-fired power plants and plans for two more. Already, classes in the town's overcrowded schools are being held in shower rooms, and the school district had to find room for 1,700 students by the end of 1976.[21] Other small towns in the western states, faced with the influx of workers for coal or shale oil facilities, have experienced similar problems.

With the likelihood of significant citizen opposition, the power industry is ready to recommend Draconian measures. One of the industry's perceived advantages of energy parks is that they could simplify the power plant siting process. It would be much easier for a utility to obtain permission to construct several plants at one site—and to proceed by building plant after plant with abandon—than to obtain necessary federal, state, and local permits at several dispersed sites. However, this advantage would vanish if the threat of an energy park mobilized citizen challenges more effectively than dispersed plants.

So the power industry wishes to reduce the threat of citizen action by establishing multi-state, regional, or federal agencies with the authority to select areas for energy park sites.[22] These agencies presumably would have ultimate veto power over the objections of towns, counties, or states unwilling to become "energy sacrifice areas" for the rest of the country. General Electric, in its recommendations for changes in the nuclear licensing procedure to facilitate nuclear centers, called for: a) elimination of construction permit hearings except upon a petition from an intervening citizens' group;* b) substitution of rule-making or legislative hearings for adjudicatory hearings (this would allow citizens the chance to make token state-

* Refer to Chapter XIX for a discussion of the intervention process and its pitfalls.

ments, but would deny the procedural rights of subpoena or cross-examination); c) restriction of the right of intervention by citizens.[23] These changes would all but eliminate citizen participation in the nuclear licensing process.

Major changes in the utility industry would also be necessary. Each plant would cost at least $1 billion, and the utility would have to order several in advance and then maintain a tight schedule in bringing the plants to operating status, whether the plants were actually needed or not. Total costs over the lifetime of a proposed nuclear park are estimated at $180 billion.[24] The large capital requirements for a park might require new arrangements in the utility industry to allow for joint ownership of land and plants. Even these changes might not be enough. One section of the GE study recommended that the federal government consider funding of "demonstration" energy parks.[25] A relevant question is: how does one go about "demonstrating" an energy park without actually building the entire twenty-plant complex, with taxpayers' money? It is conceivable that a "demonstration" park could more than bankrupt the proposed $100 billion Energy Independence Authority.*

The problems of energy parks sharpen the larger issue of just what type of energy system this country should develop. Energy parks are the inevitable extension of a system which encourages the wasteful use of energy and electricity. If the nation continues to expand its use of electricity, then more power plants, larger plants, and, finally, energy parks will be necessary.

But increased dependence on centralized generating stations further removes the citizen and consumer from the energy production process. Large energy corporations (which may control oil, natural gas, coal, and uranium) sell fuel to monopoly utilities. Electricity from the utility, as well as all costs, are passed on to the consumer. Decisions on where power plants are to be sited are made by the utility, with the approval of state commissions. The individual has little voice in choosing a power source, or the company that will provide it. Energy parks will be sited by federal or regional authorities, remote from voters and, quite likely, tolerating little interference from individuals. Citizens will be removed not only from

* The Energy Independence Authority was a Ford administration proposal that would give away $100 billion in federal funds for energy production projects. It is discussed in the chapter on nuclear economies.

the production of energy but from most of the decision-making process as well.

Growth in electrical generation will strain capital supplies. As more and more power plants are built, the industry must find the money to invest in the plants. If the utility industry is to grow at its own projected rates, it may need as much as $500 billion—about half of which will go to nuclear plants—during 1975–85. This is more than three times the industry's capital requirement for the previous decade.[26] Meeting these capital demands will require higher interest rates as money becomes in shorter supply, or federal subsidies, or both. The utility industry also competes with other segments of the economy: every dollar invested in a power plant is a dollar taken away from an industry that probably would provide more jobs. Per dollar invested, electric plants provide fewer jobs than virtually any other industry.[27]

A more sensible approach to energy would be to develop decentralized energy technologies and, specifically, to cut the wasteful growth in energy consumption that contributes to decisions promoting centralized systems and "streamlined" licensing procedures. A more efficient use of energy, to slow the growth rate, will allow for a more orderly development of decentralized energy alternatives, which are necessary to give citizens more control over the production of energy and over the decision-making process.

It is long past time for the utility industry to recognize that energy efficiency improvements and decentralized energy sources are in the national interest. These objectives will improve financial conditions by requiring fewer power plants and by easing strains on capital markets. As a first step in recognizing the benefits of a noncentralized energy system, the utility industry should heed the advice of the *Electrical World* article on energy parks, which concluded: "The surfeit of unanswered questions raised most certainly calls for intensive follow-on studies. Therefore, we hope that utilities will not, as an expedient to clear the way for large, multi-unit installations, be tempted to join easily in the foreseeable rush to promote the concept of the energy park."[28]

Decommissioning

Nuclear parks are a hypothetical problem for the future, since there are no energy parks presently in existence. But reactor decommissioning is a problem for the future which has already been created by reactor operation. Decommissioning is the process by which reactors, which themselves become radioactive, must be guarded or dismantled at the end of their useful lives. Nuclear plant licenses are granted for no more than forty years, for reasons both of safety and of economics. After forty years or less, a plant will probably be technically obsolete. Even if the plant were not to be replaced by one with more current safety technology, its older equipment would be difficult to maintain. A utility would want to take the plant out of service because operating and repair expenses would have become too large.

But while a nuclear plant license expires after forty years, its radioactivity does not cease. Not only does a reactor produce radioactive by-products in its fuel, which can be removed, but the steel pressure vessel and the pipes associated with the reactor themselves become radioactive. The radioactive materials in a plant will have to be excluded from contact with people or the environment until the materials are harmless. But the nuclear industry has little basis on which to predict the difficulties or costs of plant decommissioning. Only a few small research reactors—no more than one-tenth the size of modern reactors—have been decommissioned.

There are three general methods which have been advanced by the atomic industry as decommissioning procedures:

1. Mothballing. This would involve removing the nuclear fuel and "fluids"—chiefly the water that was used for coolant. The plant would then have to be locked and guarded for as long as the radioactive components inside remained hazardous. Estimates of how long the plants would have to be guarded vary, but even industry sources estimate that surveillance could be required for as long as 200 years.[29]

Mothballing will probably not be practical for reprocessing plants. These plants will contain significant levels of plutonium and other transuranium* isotopes and will present a waste disposal problem themselves. At a minimum, those components contaminated with transuranium elements will have to be removed from the plant.

* "Transuranium" refers to isotopes of atomic weight greater than uranium. These isotopes, such as plutonium, are formed as by-products of reactor operation. Many of them, like plutonium, emit alpha particles, which are highly toxic if ingested or inhaled. Several transuranics also have very long half-lives, on the order of several thousand years.

If the reprocessing plant became contaminated throughout with transuranium elements, mothballing alone would be insufficient.

2. *Entombment.* This would be one step further than mothballing. Radioactive components such as the pressure vessel, or perhaps the entire containment building, would be sealed. This could be done, for example, by filling or surrounding them with a layer of concrete. Security would still be required to prevent vandals or others from breaking into radioactive areas, but the guard force would be smaller than for mothballing.

3. *Dismantling.* This involves the removal of all but the uncontaminated foundation structures, and restoring the site to its original condition as nearly as possible. The entire plant would be broken down and disposed of as radioactive waste. This would be the only practical "solution" for some of the components of the reprocessing plant, if not for the entire plant. This method only transfers the problem, however, since the dismantled plant would then become part of the nuclear waste inventory in need of storage.

The costs of decommissioning are as sketchy as the methods. The U.S. General Accounting Office (GAO) reported: "There are not firm estimates of decommissioning costs for large-scale nuclear power plants in operation or under construction. The seven reactors that have been decommissioned were small and differed substantially in design from modern reactors."[30] A very small nuclear plant in Saxton, Pennsylvania (3 megawatts-electric), was mothballed for about $500,000 in 1973. The Elk River Power Plant in Minnesota (22 megawatts-electric) was dismantled for about $6 million.[31] These reactors should be compared in size with modern nuclear plants such as the Browns Ferry reactor, which is 1065 megawatts-electric.

The Nuclear Regulatory Commission believes the cost of decommissioning such larger plants will be much greater. It estimated that mothballing will cost $3 to 5 million, in addition to annual surveillance costs of $60,000 to $100,000. Entombment is estimated at $18 to 30 million, with annual surveillance costs of $15,000 to $25,000. Dismantling could cost $36 to 60 million.[32] But these figures are only estimates; cost predictions by the atomic industry go higher. Pacific Gas and Electric Company of California believes that the costs of dismantling each of its Diablo Canyon plants could be $70 million or more.[33] The Virginia Electric and Power Company calculates that dismantling costs of each of its North Anna plants could reach $150 million.[34] Other industry sources have even suggested that the costs of decommissioning could be as high as the initial construction cost of the nuclear segment of the plant.[35]

A recent study by the New York Public Interest Research Group (NYPIRG) indicates that even these costs may be seriously underestimated. The study was conducted by Marvin Resnikoff, professor of physics at the State University of New York at Buffalo, with the aid of four engineering students at SUNY-Buffalo. In contrast to industry statements that mothballing would take 120 to 200 years, Resnikoff's group found that at least 1.5 million years would elapse before the radiation in reactors had decayed to safe levels.

The difference between the estimates of the atomic industry and of Resnikoff's group comes from the industry's apparent oversight of the isotope nickel-59. Nickel-59 is formed as a by-product of the structural material in the reactor pressure vessel. Excess neutrons from the fission process can be absorbed by atoms in the steel pressure vessel. If a neutron is absorbed by an atom of nickel-58 or cobalt-59, both present in the stainless steel vessel, the resulting isotope can be nickel-59, which is radioactive and has a half-life of 80,000 years. The NYPIRG group calculated that it would take at least 1.5 million years for nickel-59 to decay to relatively safe radiation levels.[36] This time period makes decommissioning akin to the radioactive waste problem in terms of the time the reactor must be guarded.

Even before NYPIRG released its report, the Energy Policy Project had identified decommissioning as a major unanswered nuclear power question. The Project, a two-year study of national energy issues, was completed by a staff of economists, lawyers, writers, scientists, and engineers. Funded by the Ford Foundation, the Project was directed by S. David Freeman, former director of the Energy Policy Staff of the Office of Science and Technology, a federal agency. On the subject of decommissioning, the Project stated:

> Present plans call for industry to turn over reprocessing plants to the states in which they are located, after they are taken out of service. But the states have no idea of what they will do with them.
>
> A basic difficulty is that reprocessing plants and nuclear reactors are not designed to be decommissioned. The radioactivity hazard is generally

worse in reprocessing plants, but it can be a special problem in a reactor which has had an accident, as in the case of the Fermi breeder reactor in Detroit.

A full assessment of the decommissioning problem should be carried out—promptly—before the new reprocessing plants coming on line are fully contaminated, and before reactors proliferate throughout the country. Institutional and economic questions are at least as important as technical ones. Who should be responsible for decommissioning? How should decommissioning be paid for? How will decommissioning costs affect the economics of the nuclear fuel cycle?[37]

The Fermi reactor was a 61 megawatt-electric plutonium breeder reactor which, in October 1966, experienced an accident that melted four of its fuel assemblies. Decommissioning costs for Fermi are estimated by the Nuclear Regulatory Commission at $7.1 million. Fermi, moreover, is essentially being decommissioned by the mothballing option.[38]

Dr. Resnikoff's report may spur the Nuclear Regulatory Commission to examine the decommissioning issue in more detail. But the NRC's response to the concerns of the Energy Policy Project has been languid. The commission's regulations address decommissioning in a very general way—requiring only that when it becomes time to decommission a plant the licensee will supply plans to do so.[39] Apparently, decommissioning is not considered a problem until the day it must begin. The atomic establishment is apparently attempting to make the decommissioning issue disappear by ignoring it. But the radioactivity from burned-out nuclear plants is not likely to go away, and decommissioning will probably become one more issue to haunt the atomic industry's future.

The Back-End of the Fuel Cycle

THE "BACK-END" of the fuel cycle refers to the treatment of nuclear fuel after it is removed from the reactor. The fuel, in theory, will be transported to a reprocessing plant, where uranium and plutonium will be chemically separated from the other radioactive products of reactor operation. The uranium and plutonium are to be re-used as reactor fuel, and the remaining radioactive garbage becomes "nuclear waste," which must be disposed of or managed until it decays into stable or harmless isotopes. For reference, Figure 1 shows the fuel cycle.

The back-end of the fuel cycle is presently in disarray. There are presently *no* reprocessing plants in operation, and it is likely to be several months before one begins to process spent fuel. While there are many speculative "solutions" to the ultimate problem of nuclear waste, none has been demonstrated, and the history of attempted solutions is full of failures and false starts, raising serious questions about the ability of human institutions to manage nuclear waste for the centuries which may be required.

Reprocessing

Uranium fuel is depleted in a reactor as the fuel fissions and produces heat. The spent fuel is removed and replaced with fresh fuel rod assemblies from the fabrication plant. Once a year, about one-third of the fuel—roughly thirty metric tons—is replaced at each reactor. After removal, the spent fuel is supposed to be sent to a chemical reprocessing plant.

However, as noted, no such plants are operating at this time,

Figure 1. *The Light-Water Reactor Nuclear Fuel Cycle*

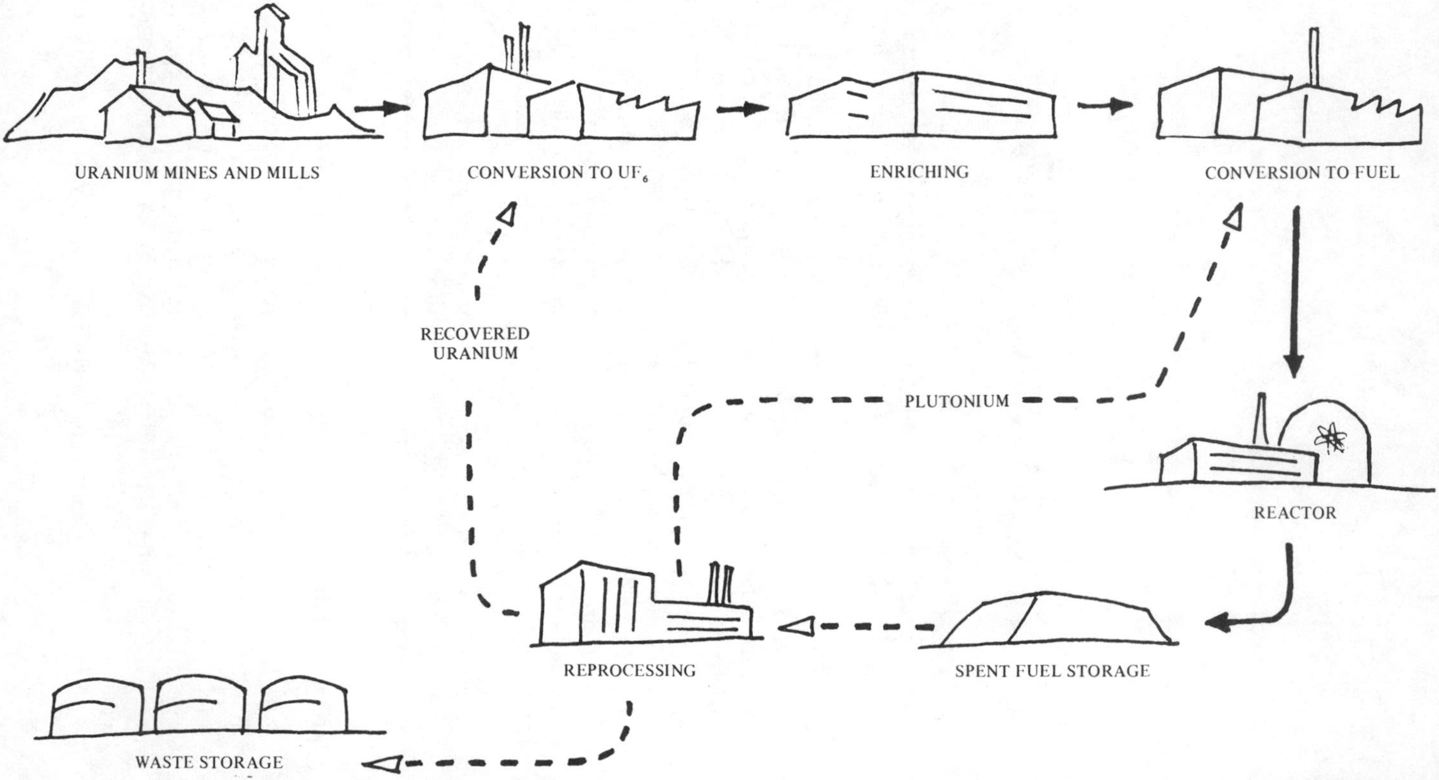

which means that the nuclear fuel "cycle" is not a reality. From 1966 until late 1971, Nuclear Fuel Services (NFS), a subsidiary of Getty Oil, processed spent fuel, with "sporadic interruptions," at a plant in West Valley, New York.[1] In December 1971, the plant was shut down. The official reason was that the plant's capacity was being expanded, but there are indications that the plant was closed because radioactive releases and exposure of workers were excessive. In February 1972, the Atomic Energy Commission accused NFS plant officials of "failure to make reasonable efforts to maintain the lowest levels of contamination and radiation" and "failure to adequately instruct or effectively train employees . . . in the radiation hazards involved in their job assignments." Three months later, company officials announced that the plant was beginning a long-planned shutdown for expansion.[2] In September 1976, NFS announced that it was abandoning plant operations for economic reasons.[3] NFS has also asked the State of New York to assume responsibility for the 600,000 gallons of radioactive liquid on the plant site.[4] It is estimated that the cost to the state to dispose of this material will be $480 million or more.

The NFS plant was sufficient to handle the spent fuel from a fledgling nuclear industry, but it was operated under significant government subsidies. As the atomic industry grew, it became apparent that more reprocessing capacity would be necessary. General Electric believed it had developed a process that would make its Midwest Fuel Recovery plant in Morris, Illinois, profitable. But the plant, which was originally expected to begin reprocessing fuel in 1973, developed major technical difficulties.[5] Despite General Electric's investment of six years and $64 million for design and construction, the plant has yet to work satisfactorily. GE has admitted that the plant must be redesigned and rebuilt—a task that will take at least four years and an additional $90 to $130 million. The failure of the plant apparently involves a technical process that worked in the laboratory but did not work at the full-scale plant.[6] If the Midwest plant is rebuilt, the earliest it could be processing spent fuel would be 1980. However, there appears to be a good deal of sentiment at General Electric for scrapping the plant completely, before it creates a greater financial morass.

A third reprocessing plant is being constructed near Barnwell, South Carolina, by Allied Chemical Corporation and General Atomic Company, a subsidiary of Gulf Oil Company. The Barnwell plant will probably not begin operation until 1977 at the earliest. Although the plant originally was to have been a profit-making venture, the owners have already asked the federal government for sizable subsidies.

With no place to ship spent fuel, utilities are merely piling up their used fuel assemblies at reactor sites. Each power plant maintains a water pool where fuel removed from the reactor is stored. In the pool, the radioactivity of the fission products decays, and the water provides a means for the removal of residual heat. After about 120 days, the fuel radioactivity has decayed sufficiently to allow

General Electric's defunct Midwest Fuel Recovery plant near Morris, Illinois.
(J. E. Westcott, U.S. Energy Research and Development Administration)

transfer from the storage pool to a truck for transport to a reprocessing plant.

With no reprocessing plants operating, however, utility companies must store spent fuel until their water pools are filled. With its storage pool filled, a power plant would have no choice but to shut down, since it would have no place to deposit used fuel. If no additional temporary storage space can be found for used fuel, as many as five nuclear power plants would have to shut down by 1978.[7]

The industry has already begun to take stopgap measures to buy time until some reprocessing plant can begin operation. Several utilities have asked the NRC for permission to expand the size of their water storage pools, giving more time before the pools fill up.

The uncoordinated nature of these measures has forced the NRC to announce that it will evaluate this storage problem on a comprehensive basis and will promulgate regulations, at some unspecified future date, to establish storage criteria on an industrywide basis.[8]

Radioactive releases to the environment from reprocessing plants are also a source of concern. The only experience with commercial reprocessing was at the Nuclear Fuel Services plant, and its record is not encouraging. To begin with, there are the admonitions from the AEC, mentioned above, which probably caused the plant's shutdown. The plant contaminated the air with plutonium, tritium, and krypton, as well as other radioactive isotopes. The plant contaminated the streams on its boundary (Buttermilk Creek and Cattaraugus Creek, which flows into Lake Erie) with plutonium, iodine-129, strontium, and other radioactive materials. The environmental contamination in turn led to radioactive contamination of fish and deer around the plant.[9]

The builders of the Barnwell reprocessing plant argue that the Allied-Gulf facility is to have a different technology than the NFS plant and that Barnwell will be "cleaner"—that is, will release less radioactivity to the environment. Despite these assurances, the radioactive inventory at Barnwell will be monstrous. John Gofman, M.D., and professor emeritus of medical physics at the University of California, Berkeley, estimates that the material at Barnwell would represent "approximately the radioactivity that would be left decaying for tens and hundreds of years from a large, full-scale nuclear war." Gofman estimates that the accidental release of just 1 percent of the Barnwell radioactive inventory could contaminate 33,400 square miles of land and "easily" cause $10 billion or more in damage.[10]

Even without an accident, the Barnwell plant would routinely emit large amounts of radioactivity. The radioactive isotopes of greatest concern will be krypton-85 (krypton of atomic weight 85), iodine-129, plutonium, tritium, and carbon-14. The Environmental Protection Agency in 1975 proposed stronger standards for radiation emitted from the nuclear fuel cycle. The EPA focused on krypton, iodine, and plutonium, which are isotopes chiefly emitted at reprocessing plants. The EPA recognized, however, that control technology for krypton and iodine is limited, and set a 1983 target date for reducing emissions of these isotopes.[11] Moreover, the EPA recognized that virtually no controls have been developed for tritium and carbon-14 emissions, and it set no standards for these isotopes. This is recognized as a major omission; by the EPA's own calculations, carbon-14 emissions from reprocessing plants through the year 2000 will eventually cause 12,000 cases of cancer, leukemia, and genetic diseases.[12]

If reprocessing plants, with all their dangers, do begin operation, they will leave a radioactive residue, referred to as "nuclear waste." This waste will have to be trucked away, guarded, or disposed of in some way, as the next section discusses.

Nuclear Waste

As an ethical question, the nuclear waste problem is the most difficult facing the nuclear industry. Consumption of nuclear power today will leave radioactive by-products that must be managed for hundreds of centuries. Some way must be found to either dispose of these by-products or transmute them into harmless substances. No such disposal or transmutation scheme exists today, and if none is developed, society will be burdened with storing and guarding nuclear wastes for a quarter-million years or more. To many people, this means that nuclear power is a moral problem: the power we consume today will leave radioactive garbage for thousands of generations to come.

The nuclear waste problem exists because the fission process creates radioactive products. As a uranium atom fissions, two new atoms are formed, each of lower atomic weight than uranium. (The combined atomic weight of the two atoms, in fact, will be slightly less than the atomic weight of uranium, since part of the mass in the uranium atom was changed to energy during fission.) These substances which are of lower atomic weight than uranium are called "fission products," and are themselves often radioactive. Another category of isotopes is also formed by the fission process. Substances of higher atomic number than uranium can be formed when neutrons, produced by fission, are absorbed by atoms in the fuel. An atom of uranium-238 (U-238), for example, can absorb a neutron and be changed to plutonium-239 (Pu-239). Such substances are called "transuranium" elements, or "transuranics," because their atomic numbers are greater than uranium's. They are also radioactive.

The radioactive danger to biological organisms differs for each of these two categories of waste, due to differences in the longevity of different radioisotopes. It should be remembered that each radioactive substance has a characteristic "half-life"—the time it takes one-half of the material originally present to undergo radioactive decay. After ten half-lives there would be only about one-thousandth of the original amount of any radioactive isotope. A rule of thumb has been established that ten to twenty half-lives are necessary for a substance to decay to amounts where it becomes radioactively harmless.

For the fission products, strontium-90 (Sr-90) and cesium-137 (Cs-137) are the most important biologically hazardous isotopes in the waste. Sr-90 has a half-life of 28 years, and Cs-137 has a half-life of 30 years. These isotopes must be guarded for about 600 years before they become relatively harmless. Other fission products are still active after about 800 years, at which time their relative toxicity is about fifty times that of a similar volume of natural uranium ore.[13]

Although the danger from fission products is significant, it is much less than the danger from the transuranium elements, which have extremely long half-lives. One isotope of plutonium, Pu-239, is one of the most toxic isotopes known and has a half- life of 24,000 years. Ten half-lives of Pu-239 represent nearly a quarter-million years, during which time it is capable of causing cancer, leukemia, and genetic damage.

But the problem with the transuranium elements will last even longer because the products of their radioactive decay are actually more toxic than the original substances. Americium-243, for example, with a half-life of 7300 years, decays to Pu-239, which is more toxic than Americium. This means that the decay chain of transuranics may result in a decline in toxicity for about 20,000 years, after which the waste becomes more toxic. After a million years the transuranic waste will be over 1,000 times more toxic than natural uranium, and will be nearly as toxic as the transuranics that came directly from the reactor.[14]

Figure 2 shows that each waste category—fission products and transuranics—is the dominant biological hazard at one time or another. Since the volume of fission products in spent fuel is greater when the fuel assemblies are removed from the reactor, it is the fission products which are more important for about the first 600 years. After 600 years the transuranics, which have been decaying more slowly, become the more important component of the waste, and they will remain dangerous for a million or more years.

Clearly, some method must be developed to deal with nuclear waste. It would theoretically be possible to transmute the waste isotopes to non-radioactive substances—via bombardment with nuclear particles, for example. There are also several schemes for waste "disposal," but these are only speculative. If no practical disposal method can be developed, then society will be burdened with storing and guarding the waste for geological time periods and with preventing the waste from contaminating the environment or populations.

Hannes Alfvén, a Nobel laureate in physics, summarizes the nuclear waste dilemma:

Figure 2. *Toxicity of Radioactive Wastes*

Union of Concerned Scientists, *The Nuclear Fuel Cycle*, Cambridge, Massachusetts, October 1973, p. 47.

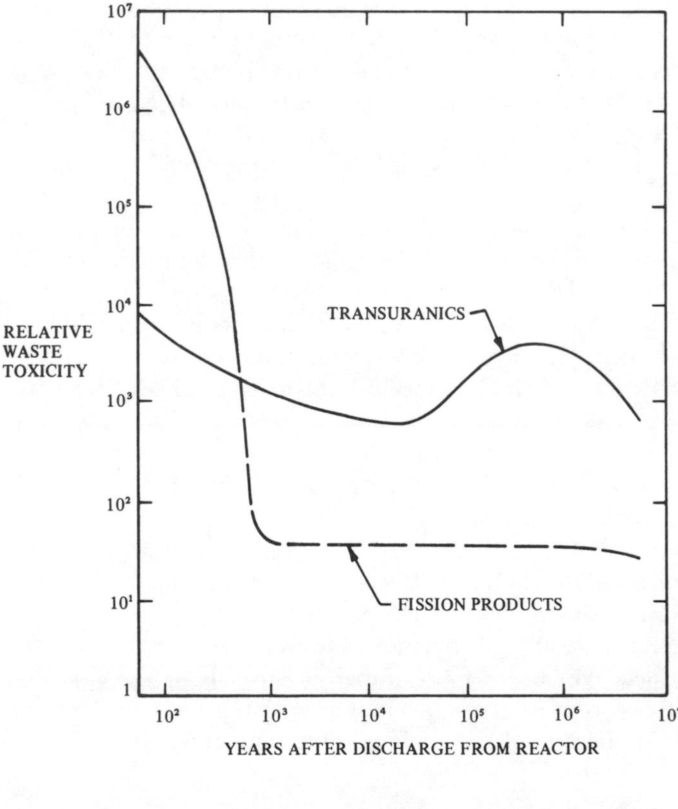

At present there does not seem to be any existing, realistic project on how to deposit readioactive waste; but there are a multitude of optimistic speculations on how to do so. The problem is how to keep radioactive waste in storage until it decays after hundreds of thousands of years. The deposit must be absolutely reliable as the quantities of poison are tremendous. It is very difficult to satisfy these requirements for the simple reason that we have had no practical experience with such a long term project. Moreover, permanently guarded storage requires a society with unprecedented stability.[15]

Storage of nuclear waste is much more than a problem of technology. Safe storage requires stable geological formations, a guarantee which is beyond the promise of technology. Safe storage also requires the development of stable human institutions to exist for thousands of years to prevent the waste from leaking and contaminating the biosphere. It should be remembered that Neanderthal man appeared "only" about 75,000 years ago.[16]

The federal government's response to the nuclear waste problem has been a mixture of hope and speculation. The Atomic Energy Commission's proposal, outlined in a September 1974 environmental impact statement, can be summarized as follows: After the spent fuel is removed from the reactor, it will be sent to a reprocessing plant for removal of the remaining uranium (and possibly plutonium) which will be re-used as fuel. The residue of fission products and the transuranics will be in liquid form at the reprocessing plant. After no more than five years, this liquid waste will be solidified and transferred to a federal storage facility until such time as an ultimate disposal method can be developed—which the AEC, at the time, believed would be matter of twenty to thirty years.[17]

Although the AEC proposal was supposed to reassure nuclear power skeptics, it simply raised more unanswered questions. There are no operating reprocessing plants; there are no commercial waste-solidification plants in operation; and the site for the federal waste storage facility has yet to be suggested. The Energy Research and Development Administration (ERDA), which inherited the waste problem from the AEC and which would be responsible for maintaining the federal storage facility, has backed off from the AEC proposal. In April 1975 ERDA announced the impact statement would be redone to include a broader analysis of nuclear waste and permanent disposal.[18] ERDA is expected to complete a new impact statement and announce its own waste policy in spring 1977.[19]

As to the technology for solidifying the waste, there are even questions over ERDA's ability to solidify its own nuclear waste, which was generated largely by the nuclear weapons program. An ERDA publication has stated:

> The ultimate disposition of the ERDA radioactive waste inventory will require the development and implementation of special technology on an unprecedented scale:
>
> The present technology for solidification by calcination of conversion to glass does not work with ERDA waste because of the chemical behavior of sodium compounds.[20]

Even if the waste were successfully solidified, the question of the federal storage facility remains. As yet, no site has been selected. The AEC contended that only twenty or thirty years of storage would be necessary until an ultimate disposal method is developed, but the commission did not explain why it chose the twenty-to-thirty-year figure, nor did it identify the anticipated method. The assertion that a method could be developed was built on blind faith; the AEC apparently expected the general public to have the same blind faith.

Several solutions have been suggested. But these are the "speculations" to which Professor Alfvén refers, and are not well-founded. One plan would place the wastes on polar ice sheets; another proposes disposal at sea; shooting the wastes into outer space is yet a third. Even the AEC, however, rejected these schemes, stating that:

> A number of concepts for disposal of high-level waste within geological formations by unconventional methods (for example, placement within polar ice sheets) are lacking in knowledge of the environmental and natural phenomena involved, technology for emplacing the waste, methods for evaluating safety, and technology for retrieval of waste if necessary.
>
> Analyses of disposal in outer space indicate that if safety technology can be developed, costs will limit disposal to the highest-hazard radionuclides within the waste.[21]

The Union of Concerned Scientists has also pointed out that space disposal would require a "zero" probability of atmospheric reentry or burnup, a situation that is "currently unfeasible."[22]

Another of the speculative solutions is "partitioning," which involves the chemical separation of the transuranic elements from fission products, and the insertion of the transuranics in a reactor for "burn up" by fision of the transuranics. This would leave only fission products, reducing the million-year problem to a thousand-year problem, although a waste problem would still exist.

While possible, partitioning at present is not close to reality. The only elements contemplated for recovery at fuel reprocessing plants are uranium and plutonium, at 99.5 percent efficiency (that is, 0.5 percent of the Pu and U is not separated from the fission products, due to limits in the technology). For partitioning to be workable, 99.95 percent of uranium, neptunium, and plutonium, and 99 percent of americium and curium would have to be separated. Further, it is likely that special reactors would have to be built to fission the transuranics.[23]

As long as fuel is reprocessed without removal of the transuranics, high-level wastes, both fission products and transuranics, will be stored together. If the wastes are stored in a solid form, partitioning would be impractical and would make the feasibility of successful partitioning in the future questionable.[24] Moreover, even the AEC acknowledged that partitioning would be viable only if "present technology could be extended to prevent the present small losses of plutonium from getting into the waste from now on."[25] It is also significant that in this statement the AEC discussed only plutonium. For partitioning to be meaningful, all the important transuranics must be separated.

Another speculative solution, already the basis for aborted AEC projects, is "geologic disposal," which requires that a stable geological formation be excavated for storage of the waste. There would, of course, be the need to monitor the waste to insure that none escaped into the environment. This proposal raises the very significant question of whether the atomic industry can guarantee geological stability of *any* formation for a quarter-million years or longer. The AEC's experience with geologic disposal is not encouraging.

In 1971, the AEC announced that it had selected a salt deposit near Lyons, Kansas, as the site for permanent disposal of radioactive waste. Because salt is soluble, its presence indicates the absence of water, which means there would be little groundwater for the waste to contaminate if it were to escape its disposal area. Salt is also an effective radiation shield. Under heat and stress, salt bends rather than breaks. During an earthquake, for example, salt storage chambers would tend to be self-healing.

The AEC was convinced that the Lyons salt domes were the

proper place to store radioactive waste and assured Congress that all the necessary studies had shown that there would be "no significant impact on the environment."[26] However, the AEC had not done its homework. Fortunately, the Kansas Geological Survey and individual citizens did theirs. Less than a half-mile away from the AEC's proposed waste dump was an active salt mine where the owners had experimented with hydraulic mining, a technique where water is pumped into the salt formation. In one such experiment, the operators had "lost" about 175,000 gallons of water. The water had not returned to the surface, and no one knew exactly where it had gone.[27] The implications for the waste dump were serious.

Because the radioactivity in the wastes generates heat, the canisters holding the wastes can reach temperatures of over one thousand degrees Fahrenheit. This heat could vaporize any water which touched the canisters and expel steam, water, salt, and waste particles throughout the mine and possibly outside it.[28] Small amounts of water could also corrode the canisters and result in a slow release of radioactivity to the environment.

The AEC withdrew its proposal in September 1971.[29] Despite the problems with the Lyons site, the AEC insisted that salt deposits were good candidates for waste dumps and began investigating other sites. ERDA, continuing the hunt in deposits in New Mexico, in 1975 found brine and toxic gases near one survey site—a situation which delayed the ERDA project for several months.[30] More recently, ERDA has announced that it may conduct geological surveys in as many as forty-five states.

All the "solutions" for the waste problem are long-term proposals. But the inability of the federal government to handle its own wastes gives little confidence that the government can manage future commercial nuclear wastes, for which the solidifaction and storage technologies are still unknown. The government wastes, generated from the nuclear weapons program, are stored in liquid form in steel tanks at various ERDA "reservations." One such reservation is in Hanford, Washington:

DR. PARKER (*Advisory Committee on Reactor Safeguards*): "*Did you try the system when they had the leaks in the tanks at Hanford?*"

MR. DEAL (*Atomic Energy Commission*): "*That is why we were there, yes. There is a funny story associated with that. We were flying around. There is quite a bit of cesium around one of those tank farm areas where it came out.*

And the fellows kept seeing this stuff around. The Hanford people were trying to get a handle on how extensive it was.

It turns out somehow the jackrabbits had gotten ahold of the nitrate solution and were eating it. They were seeing these hot pellets from jackrabbits. Some were quite warm in terms of radioactivity. We found out the extent the jackrabbits had spread around the area. They saw one quite very hot place nobody could figure out. Finally it turned out it was the carcass of a coyote."

DR. MOELLER (*ACRS*): "*So you learned about the dietary habits.*"

MR. DEAL: "*Yes.*"[31]

An abandoned salt mine near Lyons, Kansas, site of the Atomic Energy Commission's abortive attempt to store radioactive waste

U.S. Energy Research and Development Administration, Oak Ridge National Laboratory

The leaks to which Joseph Deal referred occurred at the government facilities near Hanford, Washington. The leaks show that even with government programs there has been irresponsibility. Atomic promoters have argued that the leaks are insignificant because they will "never" migrate from the soil or have any effect on the environment or human beings. But the Hanford jackrabbits have turned such reassuring statements into nonsense.

At Hanford, about 500,000 gallons of radioactive waste liquid have seeped into the soil from leaks at several tanks.[32] A single leak of 115,000 gallons from one tank was particularly disturbing:

> The official investigation revealed that the tank had been leaking for several weeks, that no automatic alarm system alerted anyone to the leak, that the management in charge of the storage facility did not review monitoring reports that would have shown the leak, that there was no preventative maintenance applied to the monitoring equipment, and that persons with responsibilities for overseeing the storage of these waste materials had no formal training to assist them in the execution of these important responsibilities.[33]

Also at Hanford, until 1973 it had been the AEC's practice to discharge liquid plutonium wastes into enclosed trenches. In 1973, there was some concern expressed by AEC officials that there was enough plutonium in one trench, number Z-9, to conceivably cause a nuclear chain reaction in the trench. Although subsequent studies showed that a chain reaction was probably not possible, the situation provided little encouragement that the government could manage nuclear waste.[34]

Nuclear power promoters would claim that the AEC's mismanagement was applicable to weapons waste, not commercial waste. The atomic industry has claimed that commercial waste will be solidified in small volumes, approximately 90 cubic feet per year for each 1000 MWe nuclear plant.[35] The industry likes to point out that the waste generated from a family's yearly consumption of nuclear electricity will be no larger than a few aspirin tablets.

This statement fails to point out that volume is a relatively unimportant measure of hazard; these "aspirins" are extremely toxic. The amount of strontium-90 alone in the waste generated by one family's annual consumption of nuclear plant electricity is enough to contaminate one billion gallons of water beyond the NRC's maximum allowable concentrations in drinking water.[36] Further, the 90 cubic feet figure represents only "high-level" waste. Nuclear plant waste at lower levels of contamination can represent 8000 cubic feet or more per plant per year.[37] Also important is the fact that the annual high-level waste generated by each large nuclear power plant will become 10,000 gallons of liquid waste at a reprocessing plant. Only after solidification by a questionable technology will this liquid become 90 cubic feet of solid material.[38]

It is also true that waste from commercial nuclear plants presently is small compared to the weapons program. But the amount of commercial waste that must be handled will grow rapidly with the proposed growth of the atomic power industry. By the year 2000, accumulated wastes could be 25,000 metric tons—40 to 60 million gallons in liquid form.[39]

Radioactive waste storage tanks under construction at the federal government's Hanford, Washington, site. Each tank is designed to hold one million gallons of liquid radioactive waste.

Battelle Northwest Laboratories

In summary, nuclear waste represents a serious unresolved hazard which the industry, despite public relations efforts to proclaim it a non-problem, cannot banish. There are presently 600,000 gallons of high-level "commercial" nuclear waste in storage at Nuclear Fuel Services, and the total will grow rapidly.[40] Although the industry promises all types of solutions to waste storage, the solutions are at best speculative. When the proposed solutions have been tested, even on a small scale, they have been fraught with practical problems. The AEC's mismanagement raises questions about any waste storage program. Moreover, it is unlikely that any method can be developed which can realistically guarantee the stability of geological formations as well as human institutions for the quarter-million years or more which may be necessary. The waste problem is so serious that many scientists and citizens believe a nuclear moratorium is dictated by that problem alone. Nuclear electricity generated now could burden thousands of future generations with its lethal byproducts.

Transportation

Transportation is the thread that binds together all the other segments of the nuclear fuel cycle. The activities of greatest concern are the shipment of highly radioactive spent fuel from the reactor and subsequent shipments of waste from the reprocessing plant. The irradiated fuel is shipped in large steel casks designed and constructed to prevent accidental release of radioactivity. But, as with the controversy on reactor safety, there is also a controversy on transportation safety.

One group which did not agree with the AEC's evaluation of hazards from nuclear material transport was the Public Interest Research Group in Michigan (PIRGIM). In a 1974 study, PIRGIM made extensive criticisms of the AEC's transportation regulations. For example, PIRGIM noted that companies designing the shipping casks are required to bring the designs to AEC field offices for approval. But a report by the General Accounting Office (GAO) examined four AEC field offices and concluded that three did not have the expertise to properly evaluate the proposed cask designs.[41]

The same GAO report found that the AEC's requirements for reporting leaks of radioactive material from the casks were so vague that many leaks went unreported. Between 1969 and 1972, there were sixty-four unreported instances in which the containers or the vehicles carrying them were contaminated beyond specified levels. These unreported instances were discovered only by the GAO's special investigation.[42]

Moreover, the casks are by no means fail-safe. The AEC admitted that an accident at speeds over fifty miles per hour could rupture the fuel rods, resulting in a radioactive release.[43] PIRGIM pointed out that the AEC did not test the casks for impalement at substantial velocity, or for broadside impacts at high velocity against curved objects such as bridge abutments.[44] The casks are supposedly built to withstand temperatures up to 1475° for half an hour. But PIRGIM pointed out that eleven substances—such as benzene, propane and toluene—which are shipped in trucks on the highways can reach temperatures as high as 4275° when they catch fire.[45]

Nor were the AEC's assurances on transportation accepted by the Association of American Railroads (AAR), an industry group whose member companies include almost all railroads in the United States. During 1974 hearings before the Senate Commerce Committee, the Association pointed out that rail accidents can and do take place in which more energy or longer fires are involved than assumed in the NRC certification and testing process.[46] The Association recommended the following operating procedures for nuclear fuel: "Shipments of casks containing irradiated spent fuel cores should move in special trains containing no other freight, not faster than 35 MPH. When a train handling these shipments meets, passes, or is passed by another train, one train should stand while the other moves past not faster than 35 MPH."[47]

Another point of concern for the railroads is the acknowledged poor condition of their tracks in many locations, along with the poor financial condition of many railroads which inhibits repair of track and equipment. In December 1975, twenty-six railroad companies requested from the Interstate Commerce Commission (ICC) the right to insist that spent fuel cores be transported on special trains. Two federal agencies, ERDA and the Tennessee Valley Authority, which builds and operates nuclear plants, have asked the ICC to deny the railroads' request. ERDA, which operates research reactors, and TVA both require rail carriers for their own nuclear waste and would be affected if the ICC granted the request.[48]

The NRC, the one agency which should welcome the railroads' request for improved safety, has taken no action to support the twenty-six companies against ERDA and TVA. Nor has the NRC taken any steps to adopt the recommendations of the Association of American Railroads.

In at least one instance, Congress has declined to accept the assurances of the Nuclear Regulatory Commission on transportation. On February 24, 1975, a shipment of powdered plutonium on a cargo plane entered Kennedy Airport in New York City, and was subsequently trucked through Queens and Manhattan on its way to a Westinghouse facility in western Pennsylvania.[49] The New York branch of Friends of the Earth, the New York Public Interest Research Group, and the office of Congressman Les Aspin (D., Wis.) all made inquiries and determined that the NRC had authorized four plutonium shipments to be flown into Kennedy between July 1974 and February 1975, although there had never been an assessment of the health hazards or environmental contamination that might have resulted from an air crash with plutonium. The NRC licensing of these plutonium shipments had been contrary to the recommendations of a select panel to the Joint Committee on Atomic Energy. The panel had recommended that air shipments of plutonium be banned because of the danger of an aircraft accident and resulting contamination.[50]

Analyses performed by independent scientists have indicated that the consequences of a plutonium air crash could be severe indeed. In material prepared for the New York State Attorney General, Marvin Resnikoff, professor of physics at the State University of New York at Buffalo, calculated that the release of 2.8 percent of the plutonium in a single shipment could kill 30,000 persons from exposure at the airport, and 46,000 members of the general population could develop lung cancer. John Gofman, professor emeritus of medical physics at the University of California at Berkeley, stated that for the dispersal even of one-hundredth the amount of plutonium in Dr. Resnikoff's analysis, the evacuation of New York City "would require serious contemplation."[51]

Congressman James H. Scheuer's (D., N.Y.) concern about plutonium shipments led him to propose an amendment prohibiting air transport of plutonium until the Nuclear Regulatory Commission certified ". . . that a safe container has been developed and tested which will not rupture under crash and blast-testing equivalent to the crash and explosion of a high-flying aircraft."[52] Congressman Scheuer's amendment passed both houses of Congress and was signed into law. The NRC has not yet certified a crash-proof container, so air shipments of plutonium have ceased, for the time being.

Nor did the city of New York accept the NRC's assurances on transportation. On July 9, 1975, Dr. Leonard Solon, director of the city's Bureau of Radiation Control, attempted unsuccessfully to prevent a shipment of highly enriched uranium from entering the city. Solon was concerned because the highly enriched uranium was weapons-grade material, and he viewed the possibility of theft or sabotage of the shipment as an unacceptable hazard to the people of New York. Solon met the truck carrying the uranium at the George Washington Bridge at 4:00 A.M. *Although he was unsuccessful in stopping the truck, he accompanied the shipment with a police escort to Kennedy International Airport and ordered that the nuclear material be taken out of the city as soon as possible.*

As a result of the July 9 incident, the plutonium air shipment incidents, and a general concern for the transportation of radioactive material through highly populated areas, the New York City health commissioner, Dr. Lowell E. Bellin, declared an unconditional ban on the shipment of nuclear material through the city. Bellin stated: "The incident at the [George Washington] Bridge points to the compelling need for a consistent posture by the city in order to protect the health and welfare of New Yorkers from the incipient danger from radioactive materials, which are being transported through the city, at increasingly frequent intervals."[53]

Bellin later proposed an amendment to the city's Health Code which would allow the transportation of medically related radioactive material but severely restrict the transportation of plutonium, highly enriched uranium, or spent fuel rods into or out of the city. Bellin held hearings on his amendments, and federal agencies produced a troop of technical and legal witnesses to oppose the amendments.

The federal witnesses argued that the U.S. government had "preempted" the regulation of transportation of nuclear material, and that the state and city of New York had no authority in the matter. Not satisfied with this response, Bellin tried to make clear that the city was determined to protect its streets from radioactive material. At one point, he replied to a Nuclear Regulatory Commission witness with, "There are those of us who are prepared to secede." When the laughter had died down, Bellin continued, "We have had firm offers from Canada and Mexico."[54]

The federal government was unsuccessful in convincing Dr. Bellin to withdraw his amendments and subsequently took New York City to court in January 1976.[55] New York City had thus joined with public interest groups, the railroads, and the U.S. Congress in expressing its skepticism over the safety of nuclear material transportation. The NRC, in attempting to prohibit the Health Code amendments, had placed itself in the ironic position of opposing an action by the city of New York to protect its citizens from the dangers of radioactive material.

X

Worker Safety

Albert Moon is fifty-two years old. He suffers shortness of breath and extreme fatigue, and requires the use of an oxygen tank for several hours each day. Moon worked for twenty years as a machinist at the Dow Chemical Company's Rocky Flats nuclear weapons plant outside Denver. While working at the plant he was routinely exposed to radioactive uranium dust.

In 1972, Moon had part of his left lung removed because he had contracted cancer. Ironically, Dow would not permit him to return to his job because of his disability. After eighteen months of attempting to obtain disability benefits from Dow, Moon filed a claim for benefits with the Colorado Workman's Compensation Board. The Dow Company denied liability for Moon, maintaining that his health problems were not work-related because Moon was a heavy smoker. In 1976, the Compensation Board denied Moon's claim on the grounds of "insufficient medical evidence" that his cancer was work-related. With Dow's refusal to grant a pension, and Colorado's refusal to grant compensation, Moon is eligible only for social security payments . . . hardly adequate in view of the extended medical care he requires.[1]

ALBERT MOON'S STORY typifies the hazards that workers in the atomic industries face. Moon has suffered permanent damage and cannot receive compensation because he cannot "prove" that his injury was radiation-caused. Other workers who have received high exposures may be in fine shape now, but they know that radiation can cause cancer and recognize that cancer has a long latent period. So they can do little but wait and hope. If they do come down with cancer, they face the same barriers as Albert Moon did when he tried to obtain compensation.

Although Dow Chemical, at its Rocky Flats plant, produced materials for nuclear weapons, workers in the nuclear power program come in contact with the same substances—uranium and plutonium—that are used in weapons. In addition, workers in the back-end of the fuel cycle will face all the radioactive reactor by-products. As with other environmental pollutants, the worst radioactive contamination occurs in the environment of the workplace, where the worker is exposed to greater concentrations of pollutants than the general population. Nuclear radiation, like other pollutants, threatens to make man physiologically obsolete. In small amounts, radiation cannot be felt, heard, smelled, or tasted until it is too late. The normal human senses thus offer no warning of radiation dangers.

As with other pollutants, exposure to radioactive materials can cause cancer and the illness may not develop until five to thirty years after the initial contact. Furthermore, if a cancer develops, the tumor offers no indication of which carcinogen (cancer-causing substance) might have been responsible. This places the burden of proof on the worker. Once exposed, the victim must prove that the resulting cancer was work-related if any compensation is to be paid. As in the case of Albert Moon, the victim's employer will typically devote a substantial amount of time, money, and legal talent to discrediting a worker's claim of job-related disability. Corporations want no precedents established.

Promoters of atomic power like to portray an image of a "clean" industry. The smokestacks of nuclear plants don't emit smoke (although they do emit radioactive materials on a routine basis). Employees at nuclear plants are depicted as highly-trained engineers or technicians in white lab coats. This picture may be reasonably accurate at the power plant during routine operation, but during shutdown maintenance operations (for refueling or repairs) workers can receive significant radiation exposures. Nuclear reactor plant accidents have also claimed the lives of at least five U.S. workers—three men at the SL-1 research reactor in Idaho Falls were killed in 1961 when the reactor underwent a steam explosion, and two workers were killed at Virginia Electric and Power's Surry plant in 1972 when a valve backed out of a pipe and scalded them with steam.

The dirty work in the nuclear industry, however, is done outside the power plants, in the other operations that make up the nuclear fuel cycle. The worst exposures have been and will be received in the mines, mills, fabrication and reprocessing plants.

Historically, radiation standards have been continually made more stringent as more information has become available. This is illustrated by the proposals of the International Commission on Radiological Protection (ICRP) and the National Council on Radiation Protection (NCRP), the non-governmental organizations established to make recommendations on radiation standards. In 1925, the ICRP's first recommended limits allowed up to 100 rem per year, a standard which might have prevented immediate death, but which did not recognize radiation's ability to cause cancer or genetic effects. In 1934, the NCRP recommended 36 rem per year, which the NRCP in 1947 lowered to 15 rem per year for workers.[2] In 1958, the recommended average yearly occupational dose was reduced to 5 rem per year, which is presently in effect. However, both the historical record and the continuing controversy on radiation standards suggest that further reductions may be in order.

Unless a worker drops dead at the workbench the atomic industry will usually argue that death was due to "natural causes." But nuclear power's occupational hazards are manifested more in long-term cancer than in immediate lethality. The long latency period of radiation-induced illness means that the occupational toll of atomic power might not become apparent for several years. But even the recognized dangers are considerable.

The Karen Silkwood Case

On the night of November 13, 1974, a small white Honda automobile careened off the road and crashed outside Crescent, Oklahoma, thirty miles from Oklahoma City. The driver, twenty-eight-year-old Karen Silkwood, died almost instantly. Silkwood was a plutonium technician on her way to Oklahoma City to meet with a reporter from the *New York Times* and with a union official from the Washington, D.C., office of the Oil, Chemical, and Atomic Workers International Union. At the meeting, Karen was to have handed over records documenting alleged irregularities and falsification of quality control data at the Kerr-McGee Corporation's nuclear facility near Crescent. The documents, carried in a manila folder, were missing when union officials reached the scene of the accident and to this day have not been found.[3]

Karen Silkwood's death set in motion a chain of events that has yet to be completed, including investigations by the Atomic Energy Commission and the FBI which have raised more questions than they have answered. The case has also raised questions of employee intimidation by Kerr-McGee. The history of the Silkwood case, described by several observers as "bizarre," is instructive in what it reveals about the problems of workers in the nuclear industry.

An interview with William and Merle Silkwood, Karen's parents, was published in the November 1975 issue of PIRG News, a monthly newsletter prepared by the Public Citizen Action Group. The following is an excerpt from that interview:

MR. SILKWOOD: *Most people don't answer our letters, even the FBI. We have no way of getting information. There just isn't anything being done about it, but there seems to be pressure building nationally to get this thing out in the open.*

QUESTION: *Have all the documents in the case been released?*

MRS. SILKWOOD: *No. We can't even get an answer from the FBI.*

Q.: *Has anyone tried using the Freedom of Information Act [FOIA]?*

MRS. S.: *National Public Radio has already filed suit to get this information. Who knows how long it will take or if they can ever get it? All the FBI and the AEC have to do is say the information is confidential or restricted. Then they're out of it. They don't have to go any further. So it looks like the FOIA isn't what it's supposed to be. This case has been covered up ever since Karen was killed. Both the Highway Patrol and Kerr-McGee have made statements they've had to retract. I'm sure the Atomic Energy Commission [AEC] finally told Kerr-McGee to keep quiet and let the AEC handle it.*

Q.: *Do you think the AEC has done all it can to find out exactly what happened?*

MR. S.: *If they did, they didn't show it in their report. Just like any other report, they'll let you see just so much of it. They may say they interviewed fifty or sixty people, but you don't know who they interviewed or what questions they asked. That's the part you don't see. If they don't ask the questions about the cover-up, then they haven't done their job. Their report looks good until you dig into it.*

Q.: *Do you think public awareness is increasing about Karen's death and about the nuclear industry's apparent lack of concern for safety?*

MRS. S.: *Yes. I think the public is becoming more aware. In fact, at NOW's [National Organization for Women] recent conference in Philadelphia, most of the five thousand attendees signed petitions calling for a Senate investigation.*

MR. S.: *Of course, not only Kerr-McGee is responsible, but the AEC is too. According to the AEC, no nuclear plant worker has died of radiation. That's a damn lie. I know of three people who have died within three months of an accident. I don't know how they can make that broad statement, but they do. They come out and tell the press that no one has died as a direct result of working in one of their plants. Look at all the people who die of cancer ten or fifteen years after being exposed to radioactivity by working in a nuclear plant.*

Q.: *The plant is being shut down. By January 1976 it will no longer manufacture plutonium fuel rods. Will the plant's closing be permanent?*

MR. S.: *No, I think Kerr-McGee is waiting for things to cool off surrounding our daughter's death. Kerr-McGee really turned people against Karen before Christmas when they closed the plant down for two weeks. Of course a lot of workers blamed Karen for it. She was already dead; she had nothing to do with it. Kerr-McGee closed the plant to show people that if they didn't do what Kerr-McGee wanted them to, they could just go without jobs. And then when they reopened the plant, they required all the employees to take a lie detector test. The ones who didn't lost their jobs shortly after that.*

On November 5, 1976, the parents of Karen Silkwood filed suit against Kerr-McGee Corporation, charging the company with conspiring to prevent her from union activities and from reporting nuclear safety problems to the federal government. The suit asked for compensation and punitive damages totaling $160,000.

The Kerr-McGee Corporation is an energy giant in the southwestern United States. The company is perhaps most widely known for its holdings in oil and gas. But Kerr-McGee also has a sizable interest in nuclear power. The company owns uranium mines, mills, a uranium fuel conversion plant, and fuel fabrication plants.[4] The Crescent facilities include plants to fabricate fuel from uranium and plutonium for nuclear reactors.

The plutonium facility, where Karen Silkwood worked as a laboratory technician, existed almost exclusively to produce fuel rods for the Fast Flux Test Facility (FFTF), part of the federal government program to develop the breeder reactor. The FFTF is located at the ERDA research reservation near Richland, Washington.

Plutonium, the fuel material for the FFTF and the breeder, is an extremely toxic substance. Dean Abrahamson, a medical doctor and professor at the University of Minnesota, has called plutonium 20,000 times more toxic than cobra venom or potassium cyanide, the gas used in gas chambers.[5] A few millionths of a gram of plutonium, when inhaled, has caused cancer in laboratory animals.[6]

Because plutonium is such a deadly substance, one would expect meticulous care at the plant—including training of workers and adequate protective measures—to prevent exposure to plutonium. At Kerr-McGee, however, eighty-seven individuals were exposed to excessive levels of plutonium in twenty-four different accidents between July 1970 and December 1974.[7] Karl Z. Morgan, a supporter of nuclear power who is a recognized expert on radiation and professor of nuclear engineering at the Georgia Institute of Technology, testified in 1976 that he had never known any plant in the nuclear industry to be "so poorly operated" as the Kerr-McGee plutonium facility, with the possible exception of the Nuclear Fuel services plant in West Valley, New York.[8]

One of the workers contaminated was Karen Silkwood. Her concern about health and safety problems at Kerr-McGee led her to become an activist with the local Oil, Chemical, and Atomic Workers (OCAW) union at the plant. In September 1974, Karen and two other members of the union flew to Washington, D.C., for a meeting at the OCAW International office. There they met with Director Anthony Mazzocchi and his legislative assistant Steve Wodka. The Oklahoma delegation gave their information on health and safety problems to Mazzocchi and Wodka, who took them to AEC officials. The AEC listed the charges and promised an investigation.

Steve Wodka was also interested in Karen's suspicions that quality control records on plutonium fuel rods at Kerr-McGee were being falsified to conceal defective welds. At the suggestion of the OCAW International officials, Karen agreed to do some investigating on her own to document rumors about the falsified records. When she got back to Crescent, Karen began compiling a dossier on the fuel rods. Apparently unknown to her, other employees noticed her actions.

On November 5, a monitoring device unexpectedly detected plutonium on Karen's skin and clothing. After showering, she checked with the monitor to be sure that the plutonium had been removed. Just before leaving work to go home she checked herself once more with a monitor and found a normal reading. The following morning, shortly after she arrived for work, monitors again detected contamination and she repeated the decontamination procedures. Then, on November 7, she found contamination on her clothing a third time after she reported to work. A nasal smear indicated that she was also contaminated internally.

Since there had been no recent accident at Kerr-McGee to account for Karen's repeated contamination, she requested that the company check her apartment. They found radioactivity throughout the apartment, the source of which appeared to be packages of cheese and bologna in the refrigerator. On November 10, Kerr-McGee flew Karen, her roommate, and Drew Stephens, a friend who had been in Karen's apartment, to Los Alamos, New Mexico, for further testing of internal plutonium contamination.

The Los Alamos testing, which went on for two days, found that the three contaminated people were in no immediate danger; the doses to Karen's friend and roommate were considered biologically insignificant. Karen and her friends returned to Oklahoma. She had promised Steve Wodka that she would deliver the information on falsification of records after she returned from Los Alamos.

In the early evening of November 13, Karen left a restaurant near the Kerr-McGee plant and started driving to Oklahoma City. A fellow union member later swore in an affidavit that Karen was carrying a manila folder about an inch thick with papers. Seven miles and ten minutes down the road from the restaurant, Karen Silkwood was dead. By the time Steve Wodka, Drew Stephens, and David Burnham, the *New York Times* reporter—who had all been waiting for Karen—got to the scene of the accident, the car had been towed back to town. When the three reached the car the next morning, the manila folder was not present.

The Oklahoma Highway Patrol concluded that Karen had fallen asleep at the wheel. An autopsy by the state of Oklahoma found that Karen's body contained "more than a therapeutic dose" of the tranquilizer she had been taking for the past several days to calm her nerves through the ordeal of plutonium contamination.[9] Conceivably, the tranquilizer could have put her to sleep. But the OCAW was bothered by the missing manila folder, and the union called in an outside expert to investigate the crash.

The OCAW's investigator was A. O. Pipkin, a former policeman who had investigated over 2,000 auto accidents and was a recognized expert in traffic accident reconstruction. Pipkin disagreed with the Oklahoma police. First, he found that Karen's Honda had drifted off the road to the left. Because the road was peaked in the middle, the car should have left the right side of the road if the driver had been asleep. Next, he noticed that the steering wheel was bent forward, indicating that the driver had locked elbows before the crash. Only an awakened driver could have done that. Finally, Pipkin saw dents in the left rear fender and bumper. His laboratory analysis indicated that the dents came from another car. Pipkin's conclusion: Karen Silkwood was struck from behind by another car which either forced her off the road or startled her into leaving the road.[10]

After Karen's death, events at the Kerr-McGee plant continued to unfold:

From November 8 to December 4, 1974, the Nuclear Regulatory Commission (NRC) investigated the plutonium contamination of Karen and her apartment.

From November 21 to December 6, the NRC investigated the union's charges on health and safety violations.

During these investigations, Kerr-McGee found uranium pellets and pellet fragments on the ground outside the uranium pellet manufacturing building, but within the Kerr-McGee perimeter fence. The NRC added the uranium pellet incident to its investigation.

On December 10 the Energy Research and Development Administration (ERDA) set up a task force to review all quality assurance and inspection procedures related to Kerr-McGee's fabrication of plutonium fuel.[11]

In early January 1975, Kerr-McGee conducted lie detector tests for all of its plutonium employees. The test consisted of such questions as: Are you a member of the union? Do you take or use narcotics? Did you ever talk to Karen Silkwood? Have you ever done anything detrimental to Kerr-McGee? Have you talked to the press or media? Workers who refused to take the polygraph test or who did not "pass" the test were fired or demoted. The union complained to the National Labor Relations Board, charging Kerr-McGee in January 1975 with harassment and intimidation of employees who had complained to the NRC about safety violations. (One employee has since been awarded compensation as the result of an arbitrator's ruling that he had been unjustly dismissed.)[12]

Meanwhile, union complaints in November 1975 prompted the

Department of Justice to order an FBI investigation into events surrounding Karen Silkwood's death and the contamination in her apartment—which signified unauthorized removal of plutonium from the Kerr-McGee plant. The FBI also considered the union's charges of harassment.[13]

On January 6, 1975, the NRC released its report on the contamination of the apartment. This investigation found that two urine samples submitted by Karen had been tampered with: plutonium had been added to them. Who had tampered with the samples could not be determined. Nor could the NRC determine how Karen and her apartment had been contaminated with plutonium in the first place.[14]

A day later the NRC released its investigation of health and safety conditions at Kerr-McGee, concluding that twenty of the thirty-nine allegations made by the OCAW were substantiated in whole or part. The remaining allegations either could not be confirmed or were said to be "outside" the scope of the investigation. Three of the substantiated allegations represented violations of NRC license requirements. Kerr-McGee claimed to have corrected those violations and the NRC took no further action against the company.[15]

ERDA also released its investigation into quality-control procedures at Kerr-McGee on January 7. ERDA conducted an investigation because the plutonium fuel rods made at the Cimarron facility were to be used in the FFTF, a reactor under ERDA's jurisdiction. ERDA concluded that photographs of fuel rod welds had been touched up with felt-tip pen markings on the negatives. The worker who used the pen told ERDA that it was done to improve the quality of the photographs, not to hide defects. ERDA accepted this explanation.[16]

Later in January the NRC released its report on the uranium pellets. The NRC could not determine how the pellets had gotten outside the manufacturing plant but implied that an employee had thrown them there to embarrass the company.[17] On May 2, 1975, the FBI unexpectedly announced that it had closed its books on the Silkwood case and concluded that her death was accidental.[18] The agency offered no conclusions on the illegal removal of plutonium from the plant or on worker intimidation by Kerr-McGee.

Kerr-McGee's response to the Silkwood affair is also noteworthy. When five employees were contaminated at the plutonium plant on December 17, 1974, the company closed the plant with the announcement that the accidents were "contrived."[19] All workers were laid off for ten days, and when they returned the polygraph tests were begun.[20] Closing the plant three days before Christmas also reminded the workers that continuing bad publicity could shut down the plant and leave the workers without jobs. In December 1975 Kerr-McGee quietly announced that it would shut down its plutonium plant because it had completed its contract with ERDA and had no more buyers for plutonium fuel rods.[21]

Although the federal agencies consider most of the Silkwood investigations closed, difficult questions remain: What happened to the manila folder? How was plutonium removed from the plant to Karen's apartment? What was wrong with the Kerr-McGee security procedures intended to prevent or detect plutonium removal? What was the real cause of Karen's "accident"? These and other questions led to demands that the Silkwood case be reopened.

The National Organization for Women (NOW) on August 26, 1975, petitioned the Justice Department to reopen its investigation. NOW stated that it made its request on the fifty-fifth anniversary of women's suffrage in the U.S. because "there is no better example of violence against women than the Silkwood case."[22] NOW has been successful in getting the House Small Business Subcommittee on Energy and the Environment to hold hearings on the case, but the Justice Department has not ordered the FBI to reopen its investigation.

National Public Radio (NPR) filed a Freedom of Information suit on September 29, 1975, against the Justice Department for records on the Silkwood case. NPR later received documents from the NRC which were disturbing. The NRC health and safety inspector with responsibility for the Kerr-McGee plant stated in a March 20, 1975, memo that he thought Kerr-McGee had a "good safety attitude." The same NRC inspector believed that the OCAW allegations were "motivated as bargaining tools and not as a real concern for safety."[23] This attitude is not supported by the company's record of eighty-seven contaminated workers in four and a half years. The NRC's attitude, along with all the unanswered questions that remain, indicate that the Silkwood affair can by no means be considered a closed case.

Other Plutonium Factories

Kerr-McGee is not the only company which has had difficulty handling plutonium. Other plants have not received the publicity that Kerr-McGee has, but they have had safety records which are equally suspect. In September 1974, Robert Gillette of *Science* magazine reviewed the records of commercial firms which had handled plutonium and concluded that none of them was in command of the technology:

> The record reveals a dismal repetition of leaks in glove boxes; of inoperative radiation monitors; of employees who failed to follow instructions; of managers accused by the AEC of ineptness and failing to provide safety supervision or training to employees; of numerous violations of federal regulations and license requirements; of plutonium spills tracked through corridors, and, in half a dozen cases, beyond plant boundaries to automobiles, homes, at least one restaurant, and in one instance to a county sheriff's office in New York.[24]

Gulf United Nuclear Fuels produced small amounts of plutonium in a laboratory at Pawling, New York between 1970 and 1972 until a fire and explosion injured one worker and contaminated two others.[25] The accident, on December 21, 1972, resulted in extensive plutonium contamination within the facility, a breach in the exhaust system in the plutonium-handling room area, and the release of an undetermined quantity of plutonium from the building through blown-out windows. Subsequent AEC investigations found a multitude of violations and safety problems, including failure of the company to require personnel to be monitored before leaving a contamination area.[26] The Gulf facility never reopened after the fire.

The Nuclear Materials and Equipment Corporation (NUMEC), now a subsidiary of reactor manufacturer Babcock & Wilcox, operates a plutonium production plant near Leechburg, Pennsylvania. Although the working crew numbers only one hundred or so, thirty persons were overexposed to airborne plutonium in at least thirteen incidents from late 1969 to September 1974. Six of these incidents resulted from repeated leaks of plutonium from one piece of equipment over a thirty-day period in 1973.[27]

On August 12, 1974, the AEC fined NUMEC $12,170 for sixteen separate violations related to health, safety, and security.[28] In a letter to NUMEC citing the reasons for the fine, the AEC stated:

"Our review of the NUMEC enforcement history for the calendar year 1973 and the violations noted in inspections during calendar year 1974 indicates a history of repeated violations and unfulfilled commitments to correct violations."[29] On June 5, 1974, the AEC had informed NUMEC of its intent to fine the company. As an indication of the company's problems, on June 11, a worker was contaminated by plutonium as a result of failure of a glove box.*[30] After NUMEC had paid its fine in August, yet another worker was contaminated by plutonium within fifteen days.[31]

Nuclear Fuel Services, a subsidiary of Getty Oil, has operated plants to handle plutonium in two different areas of the country. A facility in Erwin, Tennessee, is still operating, but has experienced at least fifteen separate incidents since 1969 in which more than fifty workers have been exposed to radiation above permissible limits. On October 18, 1974, the AEC cited the Erwin plant for five licensing violations—all related to health and safety. These violations were discovered only after members of the local OCAW union requested a meeting with the AEC to complain about unsafe working conditions.[32]

Another dramatic example of safety violations is the Nuclear Fuel Services (NFS) reprocessing plant at West Valley, New York. As has already been reported, the probable reason for this plant's shutdown was its failure to control releases of radioactive material and failure to instruct or train workers.[33] At NFS, West Valley, at least fifteen separate incidents between 1966 and early 1973 exposed thirty-eight persons, who either inhaled or ingested the materials, to "excessive concentrations of radioactive materials."[34]

Another disturbing practice at West Valley was the use of several thousand temporary employees to work in the "hot" (intensely radioactive) areas. These short-term workers were unemployed laborers, moonlighters, college students, and other individuals from a construction firm and a temporary labor firm in the Buffalo area. If NFS had required its permanent employees to perform the work, they would have reached their radiation exposure limits before all tasks could be completed.

* A glove box is a container with a transparent window and permanently installed gloves. Such a device allows a worker to handle plutonium, by inserting his or her hands into the gloves, without coming into contact with the material. The transparent window allows workers to see their own actions. A leak in the glove box, of course, would allow plutonium to escape into the air, where it could be inhaled.

Imported temporary workers were hired to go into high-radiation areas to perform unskilled tasks, such as wiping radioactivity off equipment or removing nuts and bolts. After a worker absorbed the radiation limit for a calendar quarter (which is the shortest time period for any AEC regulations), or had come close to the limit, the worker would be paid and sent home. Depending on the intensity of radiation, an individual could reach a quarterly limit in a few days or even a few hours.

NFS used an average of 1,400 temporary workers in each year of its operation. Robert Gillette, the *Science* journalist, concluded that the NFS temporary workers were given "an apparent minimum of instruction in safety procedures and the potential hazards of their jobs."[35] This practice at NFS and at other plants of using occasional workers blurs the distinction between workers and "the general public," who have exposure limits ten times more stringent than those for industry employees.

It is important that the problems mentioned above are relevant to commercial plants handling nuclear fuel; they do not include experiences from the weapons program, which also has had problems with plutonium. The amount of plutonium in the commercial sector at present is relatively small but will expand dramatically if the industry is successful in its efforts to have wide-scale use of plutonium licensed.

The atomic industry has argued that, among workers in the weapons sector or the commercial sector, there have been no proven occupational cancers from plutonium exposure. This claim can remain unchallenged only because there has been little follow-up investigation into the health of plutonium workers after they leave the job.

It was not until 1968 that the Atomic Energy Commission began to set up the Transuranium Registry—a medical data bank to maintain the health records of former plutonium workers. More than 9,000 workers have been identified as having worked in or around plutonium facilities, chiefly operated by the government. While the AEC for several years maintained that the Registry would establish an epidemiological study of plutonium workers, the government now claims that the Registry's only purpose was to measure plutonium depositions in workers' bodies. The Energy Research and Development Administration (ERDA) indicated in 1976 that it will be several years before the Registry gathers enough information to make meaningful comparisons between plutonium workers and the general population.[36] Meanwhile, ERDA is promoting the breeder reactor and a plutonium industry. This lack of follow-up may only be setting the stage for a tragedy similar to that with uranium miners. In that situation, unbending government officials would not acknowledge dangers to the miners until too late.

Uranium Miners

Helen Caldicott is an Australian doctor presently working, under a fellowship, as a pediatrician with the Children's Hospital in Boston. During 1974 and 1975, she organized opposition to the exportation of uranium from Australia. She described this process in Australia at Critical Mass 75, a national gathering of the citizens' movement to stop nuclear power held in November 1975 in Washington, D.C.:

In 1972, there was a widespread public response when it was revealed that radioactive rain had fallen on some parts of Australia as a direct result of atmospheric testing of nuclear weapons, by the French, in the Pacific Ocean. At that time the media, the people, and the government of Australia responded to a case based on the medical and biological dangers of radiation. A Gallup Poll showed that more than three-fourth of the Australian people were opposed to the testing, and both major political parties were obliged to include opposition to French testing in statements of policy. However, three years later, when the Australian government announced plans to promote uranium sales on the world market, neither politicians nor the media would listen to the same biological arguments against radioactive pollution caused by the nuclear power industry. Clearly the government was motivated by a potentially huge increase in national wealth. Initially the response of the media was baffling. They had no great love for the government of the day, but it was learnt that some of the media companies have a significant financial interest in uranium mining.

I approached the seventy-six trade unions in the state of South Australia, asking them to give me the opportunity to present to their members the biological arguments against the mining and exportation of Australia's uranium resources. The response was spectacular. Many of the trade unions passed resolutions against the uranium industry and sent strongly-worded protests to the government. Several of the more influential unions made even more dramatic stands. The Railway Workers' Union and the Waterside Workers' Union decided not to allow their members to handle uranium. The miners decided not to work uranium mines. Finally, the parent body of the Australian Council of Trade Unions put a ban on the handling of uranium until there had been an environmental impact inquiry into the effects of the industry.

The medical profession in Australia had remained silent. I wrote to the South Australia branch of the Australian Medical Association (AMA), asking them as physicians to consider the medical implications of nuclear power. Within two weeks, they passed a resolution expressing concern at the public health ramifications of the government uranium policy, and they referred the issue to the AMA Federal Council. I then approached the Royal Australian College of Physicians, which established an interdisciplinary committee to investigate the public health aspects of uranium mining. . . .

I believe that in Australia we have demonstrated that there are ways to circumvent a self-censured press and an unresponsive government.

An underground uranium mine on the Colorado Plateau.
(U.S. Atomic Energy Commission, Grand Junction Operations Office)

Unfortunately, many uranium miners in the United States did not have a Helen Caldicott to warn them of the dangers until it was too late. During the period 1946–68, about 6,000 underground uranium miners were needlessly and significantly exposed to radioactive gases. C. C. Johnson, an official of the U.S. Public Health Service, estimated in 1969 that 600 to 1,100 lung cancer deaths, in excess of what would statistically occur among a similar sample of the general public, could occur in this group of miners.[37] The miners were affected while digging uranium chiefly for the nuclear weapons program. Despite the fact that the mine owners were AEC contractors, the AEC maintained that it had no jurisdiction over the mines. The Atomic Energy Act of 1954 gave the AEC regulatory authority over uranium ore "after removal from its place of deposit in nature,"[38] which the AEC interpreted as the mine rather than the body of the ore itself.[39] This interpretation of the law resulted in an AEC policy of "hands off" miners, which meant that there were essentially no standards for uranium miners until 1967. Consequently, many miners were needlessly exposed to high concentrations of radiation.

The danger to uranium miners results from the inhalation of radon gas, a decay product of uranium which is released when the ore is mined. When inhaled, the gas and its own decay products (called "radon daughters") can inflict strong radiation doses to the lungs—resulting in cancer several years later, as a series of long-term studies has shown. Poorly ventilated underground mines are the primary source of radon contamination for miners.

By 1944, several studies had established that 50 to 75 percent of the miners in the Schneeberg (Germany) and Joachimstal (Czechoslovakia) mines died of lung cancer, and that the deaths were related to high radon gas concentrations in the mines.[40] These shocking results prompted the AEC to perform its own surveys of mines in the United States—and it found radon gas levels to be higher than in the European mines. These findings apparently were never announced, and in 1948 the AEC issued its first price schedule for uranium as part of an effort to induce uranium prospecting and mining.[41]

In 1952, U.S. Public Health Service (PHS) studies revealed that radon gas levels in uranium mines in Colorado were comparable to the levels in the Schneeberg and Joachimstal mines.[42] The International Labor Office (ILO), an organization in partnership with the

United Nations founded to improve labor conditions and promote social justice, reported the results of PHS surveys to the 1955 Atoms for Peace Conference,* the first to be held. Stating that the results were "cause for considerable disquiet," the ILO recommended improved mine ventilation as corrective action.[43] In 1957, the Public Health Service released yet another survey which reported that 65 percent of miners in the Colorado Plateau area had been exposed to radioactive levels comparable to those in European mines.[44] In spite of this information, the AEC continued to leave uranium mine regulation to individual states. This amounted to no regulation at all.

The Congressional Joint Committee on Atomic Energy (JCAE), after a series of hearings in 1959 that included an investigation of radon gas inhalation in the mines, concluded that regulation and enforcement were needed. The committee pointed out that all AEC contractors "fall under the Walsh-Healy Public Contracts Act [and therefore] the AEC had direct control over their procedures and methods of operation." "In these cases," the committee stated, "a simple directive to the contractor results in the necessary changes being made, with the cost of the changes being borne by the AEC."[45] At the same hearings, Dr. Duncan A. Holaday of the U.S. Public Health Service testified that the states had insufficient numbers of technical people to survey the mines or to develop an effective enforcement program.[46] But the AEC continued to do nothing.

In 1967, the U.S. Department of Labor attempted to correct several years of AEC and state inaction by setting a radon–radon daughter atmospheric limit of 0.3. working levels (WL). A miner exposed to 0.3 WL for a forty-hour week, twelve months a year, would receive a whole-body radiation dose of roughly 5 rems. The limit of 0.3 WL was, at the time, the limit for occupational exposure endorsed by international radiation standard-setting bodies.

The Joint Committee, by now the AEC's partner in inaction, held hearings to criticize the Department of Labor and to delay imposition of the standard.[47] Through the JCAE's efforts, a standard of 12 working-level months (WLM) per year (equivalent to exposure at 1 Working Level for forty hours a week for twelve months) was

*The Atoms for Peace Conferences were held at the urging of the United States government, under the auspices of the United Nations. At these international gatherings, participants from several nations discussed ways in which atomic energy could be used for "peaceful" purposes.

put into effect in 1967, marking the first time that a federal standard had been established. In 1971, the permissible maximum was reduced to 4 WLM per year, which remains in effect today.[48] Responsibility for enforcing the standard lies with the Bureau of Mines, part of the Department of the Interior.

The above information is important for more than its historical interest; the sequence of events is an example of the dangers of the nuclear fuel cycle and of the failure, by any definition, of the nuclear industry to regulate itself. It is also important to note that the uranium miners employed during the 1950s continue to die from cancer while their widows and families receive little financial support. Colorado is the only state which compensates uranium miners or their families for lung cancer although uranium ore was also mined extensively in Utah, Arizona, and New Mexico during the pre–4 WLM standard era.

Are the existing standards tough enough? Questions about their adequacy are beginning to surface. Victor Archer, of the U.S. Public Health Service in Salt Lake City, recently updated a study on a group of 3,366 uranium miners, of whom 745 have died. Lung cancer caused 144 of these deaths—which represents an excess of nearly 400 percent over the lung cancers which, statistically, would be expected to occur.[49] It appears that the Public Health Service 1969 estimate by C. C. Johnson that excess lung cancers would affect one-tenth or more of the miners is still accurate. Dr. Archer further indicated that thirty years of exposure at the present 4 WLM per year standards would increase by 45 percent the chances that a person would contract cancer.[50] Archer commented: "The epidemic of respiratory cancers among United States uranium miners is continuing, even though radon daughter levels have been low in recent years. A new epidemic of death from respiratory insufficiency has begun among them."[51]

Radioactive Drinking Water

More recently, another problem at mines and mills has surfaced—the drinking water for the workers and even their families may be radioactive, a situation called "intolerable" by a 1975 internal memo of the Environmental Protection Agency.[52]

In May 1975, an attorney at the Freedom of Information Clearinghouse* in Washington, D.C., received a phone call from an unidentified New Mexico state official who wanted advice on how documents relating to uranium mines could be kept secret, because release of the documents would upset people. When the official refused to divulge more, the attorney explained that the Clearinghouse was concerned about freedom of information, not its suppression. At that point the unidentified official hung up, but the attorney relayed his conversation to the Public Interest Research Group (PIRG).

After two specific Freedom of Information requests, PIRG obtained documents on surveys of water near uranium mines and mills performed for the state of New Mexico by the U.S. Environmental Protection Agency (EPA). The documents contained some rather alarming information.

According to the EPA documents, the New Mexico Environmental Improvement Agency in September 1974 requested the EPA to conduct a survey to evaluate the quality of surface and ground water in the Grants Mineral Belt of western New Mexico.[53] The EPA completed its preliminary report in June 1975 and a more extensive report in July of that year.[54] Neither of these reports was released to the public until PIRG made its request.

Large amounts of underground water in the Grants area had to be pumped out of the uranium mines to prevent flooding. When removed, the water is either discharged directly out of the mines or run through an ion exchanger to remove residual uranium from the water. Some of the water passing through the ion exchanger is used for drinking water at both the mines and at uranium mills located near the mines.[55] It was this water which was contaminated.

The EPA surveyed six drinking water supplies and found all six excessively contaminated with alpha radiation and radioactive uranium. Alpha radiation levels were 200 times those allowed by drinking water standards proposed by EPA; radium levels were 8 times the allowed levels. Even more serious was that the water from the

*The Freedom of Information Clearinghouse was established in 1972 as part of the Center for the Study of Responsive Law. The Clearinghouse gives legal and technical assistance to public interest groups, citizens, and the press in the effective use of laws granting them access to government-held information. The Freedom of Information Act, which became law in 1967, is the primary basis for the public's right of access to federal records.

mines also supplied the drinking water for mobile home camps near the mines, where the spouses and children of some of the mine and mill crews lived. The one drinking water sample that was taken from a mobile home contained alpha radiation 70 times higher than the proposed standard.[56]

One EPA report stated: "All industry potable water supply systems surveyed exceeded existing selenium and planned gross alpha limits for potable water. All but one exceeded existing and planned radium limits. Such water is supplied to families of miners at the United Nuclear Corporation Churchrock mine. These conditions are considered intolerable as they bear on the long-term health of those using the supplies."[57] Selenium is a non-radioactive substance which is chemically similar to sulfur. Its toxicity is described as similar to that of arsenic.[58] Another EPA report warned that radium in drinking water could produce leukemia.[59]

The mining companies identified with radioactive drinking water were Kerr-McGee (the same company that had employed Karen Silkwood) and the United Nuclear Corporation. Numerous environmental violations from water discharged into streams were also discovered. Mines and mills of Kerr-McGee, United Nuclear, Homestake Partners, and the Anaconda Company were found to have violated either EPA regulations, AEC regulations, or New Mexico water quality standards. Particularly outrageous was the fact that some of the drinking water supplies did not even meet EPA and AEC requirements for waste discharge into streams. Not only was the drinking water unfit to drink, it was unfit even as *waste* water.[60]

The situation in the Grants Mineral Belt area was easily remedied without plant shutdowns or worker layoffs: bottled drinking water was shipped in, until water treatment facilities were installed or improved, with the operations allowed to continue. As simple as this procedure was, it did not begin until PIRG released its information on the problem. Although the problem seems to have been corrected, at least temporarily, neither the EPA nor the state of New Mexico has determined how widespread the radioactive water might be at other mines or in other states.

The drinking water problem in the Grants area posed no such catastrophic problem to the population as a nuclear reactor accident might have. But it clearly demonstrates the ability of the nuclear fuel cycle to contaminate workers, to degrade the environment, and—by contaminating streams and drinking water supplies—to generate latent hazards, which may lurk undetected for years, for the general population. The prolonged history of this situation demonstrates the startling inertia of government agencies in enforcing minimum health standards, the negligence of large corporations in making even the simplest reforms, and the insidious danger to workers in the atomic industry.

The Plutonium Breeder Reactor

Edward J. Gleason was a dock worker living in Cliffwood Beach, New Jersey. On January 8, 1963, while he was handling a shipment at the Eazor Express Trucking Terminal in Jersey City, Gleason noticed that one of the boxes in the shipment was leaking. He had handled leaky shipments before, so without thought he simply tilted the box onto a handcart and took it to the loading dock. When the leak began forming a puddle, Gleason turned the box over; as he grabbed it with his bare left hand, the liquid came into contact with his skin. The dripping ceased and, at the suggestion of the terminal manager, Gleason covered the puddle with sawdust. The shipment, originating from the Nuclear Materials and Equipment Corporation (NUMEC) plant in Apollo, Pennsylvania, had been improperly packaged, improperly transported, and improperly labeled. It was not until much later that Gleason learned that the box he had handled contained a glass jug of a solution of chemicals contaminated with plutonium.

Three years later Edward Gleason developed cancer on his left hand, which finally required amputation. Doctors then had to amputate his arm and shoulder in successive attempts to arrest the cancer. Cobalt treatments were initiated, but the cancer continued to spread, and in February 1973 he died. The medical evidence is "overwhelming" that Edward Gleason was killed by plutonium.[1]

THE FUTURE of the atomic industry is unavoidably bound to plutonium, the agent which very likely caused Edward Gleason's death. Without question, plutonium is an unforgivingly deadly element. Glenn T. Seaborg, the co-discoverer of plutonium and a former chairman of the Atomic Energy Commission, has called "fiendishly toxic" an understatement in describing plutonium's hazards.[2] But Seaborg also envisions a "plutonium economy"[3] in which reactors, powered by this deadly fuel, actually breed more plutonium than they consume. The growth of nuclear power depends inevitably on the breeder reactor, because amounts of fissionable uranium-235 are limited. The breeder reactor would produce its own plutonium fuel from uranium-238, which is much more abundant in nature, and would greatly extend the useful life of uranium reserves.

On the other hand, many scientists, as well as members of the public and their elected representatives, do not share Seaborg's enthusiasm. These skeptics point out that the breeder reactor has its own set of serious, unresolved safety problems, and that the breeder could be even more dangerous than present-day reactors. A plutonium economy will have far more risks than benefits. In addition to its extreme toxicity, plutonium is the raw materials for nuclear bombs. Its use raises the specter of sabotage or theft and the fashioning of an illicit weapon by a terrorist or sub-national group. The plutonium economy could not be maintained in a unstable and imperfect world.

The Breeder Reactor

The principles of the breeder's operation have been explained in "Nuclear Power Overview," at the beginning of this book. As plutonium fissions, the "fast" neutrons that are generated produce plutonium if a U-238 atom absorbs a neutron. The breeder thus produces plutonium as well as heat. The coolant is sodium, which transfers heat efficiently without inhibiting the production of fast neutrons.

Any simple description of the breeder's operation tends to obscure the major uncertainties about the reactor's safety. The breeder can explode, causing a breach of the containment building surrounding the reactor, and releasing plutonium and other radioactive substances. Additional, non-nuclear explosions could result from the sodium coolant if it were to come in contact with the air or with water. The steam generator for the breeder would have sodium flowing through the tubes used for heat exchange and water flowing over the tubes. Steam generator tube leaks are a rather common occurrence with light-water reactors—because there are literally thousands of tubes, and water corrosion will inevitably cause leaks in some. The implications of a sodium-water reaction from leaking breeder steam generator tubes cannot be taken lightly, therefore. In fact, it is widely believed that a Soviet breeder on the Caspian sea experienced a sodium-water reaction that was violent enough to be detected by a U.S. reconnaissance satellite.[4] Two of the station's three steam generators have remained out of commission since that October 1973 accident.[5]

The breeder can also experience an accident called the Hypothetical Core Disruptive Accident (HCDA), a euphemism for an accident in which the plutonium fuel rearranges itself in such a manner as to cause a low-order nuclear explosion (which is not possible with light-water reactors.) Such an accident could be initiated by loss of sodium flow (caused, for example, by failure of the sodium pumps) accompanied by failure of the reactor shutdown systems to operate. Were such an accident to occur, heat would build up rapidly and the sodium would begin to boil. In normal operation, the sodium actually dampens the nuclear chain reaction somewhat. Boiling of the sodium, which would make it less dense, would therefore accelerate the nuclear reaction. The fuel would begin to melt and rearrange itself. If a significant fraction of the plutonium core material were drawn together, a nuclear explosion could occur.[6]

There are other phenomena which could accompany a breeder meltdown, creating problems even without a nuclear explosion. For example, only a partial meltdown of the fuel core is necessary to initiate a serious reaction between sodium and the fuel.[7] Three British researchers have concluded: "Fuel melting could also arise in fast reactors under fault conditions, and our calculations show that pressures on the order of 15 kbar could be produced from large scale events involving sodium and molten uranium oxide."[8] One kbar is 15,000 pounds pressure, so the interaction could generate 225,000 pounds pressure—far more than the design capability of either the breeder piping system or the containment building around the reactor. Were the meltdown accident to proceed without a nuclear explosion or fuel-coolant interaction, sodium could react with the concrete in the containment building to generate hydrogen, which itself is explosive or flammable.[9]

Although many scientists would agree that the breeder can undergo a nuclear explosion, recent AEC statements on the subject attempted to minimize the magnitude of the explosion: "It is absolutely impossible for any nuclear incident to lead to explosions of the magnitude associated with nuclear weapons; that is, many thousands of tons of explosive equivalent."[10] Of course, there is a great deal of room between no explosion and the "many thousands of tons" upper boundary set by the AEC. The authors of a 1968 AEC document were less reluctant to make estimates of the explosive potential: "Nuclear bursts: For both EBR-II [Experimental Breeder Reactor Number Two] and Fermi [the first demonstration breeder] explosive energy releases from nuclear bursts as well as sodium-air reactions were considered as design basis accidents. . . . The meltdown-reassembly or 'Bethe-Tait' accidents were calculated to produce energy releases roughly equivalent to the detonation of 300–500 pounds of TNT."[11]

One fact which is intriguing is the lengths to which federal officials will go in their efforts to avoid the phrase "nuclear explosion" in discussing the hazards of the breeder. This was apparent even in 1968, and perhaps is the reason for the creation of the term "HCDA." George L. Weil, who is now a private consultant, was once a research associate of Enrico Fermi on the Manhattan Project. After the formation of the Atomic Energy Commission, Weil was appointed chief of the Reactor Branch in the Division of Research and later served simultaneously as chief of the Civilian Power Branch and assistant director of Reactor Development within the AEC.[12] He commented in 1973 on the apparent inability of the AEC to use plain language when talking about the Liquid Metal Fast Breeder Reactor (LMFBR—"Liquid Metal" refers to the sodium coolant):

> No matter how it is phrased, "nuclear explosive energy," "rapid reassembly of the fuel into a supercritical configuration and a destructive nuclear excursion," "rapid core meltdown followed by compaction into a supercritical mass," or "compaction of the fuel into a more reactive configuration resulting in a disruptive energy release," the meaning is clear: LMFBRs are subject to "superprompt critical conditions," and, as the AEC well knows, this technical terminology translated into layman's language is an "atomic bomb."[13]

Other scientists believe that the explosive potential of the breeder is much greater than the earlier AEC figure of 300 to 500 pounds of TNT. Calculations of 6,700 pounds and 20,000 pounds of TNT equivalent have been made as estimates. A containment vessel that would not be economically prohibitive would only be able to withstand the explosive force of 1,000 pounds TNT equivalent or less.[14] The Natural Resources Defense Council (NRDC), a national environmental group, has this summary of the explosion question: "There are major uncertainties in defining the explosive potential of the breeder, which are all the more worrisome considering the several tons of plutonium in it." [15]

The discussion of the breeder's explosive potential is more than an esoteric argument between the AEC and other scientists. The concern, of course, is that an accident could rupture the breeder containment and cause radioactive material to be spread over large areas. The consequences of a light water reactor accident, which have already been reviewed, are serious enough. An accident with the breeder could be even more serious, particularly if plutonium were released. Plutonium's extreme toxicity has been mentioned before—inhalation of a few millionths of a gram has caused cancer in laboratory animals.[16]

The promoters of the breeder reactor, while they might concede the scientific controversies, attempt to distract the public with assuring but unconvincing statements that the hypothetical breeder accidents are highly unlikely. Such statements, however, ignore the problems experienced by breeders which have already been operated. There is the Russian breeder, which experienced sodium-water reactions, and possibly an explosion. The Experimental Breeder-I, an experimental reactor in the eastern Idaho desert, suffered a partial melting of its fuel in November 1955 and never operated again.[17] Industry also conveniently omits discussion of the Fermi reactor, this country's only commercial breeder reactor ever operated, sited near Detroit, Michigan. In October 1969 the Fermi reactor experienced a fuel melting accident which was more serious than the accident which had been postulated as the "maximum credible accident" for the plant.[18] Luckily, the Fermi accident did not release any radioactivity to the general population, but its aftermath was so serious that one engineer at the plant was prompted to remark, "We almost lost Detroit."[19]A 1957 university of Michigan study had concluded that a reactor accident at Fermi could kill over 60,000 people.

It is clear, in summary, that the breeder has safety problems which are unresolved and which are more serious than those with light-water reactors. If these problems are in fact amenable to technical solutions, they should be resolved before the breeder is allowed to produce power, not after.

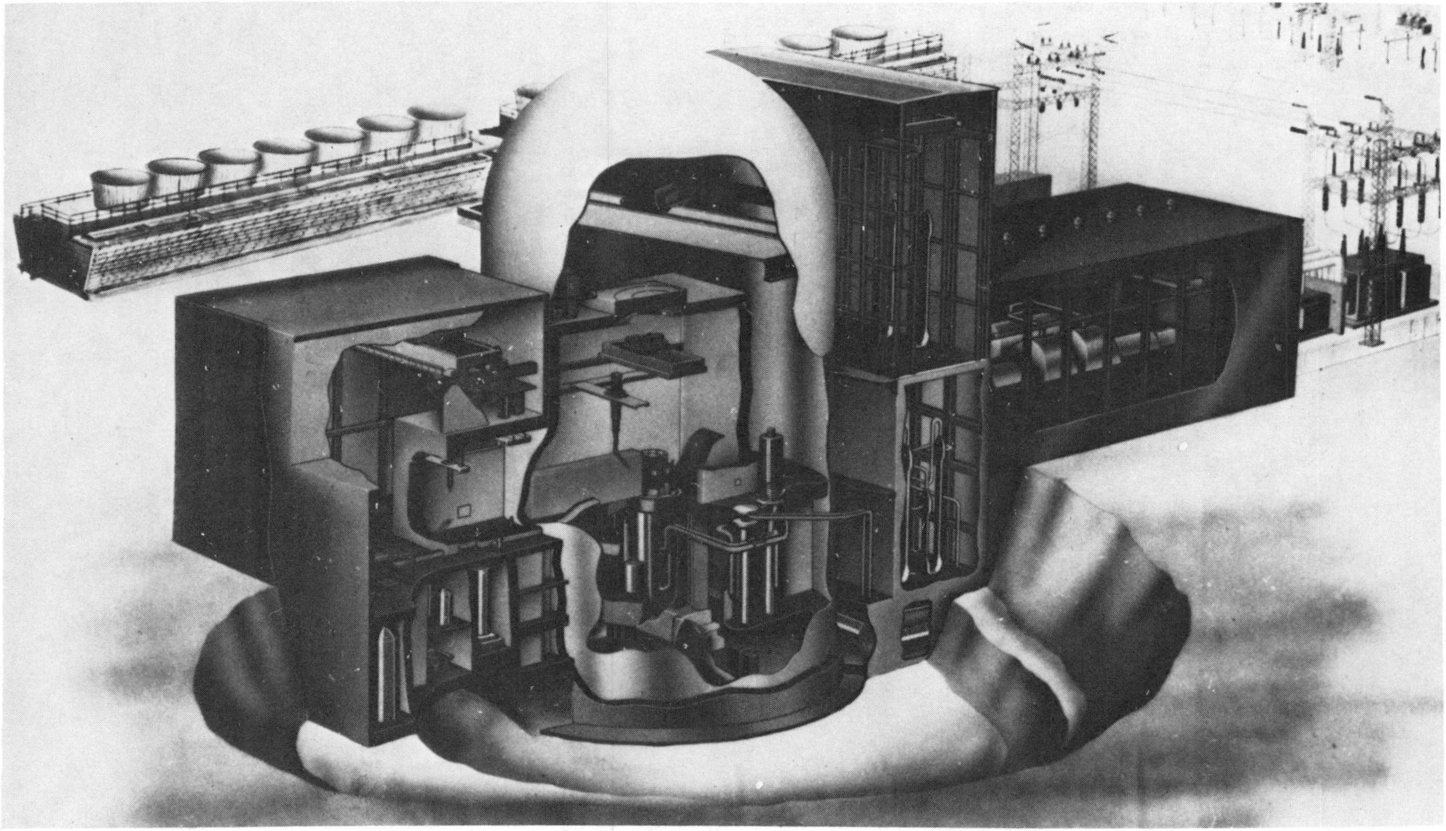

Figure 1. *Artist's Conception Showing a Cutaway View of the Liquid Metal Fast Breeder Reactor.*
(U.S. Atomic Energy Commission)

Breeder Economic Problems

Beyond the questions of safety, there are also significant questions as to the breeder's economic viability. The analyses that led breeder proponents to promote its development depend critically on assumptions that are proving increasingly unrealistic. When reanalyzed using more reasonable assumptions, the economic benefit of the breeder is found to be practically zero. Already the breeder program has experienced major cost overruns, is expected to incur additional overruns in the future, and is threatening to become an economic white elephant gobbling up massive taxpayer funds necessary for other energy research programs.

Between 1968 and 1974 the Atomic Energy Commission performed three cost-benefit analyses of the Liquid Metal Fast Breeder Reactor Program (LMFBR). In each of these, the AEC's conclusion was that the breeder program would produce economic benefits far greater than the cost of the program.[20] Each analysis made assumptions about the demand for electricity, the supply of uranium, and the capital costs of the breeder which were highly favorable to the breeder but which, if altered, would significantly change the outcome of the cost-benefit analysis.[21]

The AEC's cost-benefit scenarios generally projected electrical power demands that would continue to grow at historically rapid rates. With this rate of growth, it would obviously be necessary to build power plants rapidly in order to meet the demand—with nuclear power contributing a significant portion of the new capacity. This would require nuclear power to grow faster than the growth in electrical demand. The AEC also assumed that uranium supplies were very limited. With greater nuclear power demands, the breeder reactor becomes necessary because it requires much less uranium for its operation. To meet the AEC's expected electrical demand, the breeder reactor would have to be brought into commercial operation in a relatively short period of time. Thus, the AEC created a crisis scenario in which future electricity use was so large that uranium reserves would be quickly depleted, requiring the rapid development of the breeder. The AEC's development program was designed to make the breeder a commercial reality by 1987. The "base case" economic analysis concluded in 1974 that by the year 2020 the benefits from the breeder would be $19.4 billion in reduced electrical costs, against $4.7 billion in costs for research and development.[22] Clearly, the AEC concluded, the economic benefits of the breeder will outweigh the costs.

The Natural Resources Defense Council was dissatisfied with the AEC's analysis, so they performed their own, making only three changes in the AEC's assumptions: in electrical demand, uranium supply, and cost difference between the breeder and light-water reactors. NRDC demonstrated that the AEC's assumptions in these areas were crucial to the cost-benefit analysis, and that if more realistic assumptions were made, the breeder's estimated net benefits become negative.

NRDC first pointed out that the AEC projections resulted in estimated electrical consumption in the year 2020 that was over *fifteen* times the consumption in 1974, "a result widely regarded as completely unrealistic." NRDC referred to more sophisticated studies, including those performed by two Cornell University researchers who concluded that the demand will be less than half the AEC projections. Giving the AEC the benefit of the doubt, NRDC set the electrical demand assumption at 50 percent (which even then probably overestimates demand) of the AEC projections for the year 2020.*[23]

NRDC next investigated studies of uranium supplies and found that other scientists believed uranium reserves are greater than assumed by the AEC. The AEC included an "optimistic" estimate for uranium in its cost-benefit analysis, and NRDC concluded that this estimate was actually more realistic than the AEC "base-case" estimate. NRDC then challenged the AEC's assumption that by the year 2000 the breeder's cost of construction will be equal to those of light-water reactors. NRDC pointed out that this is extremely questionable in light of the trends which have already been observed in LWR and in LMFBR capital costs, including the cost overruns which have already occurred in the breeder program. NRDC concluded that by the year 2000 the breeder's construction costs will be at least $100 per kilowatt-electric of plant capacity greater than those for LWRs.[24]

Changing only the three assumptions on electrical demand, ura-

* Since NRDC's analysis in 1975, it has become clear that they were extremely generous in setting electrical demand at 50 percent of the AEC's overblown estimates. The "Conservation" section of Chapter XIV shows that the country can cut its *present* consumption of energy and electricity with no loss in well-being.

nium supply, and capital cost difference, the cost-benefit analysis changes dramatically—the benefits of the breeder program by 2020 will be $0.4 billion, less than one-tenth the costs of research for the program.[25] The NRDC analysis clearly showed that, even without considering the breeder's safety problems, early development of the breeder reactor is economically foolhardy.

The NRDC analysis was confirmed by later analyses performed for the Congressional Joint Economic Committee (JEC) and the American Enterprise Institute, a private organization which performs economic research. The study prepared for the JEC validated the NRDC assumptions on electrical demand, uranium reserves, and capital cost differences.[26] The study prepared by Brian G. Chow for the American Enterprise Institute concluded that the country "can and should" delay the plutonium breeder for five to ten years, until future energy demand and uranium resources are better assessed.[27]

The AEC and its successor, the Energy Research and Development Administration, grudgingly acknowledged that NRDC's assessments were more sensible than their own earlier assumptions. While the AEC, in its proposed final environmental statement on the breeder, did not change the "base case" economic analysis, it did acknowledge in its summary statement that projections of electrical demand were much too high. The AEC stated that by the end of September 1974 the equivalent of 12 large power plants had been cancelled and the equivalent of 109 plants had been deferred by about two years each. In conceding the slowdown in the growth rate of electrical consumption witnessed during 1974, the AEC agreed with NRDC that the projection in electrical demand by 2020 could be reduced by 50 percent.[28]

The AEC also abandoned the date of 1987 for commercial introduction of the breeder: "Recent evaluations of the LMFBR development program by the AEC, taking into consideration cancellations and deferrals of generating capacity by electric utilities that had occurred by the end of September 1974, suggest that the commercial LMFBR introduction would probably occur in the early 1990's."[29] ERDA, the government agency which inherited the breeder program from the AEC, made even more concessions, acknowledging that U.S. uranium reserves are plentiful enough so that no breeder reactors would be required until after the year 2000.[30] ERDA's program calls for the breeder to make "an initial contribution beginning before 2000," but it is only under ERDA's "Intensive Electrification" scenario that the breeder makes *any* commercial contribution by the year 2000.[31]

Important to breeder economics are the cost overruns which have already occurred. In the mid-1960s, cost estimates for the LMFBR program were about $2 billion.[32] That amount has already been spent, and the AEC estimated that at least another $8.1 billion would be necessary.[33] (The $8.1 billion would be $4.7 billion in 1974 dollars.) Only a small fraction of this overrun can be attributed to inflation.

Different components of the breeder program have experienced immense cost overruns. As program managers have learned more about the breeder technical systems, they have also learned that these systems will cost more. The Fast Flux Test Facility (FFTF), the research plant near Richland, Washington, is a significant part of the LMFBR program. Between 1970 and 1974, estimated costs of the FFTF intermediate heat exchangers more than doubled—from $3.2 million to $7.2 million. The estimated costs of primary sodium pumps for the FFTF rose from $1.8 million to $10.5 million during the same four-year period.[34]

The initial authorization for the FFTF was $87.5 million in1967. Seven years later, the General Accounting Office estimated the cost of the FFTF as more than $933 million,[35] at a time when the design was about 75 percent complete and construction 30 percent complete.[36] More recent estimates, in May 1975, put the cost of the program at more than a billion dollars.[37] The Clinch River Breeder Reactor (CRBR), ERDA's demonstration plant, has experienced similar funding difficulties. The first official cost estimate, in 1971, was about $400 million.[38] In 1973 Congress was told that the CRBR would cost $700 million. By May 1975 the estimate was over $1.7 billion.[39]

Moreover, there are further costs to the breeder program. After the CRBR (if the breeder program has not destroyed itself by its insatiable appetite for federal funds), ERDA will construct a commercial-size facility called the Prototype Large Breeder Reactor (PLBR), supposedly to demonstrate the economic feasibility of the breeder, if the CRBR could demonstrate technical feasibility. The PLBR will require a federal subsidy of at least $1 billion, and probably will need much more, if the breeder program overruns already experienced are any indication.[40]

The federal subsidies that will have to be poured into the breeder program detract heavily from the demonstration of other energy sources. Funding for the breeder program in fiscal year 1976 was over $400 million, almost 25 percent of ERDA's proposed budget for "Direct Energy" expenditures—which cover programs for research and development of energy sources, and exclude military and weapons programs. Nuclear power projects overall took almost 60 percent of ERDA's fiscal year 1976 Direct Energy Budget.[41] Nor does this figure represent the true drain of nuclear power, because the Direct Energy category did not include over $600 million for uranium enrichment, or other hidden nuclear subsidies.[42]

One of the major drawbacks of the breeder, therefore, will be that as it and other nuclear programs gobble up research money, nuclear power will become a self-fulfilling prophecy. In a world of finite budget resources, dollars wasted on the breeder will be dollars taken away from other energy options. The ERDA budget for fiscal year 1976 gave the breeder nearly double the total funding for solar energy, geothermal energy, and energy conservation combined.[43] This misappropriation could be tilted even more unfairly to the breeder in the future, because the breeder is a project with bureaucratic and corporate momentum and because of the overruns expected for the program.

It is significant that the last of the AEC's official projections for future LMFBR expenditures was $8 billion. This exceeded the estimate of a 1973 Federal Power Commission study that $6 billion would be the total research and development (although not commercial demonstration) cost for all non-nuclear technologies, including coal gasification, all the direct and indirect solar technologies, geothermal power, advanced steam cycles, Magnetohydrodynamics,* fossil fuel effluent control systems, and a variety of energy storage systems.[44] NRDC calculated that solar and geothermal technologies, along with lower growth in electrical load, could account for nearly 80 percent of even the AEC's overblown electrical demand estimates for the year 2020.[45] The AEC had assumed that the breeder would account for only 50 percent of that demand.[46]

The breeder reactor, in summary, has severe technical and

* With magnetohydrodynamics, a fuel such as oil or coal is burned to a very high temperature, producing ionized gases. The gases pass through a magnetic field to produce an electricity flow.

safety problems. It has had a history of serious cost overruns which can be expected to continue, and it can take money away from other sources of energy which can provide more than enough energy to replace the breeder. It is clear that the rational and ethical course for the future is to bypass the plutonium breeder reactor by eliminating its funding, and to develop safer and more economical energy sources.

XII

The Plutonium Fuel Cycle: The Immediate Threat

J. Gustave Speth does not flaunt his credentials. With his South Carolina drawl and his easygoing manner he could be just one of many pleasant people in the world. But his background is impressive. He has been a Rhodes Scholar and law clerk to Supreme Court Justice Hugo L. Black. As the lawyer representing Margaret Mead and the Scientists' Institute for Public Information, he won a court order forcing the Atomic Energy Commission to prepare an environmental impact statement for its plutonium breeder reactor development program. As an attorney for the environmental group Natural Resources Defense Council Speth has been specializing in nuclear power problems since 1970. His nuclear expertise, moreover, is not limited to legal matters. Ironically, when he was an undergraduate at Yale Speth worked during the summer at the Atomic Energy Commission's Savannah River Laboratory in South Carolina.

The Washington, D.C. office of the Natural Resources Defense Council (NRDC) also includes two capable technical people. Thomas Cochran, a Ph.D. in physics, has been investigating the breeder reactor since 1971. In 1974, he published The Liquid Metal Fast Breeder Reactor: An Environmental and Economic Critique, *a book which is the cornerstone of technical criticism of the plutonium breeder.*

NRDC's expert on the biological hazards of plutonium is Arthur Tamplin, the biophysicist who was John Gofman's colleague and who with Gofman challenged the atomic establishment's standards for radiation exposure. With members such as these on its staff, NRDC is a very effective environmental group. Because the nuclear staff focuses its efforts on plutonium issues—the breeder and "plutonium recycle"—they have become a respected and highly credible force.

The efforts of the NRDC nuclear staff are aided by Anthony Z. Rois-

man, who is representing them as attorney in some of their legal battles. Roisman is a Washington lawyer who has represented several citizen intervenor groups before the NRC hearings, including the hearings covering the ECCS controversy. These four individuals have played a major role in the citizen effort to halt the atomic industry development of "the plutonium economy." In fact, NRDC and Anthony Roisman in May 1976 were responsible for the court order which prohibited the Nuclear Regulatory Commission from issuing licenses for the use and processing of plutonium. These licenses would have prematurely and needlessly introduced the dangers of the plutonium economy.

ALTHOUGH THE BREEDER is not likely to be in commercial use before 1990, the atomic industry is already planning to use plutonium as fuel in light-water reactors. The rationale for this industry scheme is to "extend" the supplies of uranium, but the use of plutonium as fuel would serve as a stalking-horse for the breeder, by causing the construction and operation of the reprocessing and fabrication plants that would create the "plutonium economy." The implementation of "plutonium recycle," as this scheme would be called, would also subject society to the dangers of plutonium on a wide scale.

With plutonium the nuclear fuel cycle would be changed from its present status. At the reprocessing plant, plutonium as well as uranium would be removed from spent reactor fuel. The plutonium and uranium recovered would be shipped to a fuel fabrication plant. (The uranium would still have to first go to an enrichment plant.) At the fuel fabrication plant a "mixed oxide" fuel of uranium oxide and plutonium oxide would be placed in fuel rods, which would then be shipped to reactors. There would still be much more uranium in reactor fuel than plutonium. The use of plutonium in LWRs is estimated to reduce uranium requirements through the year 2000 by about 20 percent.[1]

The dangers of plutonium reactor fuel will be considerable. The extreme toxicity of plutonium has already been discussed. Plutonium fuel would not solve the nuclear waste problem, because other transuranics would remain in the waste, even if all the plutonium could be extracted. In fact, LWRs with plutonium as fuel will generate a comparable amount of high-level waste as reactors without plutonium.[2]

Then there is the problem of safeguarding plutonium. Many

persons, including scientists within the nuclear establishment, fear that plutonium could be stolen by a group of terrorists or saboteurs and fashioned into an illicit weapon. A group with a nuclear weapon could then engage in the worst kind of blackmail or terror. Even if a group could not fashion a nuclear bomb, havoc and numerous casualties would be easily achieved by dispersing plutonium in a populated area.

The Safeguards Threat

Theodore Taylor is a nuclear physicist who worked for seven years at the AEC's Los Alamos Laboratory, in New Mexico, designing nuclear explosives.[3] Taylor believes that a person possessing about ten kilograms (twenty-two pounds) of plutonium oxide could, within several weeks, design and build a "crude fission weapon." By "crude fission weapon," Taylor means a weapon that could explode with the force of 100 tons or more of TNT, and would be small enough to be carried in an automobile. The person who manufactured this bomb would need little in the way of extraordinary skills: "Such a person would require a working knowledge of nuclear physics or engineering, be innovative and skilled with his hands, and willing to take a significant risk of killing himself during the construction operations."[4] If one person of Taylor's description could fashion a weapon, then it certainly is likely that a group of desperate or fanatical individuals could fashion a more effective weapon in a shorter period of time.

Theodore Taylor's scenario, moreover, is not mere fantasy. A television documentary entitled "The Plutonium Connection" has been aired nationally on educational networks. In the program, a twenty-year-old chemistry student was asked to design a nuclear weapon in five weeks. The bomb was to have been built with stolen plutonium and parts from any hardware store, using information that was publicly available. The bomb that the student designed was reviewed by a weapons expert from the Swedish Defense Ministry, who stated that it had "a fair chance of working" and could explode with the force of up to 100 tons of TNT. The Swedish Defense Ministry expert called the student's work a "shocking" report.[5] The only item that was required for the weapon was stolen plutonium.

John Darcy and Joe Shapiro were developing their own private doubts. The two men were security guards at the Three Mile Island (TMI) nuclear power plant, located near Harrisburg, Pennsylvania, and operated by the Metropolitan Edison Company. Darcy and Shapiro were conscientious, took pride in their duties, and considered themselves professional security men. They were not so sure about some of their co-workers, and the two men became more and more disturbed about the lax manner in which security at TMI was handled. When they took their complaints to their supervisors, no action was taken. So Darcy and Shapiro decided to go public.

In June 1975, the two men made a public statement in which they detailed the security problems at TMI and called for an investigation of security at all nuclear power plants. As examples of the poor security at TMI, the guards recounted incidents of malfunctioning electronic security systems, insufficient screening and training of security personnel, cover-up of an incident in which one guard company breached security to embarrass another company, and lax security at entrance gates.[6]

The NRC responded to these charges by beginning an investigation, which was completed by October 1975. The NRC confirmed some of the guards' allegations, but concluded that all the specific problems cited by the guards had since been corrected. William A. Anders, then chairman of the NRC, stated that the NRC was "fully satisfied that at this time, the plant's physical protection program meets all applicable NRC requirements."[7]

On January 27, 1976, an intruder shattered the NRC's soothing reassurances by breaching security measures at TMI Reactor Unit 1. The intruder drove his car past security guards at the main gate, gained access to a restricted area of the reactor by scaling a fence, and eluded guards for an hour and twenty minutes. Having confounded the guard force, the intruder exited the plant in the same way he had come in—by driving out the main gate. It was later determined that the intruder was an electrician at TMI Unit 2, which was under construction.[8] *Luckily, he did no damage to Unit 1, which was an operating reactor full of radioactive material.*

The intrusion demonstrated the same lack of professionalism and poor security of which John Darcy and Joe Shapiro had warned, seven months earlier. The NRC's response was to fine Metropolitan Edison $8,000—a mere slap on the wrist for a company that would have to pay $200,000 a day for replacement power if it had to shut down its reactor.[9]

John Darcy and Joe Shapiro had found new jobs after their June 1975 press conference. Perhaps the most rewarding moments of their lives came in the days after the press conference—when people they had never known in the Harrisburg area called to tell them how much they admired them for being honest men. The TMI intruder had vindicated Darcy and Shapiro. The same intruder had damaged the credibility of the NRC.

John Darcy and Joseph Shapiro had been hired to protect a nuclear reactor against sabotage and the resulting release of radioactive material. The plutonium safeguards problem would not apply at a reactor, because the plutonium is encased in metal fuel rods and would have to be chemically separated from other radioactive material. But the same lax security at Three Mile Island would be of concern at plutonium processing plants, where raw plutonium would be available. A number of critical reviews and memos from federal agencies have shown that the atomic industry's security problems are not limited to reactors—security at plants which handle weapons-grade material is also a serious problem. These reports give little confidence that plutonium will not be stolen or, in euphemistic jargon, "diverted."

In late 1973 the General Accounting Office (GAO) released an evaluation of security at AEC-licensed facilities. These facilities were authorized to possess "special nuclear material" (SNM), which includes weapons-grade nuclear material—plutonium or highly-enriched uranium. The GAO found discrepancies such as the following at the plants it inspected:

1. Fences had weaknesses and holes.

2. Guards were unqualified to fire their small arms, could not see all areas of the plant from the points at which they stood guard, and did not vary the time or routes of their patrols.

3. Locks on doors were broken; alarms were inadequate, inoperative, or easily circumvented.

4. At one facility, the inspectors actuated an alarm and waited thirty minutes for someone to respond. No one did.

5. At another facility, at the request of the inspectors the facility failed to make a periodic check-in with local police. The police responded as they were supposed to, dispatching a patrol car to the facility. The patrol car went to the wrong facility, fourteen miles away. Particularly disturbing was the fact that each facility inspected by the GAO had been visited by AEC inspectors within the previous year, after which the inspectors always reported that the plant "met" or "exceeded" AEC safeguards requirements.[10]

Another GAO report investigated the security measures used in transporting SNM. This inspection also found serious problems:

1. In one case, a flatbed truck with an open cargo compartment was used to ship special nuclear material.

2. Drivers of trucks carrying special nuclear material were alone and unarmed.

3. The trucks had no alarm or communications system.

4. One truck had no preplanned route and no call-in points.[11]

Following these alarming findings, the AEC upgraded its safety regulations. But even these were inadequate, as a special AEC Task Force Report would find. In April 1974, Senator Abraham Ribicoff (D., Conn.) released the "Rosenbaum Report," named after David M. Rosenbaum, an AEC consultant who was asked to direct a special study on safeguards measures. The report had been commissioned by the AEC but had been kept secret until released by Senator Ribicoff. The report stated that the harm to the public from the explosion of an illicitly made nuclear weapon would be greater than from any power plant accident, including a meltdown. The report's view of the AEC safeguards regulations were: "Even though safeguard regulations have just been revised and strengthened, we feel that the new regulations are inadequate and that immediate steps should be taken to greatly strengthen the protection of special nuclear materials."[12]

Following the release of the Rosenbaum Report, the AEC again was forced to upgrade its safeguards regulations. But even these would not be adequate for the use of plutonium as reactor fuel, as the AEC admitted in its environmental impact statement on plutonium.[13] Moreover, there still is doubt within the government that safeguard measures are even adequate for the amounts of special nuclear material presently handled at commercial plants. In a January 1976 memo, Carl H. Builder, the NRC's director of safeguards, revealed his private doubts about the present system.

> "If safeguards are not adequate against the lowest levels of design threat that have been suggested, then we must logically conclude that such safeguards are inadequate, guite apart from the uncertainty we may accept about what constitute adequate safeguards. I am concerned that some or even many of our currently licensed facilities may not have safeguards which are adequate against the lowest levels of design threat we are considering in [plutonium recycle]. . . .
>
> The logical conclusion from all this is that current safeguards must be presumed inadequate if they cannot effectively counter internal threats of one person or external threats of three persons. It does not, of course, say that they are adequate if they can effectively counter these same threats.[14]

The Builder memorandum thus revealed that even without plutonium recycle, the threat of special nuclear material diversion is already significant. The Natural Resources Defense Council, which obtained a copy of the Builder memorandum and made it public, petitioned the NRC to take immediate and decisive action to strengthen safeguards at existing plants. The NRC's response was deplorable. The agency replied that there in fact was no safeguards problem, in contradiction of the Builder memo, and no need to grant the NRDC petition.[15] The NRC's response, in short, did a disservice to the public health and safety and belittled the private concerns of its own Director of Safeguards.

Missing Material

Previous chapters have reviewed the occupational and environmental hazards of plants handling plutonium, but their dismal safety record has been accompanied by shoddy safeguards as well. In addition to the NRC's failure to regulate the industry, and in addition to problems with the industry's physical security, the industry has also had difficulty in accounting for special nuclear material.

The atomic industries—for both nuclear power and nuclear weapons programs—have been afflicted with the problem of "materials unaccounted for" (MUF). The safeguards system for SNM includes physical security methods that are supposed to prevent SNM theft, and materials accounting methods which are supposed to detect any theft that occurs. Any civilian facility in the U.S. which possesses one kilogram (2.2 pounds) or more of plutonium or other SNM must be licensed by the NRC to do so, and must have a system of accounting for the SNM. In most cases, an SNM inventory must be performed every two months. After this inventory, the licensee must calculate a MUF. This is done by taking the amount of SNM on hand at the previous inventory, adjusting for receipts and shipments, and comparing with the latest inventory. Any discrepancy which results from this comparison is a MUF.

There are limits to the accuracy in SNM accounting methods, of course. There can be errors in the accuracy, sensitivity, and calibration of weighing and measuring devices; differences in the way two persons might read the same instrument; or even errors in the way employees report or record data. So the NRC requires that a factor called "limits of error of material unaccounted for" (LEMUF) be calculated, using statistical methods. LEMUF then represents the limits of accuracy to which materials can be accounted. Presently recognized LEMUFs in accounting for plutonium are 1.0 percent in reprocessing plants and 0.5 percent in other plants.[16] Any MUF which deviates significantly from a plant's LEMUF is an indication that SNM could have been lost or stolen.

One acknowledged instance of an excessive MUF occurred at the Nuclear Materials and Equipment Corporation (NUMEC) plant in Apollo, Pennsylvania. In the fall of 1966, NUMEC came up 100 kilograms short in an inventory of highly enriched uranium, which is SNM. The AEC ordered NUMEC to close down and look for the missing material. But at the end of the investigation, 67 kilograms (148 pounds) were still missing and NUMEC was forced to pay the AEC $834,000 for the missing material. The whereabouts of the missing material has never been determined.[17]

In addition to paying for the missing SNM, NUMEC's dubious record includes two sizable fines. In August 1974, the AEC fined NUMEC $12,170 for various health, safety, and safeguard violations at its plants in Apollo and Leechburg, Pennsylvania, which process uranium and plutonium for use in experimental reactors and for national security programs.[18] The NUMEC plants were later sold to Babcock & Wilcox, a company which manufactures reactor equipment. Even under their new owner, the plants continued to operate poorly. In April 1976, the NRC fined Babcock & Wilcox $26,500 for violations of physical security requirements at Apollo and Leechburg.[19]

Even the federal government has had trouble accounting for its own Special Nuclear Material. An investigation by the General Accounting Office of thirty-four facilities which handle SNM under contract to the Energy Research and Development Administration (ERDA) was quite critical. The GAO reported that ERDA's accounting methods were "of questionable validity," and pointed out that the cumulative MUF—over a period of nearly thirty years—at the ERDA facilities was "tens of tons." The GAO also investigated physical security measures at the ERDA facilities and found deficiencies similar to the problems found in the GAO's 1973 inspections: failure to install alarms, inadequate strength of guard posts, vulnerable communications systems, and failure to follow proper procedures. The GAO reported "repeated violations of ERDA's own standards."[20]

In December 1974 the *New York Times* asked the Atomic Energy Commission for information on the material unaccounted for at each of the civilian facilities licensed by the AEC to handle SNM. The AEC replied only that the questions had been turned over to the National Security Council (NSC)—which consists of the president, vice-president, secretary of state, secretary of defense, director of the CIA, and chairman of the Joint Chiefs of Staff. The NSC was requested to perform a "short-term" study to determine whether the release of such information would be detrimental to national security and international relations. Over two years after it was announced, the "short-term" study had not been released.[21]

The fact that material is unaccounted for does not definitely mean that it was stolen. But by the same token, the NRC and ERDA cannot give assurances that the missing SNM has *not* fallen into the hands of terrorist groups or foreign governments. The Rosenbaum Report underscored this point when it stated: "The uncertainties in the accumulated material balances of the atomic energy operation of the country already make it impossible to say that an explosive mass has not been diverted, and if reliance is placed solely on material balance methods that statement will have to be expanded many many fold in the near future."[22]

By 1990, the atomic industry hopes to have 82,000 kilograms of plutonium circulating annually in the light-water reactor fuel cycle.[23] Even if this material could be accounted for with 0.5 percent error, the MUF would be 410 kilograms. This would be enough for forty-one of the ten-kilogram weapons that concern Ted Taylor. In light of the industry's poor record in handling plutonium and the recognized inadequacies of government safeguards regulations, the first question that should be asked about plutonium fuel is: Are adequate safeguards measures possible?

The Civil Liberties Threat

Then there is the other side of the safeguards coin, which is the related question: If adequate plutonium safeguards *can* be implemented, will there be unacceptable effects on civil liberties?

One of the tentative AEC proposals for plutonium safeguards was the establishment of a federal plutonium police force to deter and investigate theft of special nuclear material.[24] There are some who fear that such a police force could become a law unto itself, given its potential for abuse of powers. The AEC had already found it necessary, during the outbreak of hostilities in the Middle East in 1973, to issue shoot-to-kill orders to personnel directing the production, shipment, and storage of atomic weapons.[25] What might be the reaction of a federal force attempting to prevent nuclear blackmail by an underground group in a limited period of time? Russell Ayres, a student at Harvard University Law School, assessed the situation in an article in the *Harvard Civil Rights–Civil Liberties Law Review:*

> [But] . . . it is entirely possible that the person or group making the threat may give the government only a few hours in which to respond. In such a case the government would be compelled to use every means at its disposal to locate the source of the threat. Searches based on the scantiest information or on no information at all as well as brutal interrogations of persons suspected of having knowledge of the identities and whereabouts of the threatening parties would be inevitable. Such desperate measures would go far beyond the sorts of emergency powers which have been upheld or even considered in the past, and would approach the establishment of tyranny.[26]

Other proposals have included widespread background checks on all persons with access to plutonium, material records, or vital equipment; and improved "hardening" of facilities. "Hardening," is a euphemism for more guards, better alarm systems, better monitoring systems, or stronger fences and barriers. In July 1976, in fact, a joint ERDA-NRC task force recommended that full security checks be required for "selected" employees of SNM licensees, and that facility guards be armed with semi-automatic rifles.[27]

Ayres assessed the effect that background checks and a garrison-state mentality could have on members of the atomic industry:

> Employees who express opposition to nuclear energy or nuclear industry policies or who associate themselves with dissident organizations may be regarded as untrustworthy custodians of plutonium by their employers and by the AEC. Aspects of employees' private lives, such as sexual conduct or personal finances, may be used as grounds for dismissal or demotion.[28] . . .
>
> A basic objection to theft-preventive safeguards is that they would require individuals to distort their assessment of their own role in society. For example, civilian employees of the nuclear power industry who would have to comply with stringent new security regulations might come to believe that they were more like soldiers than civilians in light of the background checks they would have to undergo to secure employment and in

light of the limitations on their off-the-job activities that they would have to observe to retain employment.[29]

Nor would background checks and surveillance be limited to those within the atomic industry. Police surveillance could very well extend to the general population. Ayres predicts: "The urgent need to prevent thefts of plutonium will lead to a loosening of standards for government conduct of covert surveillance. The government will probably take full advantage of the broad powers which the courts have allowed it in the use of informers and infiltrators. Moreover, the case for using wiretapping to uncover plots to steal plutonium is very strong."[30]

For these and other reasons, Russell Ayres concluded that an important social cost of plutonium recycle will be the "loss or diminution of basic civil liberties."[31] He recommended that "all other sources of energy," including conservation, "be proven unworkable or unacceptable" before plutonium is used as an energy source.[32]

Such concerns about civil liberties are not idle speculation. Even without plutonium, nuclear power has led to abuses of individual liberties. There have been the lie detector tests at Kerr-McGee following the death of Karen Silkwood. Kerr-McGee employees who refused to take the tests or who failed to "pass" the tests were demoted or fired. It was revealed that the Texas state police compiled dossiers on nuclear power critics in that state. One file was kept on a former Marine captain who is a Continental Airlines pilot and who was the leader of a local group opposing nuclear plants.[33] Following publicity about the dossier of this one individual, the Texas police claimed that they had destroyed all files generated by "non-criminal investigations." The police declined to state how many files had been destroyed.[34]

But perhaps the most ominous portent came from the Virginia Electric and Power Company (VEPCO). In January 1975, the company asked a representative in the Virginia state legislature to submit a bill that would give VEPCO authority to establish its own police force, with power to arrest persons anywhere in the state and gain access to confidential citizen records. The VEPCO security chief asserted, in response to press inquiries, that such powers were required by AEC regulations.[35] The AEC replied that such broad police authorities were not required. The bill was subsequently withdrawn following publicity and the protests of environmental and civil liberties groups. Such a bill might very well have been received favorably, however, if it had followed a dramatic sabotage or theft attempt.

This request by a utility for its own police force, and the other abuses, have occurred, it should be remembered, in the context of the present state of the atomic industry. If these incidents are any warning, then infringements on individual liberties—if plutonium becomes an energy source—could become institutionalized.

The Double Danger

The plutonium economy thus offers a double danger for society. The first is that civil liberties infringements which undermine society will be necessary. The second is that a thief or saboteur could slip through any security system, no matter how carefully it is enforced, thus rendering the defense of the nation impossible. Nuclear power installations and conveyances themselves become national security problems.

Apologists for the atomic industry have argued that the nuclear sabotage threat is not a credible one. They have stated that terrorists would be more likely to concentrate on smaller bombs and more limited targets, must rely on public sympathy, and have non-nuclear means of mass destruction at their disposal. Although these arguments might make sense to more rational men, they have not deterred terrorists. In the United States alone, from 1969 to 1975, there were ninety-nine reported threats and acts of violence directed against licensed nuclear facilities,[36] seventy-six threats and acts of violence directed against unlicensed nuclear facilities, and twenty-eight threats and acts of violence involving nuclear materials.[37] Some of the nuclear attacks or near-misses, domestically and abroad, are listed below:

- In September 1970, one month before the Point Beach 1 reactor in Wisconsin was to begin operation, dynamite was found at the site of this Wisconsin-Michigan Company plant.[38]
- In 1971, arson at one of Consolidated Edison's nuclear plants caused \$5 to \$10 million in damage.[39]
- In November 1972, aircraft hijackers threatened to crash a plane into the nuclear installation at Oak Ridge, Tennessee, unless they were paid \$10 million in ransom. In view of the threat, all reac-

tors at Oak Ridge were shut down and all but emergency personnel were evacuated. Luckily, the hijackers changed their minds.*

• In March 1973, a guerrilla band took temporary possession of a nuclear station in Argentina, although they did no damage.[40] Scottish nationalists have threatened an English reactor.[41]

• In 1974, Commonwealth Edison's Zion station, in northern Illinois, received bomb threats and experienced apparent sabotage. There had been repeated failures of a single valve, and other valves and switches were found in incorrect positions. The company believed these incidents were purposely caused by a disgruntled employee.[42]

• In August 1974, an incendiary device was detonated in a public area of Boston Edison's Pilgrim reactor.[43]

• In May 1975, two bombs planted by an anarchist group exploded at a nuclear power plant under construction in France. The bombs went off in a building close to the reactor, which did not yet contain fuel.[44]

• In the fall of 1975, a member of the West German parliament walked into a nuclear power plant in West Germany and presented a bazooka to the director of the company, who was leading a tour of the plant to explain its safety features. The member of parliament, Werner Twardzig, carried the bazooka past the plant's security force to dramatize the ease with which a nuclear plant could be attacked.[45]

To those who might believe that terrorist groups would not be capable of fabricating nuclear weapons, the Rosenbaum report had this reply: "In recent years the factors which make safeguards a real, imminent and vital issue have changed rapidly for the worse. Terrorist groups have increased their professional skills, intelligence networks, finances, and levels of armaments throughout the world."[46]

Then again, there are the comments of NRDC's J. Gustave Speth on the validity of nuclear terrorism:

. . . the image of the friendly terrorist trying to win public support hardly rings true when one remembers Munich and Northern Ireland, or recalls

* It should be pointed out that modern nuclear power plants could withstand only the impact of a 200,000-pound aircraft arriving at a speed on the order of 150 miles per hour (James R. Schlesinger, former AEC chairman, on *Meet the Press,* December 17, 1972). A Boeing 707 aircraft, by comparison, weighs 230,000 pounds when completely empty, and 333,600 pounds when fully loaded. Other aircraft, including the Boeing 747, are, of course, larger.

the bazooka rockets which narrowly missed an Israeli airliner containing 136 persons at Orly airport and the 1,000 persons evacuated from the Greyhound bus terminal in Los Angeles after a bomb was discovered in a locker. What terrorists do apparently seek to create are dramatic incidents of terror. Nuclear bomb threats, blackmail and explosions are unsurpassed for such objectives. There is also an apparent fascination with bombs. The FBI announced last February that in 1974 no less than 2,041 bombing incidents occurred in the U.S.[47]

There is the statement of Donald Geesaman, biophysicist and plutonium specialist at the University of Minnesota. Dr. Geesaman reminds one that if anything can go wrong, it will:

Reality is more inclusive than the *remote numerata* of the systems analysts. There are Klaus Fuchses and Lee Harvey Oswalds. There are heroin thefts at the New York Police Department and gambling hanky-panky in the highest echelons of AEC security personnel. There are Huston plans and White House plumbers; Argentinian kidnappings and Chilean coups. There are the angry sectarian confrontations by the Spanish Basques and the Canadian Separatists and the I.R.A. There is terrorist violence at Khartoum and Munich and Tel Aviv; the attack on Princess Anne and the Hearst kidnapping. There are resurgent nationalism and alienated minorities, the hopeless, the poor, the vicious, the pathological, and most important of all, the brilliant, for raw human intelligence spreads across all human classifications; and "where there's a will there's a way."[48]

Another atomic industry argument is to downgrade the threat to civil liberties from the plutonium economy. Members of the industry point out that plutonium has been handled in the weapons program and in the armed services, with no threat to the civil liberties of the general population. However, to make this argument is to provide its own refutation. Large amounts of plutonium will be handled in the commercial sector. If the safeguards of the armed forces are necessary, turning the nuclear industry into armed camps and requiring the clearance procedures of the military, then atomic power will have introduced a garrison-state to the commercial sector. Even William Anders, then chairman of the NRC, stated that "it staggers me" to consider applying the safeguards of the military to the commercial industry, when such a suggestion was made by John Simpson, an executive with the nuclear branch of Westinghouse Corporation.[49]

To minimize the civil liberties threat from plutonium is to ignore the points raised by Russell Ayres and other students of the law. It also ignores the nuclear power threats to civil liberties that have already surfaced, even without plutonium.

Lastly, industry apologists attempt to "detoxify" plutonium, arguing that botulin toxin, anthrax, or nerve gas are all much more toxic than plutonium. Such statements are designed to assure the public that plutonium somehow is not as dangerous as some persons make it out to be.

To begin with, this detoxification campaign has questionable scientific validity. Some scientists would point out that "hot particles" of plutonium are of comparable toxicity to botulin toxin.[50] But even if there are other substances that are deadlier than plutonium, they do not make plutonium any less dangerous on an absolute scale. Plutonium is still a very impressive cancer-causing agent in its own right.

It is also important to note that no one is recommending botulin or anthrax or nerve gas as energy sources spread throughout the civilian economy. Under normal circumstances, efforts are taken to limit the production of such toxins, and there are ways of neutralizing them. Biological toxins such as botulin and anthrax can be destroyed by boiling, heat, or ultraviolet light. Nerve gas can be destroyed by heat or chemical reactions.[51] Plutonium can be heated or chemically reacted, but it will still be a highly toxic radioactive substance that can be fashioned into weapons material. The atomic industry not only does not intend to limit the production of plutonium, but wants to manufacture tons of this poison and use it as the country's major energy source. To base an economy on such a virulent poison is illogical.

It is not yet too late to turn back the threat of the plutonium economy, but time is drawing short. In November 1975, the Nuclear Regulatory Commission announced that it would make a final decision on the wide-scale use of plutonium fuel, possibly by 1977.[52]

Another part of the NRC's November 1975 announcement, which would have allowed "interim licensing" of facilities to use and handle plutonium, was overturned by the U.S. Second Circuit Court as a result of a suit by the Natural Resources Defense Council.[53] The "interim licensing" provision had been particularly suspect, because it would allow the licensing of individual plutonium facilities before the NRC had ostensibly approved wide-scale use of plutonium. If several facilities were already handling plutonium, they would create a great deal of momentum, prejudicing the NRC to decide in favor of plutonium fuel when the time for the final decision was reached.

Although interim licensing was defeated, the NRC is proceeding on the plutonium fuel question, and public hearings began in the winter of 1976. The NRC's November 1975 announcement has already been interpreted as a clear signal to the industry that the NRC will approve plutonium recycle and that it will do so quickly, at the expense of a full public airing of all issues.[54]

In proceeding toward its decision on plutonium, the NRC has brushed aside the deep concerns of informed citizens. In March 1976 the annual convention of the National Council of Churches voted overwhelmingly to endorse a moratorium on plutonium processing plants, the plutonium breeder, and the use of plutonium as fuel, until the serious social and ethical problems of "the plutonium economy" can be resolved.[55] The Council's endorsement had been prompted by a panel of distinguished citizens, including fifteen Nobel laureates and twenty-six members of the National Academy of Sciences, which called the use of plutonium as reactor fuel "morally indefensible and technically objectionable." This statement pointed out that the key plutonium issues are not technical, but social and ethical, and that "every citizen should have a voice" in deciding on the use of plutonium. The panel further recommended that the threats of the plutonium economy demand an "unprecedented political response."[56] Industry lobbyists who descended on the Council of Churches convention in Atlanta to oppose the statement prior to this historic vote went away disappointed. Not only did they lose; they knew what they had lost—nuclear power had been judged as a moral issue.

Nuclear Economics

NUCLEAR POWER'S PROBLEMS are not limited to questions of safety. As if the threats to health and safety of present and future generations, occupational safety, national security, civil liberties, and to the environment were not enough, nuclear power is also afflicted with economic problems.* Utility public relations claims to the contrary, nuclear power is now an economic disaster. Nuclear power plant construction costs and uranium costs are both skyrocketing, causing the utility industry to cut back its plans for a rapidly expanding nuclear program. In addition, the financing of plants in the fuel cycle supporting the reactor is becoming increasingly uncertain. As a result, the giant corporations that make up the atomic industry are coming to Washington to ask for government bailouts. Federal subsidies to nuclear power are already large, but the future promises subsidies that will be much greater and more overt, if the atomic industry continues. The economic troubles that plague atomic power are detailed below.

Phantom Savings

Consolidated Edison, an electric utility that services the New York City area, claimed in a public relations campaign and at its annual briefing for members of Congress that its two Indian Point nuclear power plants saved its customers $95 million in 1974. Richard Ottinger, a Democratic congressman from New York, attended

* For a much more detailed treatment of nuclear power's economic problems, the reader should refer to Ronald Lanoue, *Nuclear Plants: The More They Build, The More You Pay* (Center for Study of Responsive Law, P.O. Box 19367 Washington, D.C., 1976).

the briefing. Skeptical of the claim, he asked the Council on Economic Priorities (CEP), a non-profit research organization in New York City, to perform an independent analysis of Con Ed's statement.

CEP found that the $95 million savings were illusory, for if Con Ed had built coal instead of nuclear plants at Indian Point, the coal plants would have been $26.8 million less expensive in 1974 than the nuclear plants. Charles Komanoff, an economist on the CEP staff and author of the report, put Con Ed's misleading statistics in perspective: "The dramatic cost saving reported by Con Edison for 1974 was obtained by comparing the cost of owning and producing electricity at Indian Point 1 and 2 with the cost of owning Indian Point 1 and 2 but not producing electricity there."[1]

The reason for Con Ed's inflated figure is that the utility computed only the *fuel* costs that would have accrued if the electricity had been generated with oil rather than with nuclear fuel. This comparison is misleading because fuel cost is the economic factor most favorable to nuclear power; other factors, such as the cost of construction or repair of a nuclear plant, or taxpayer subsidies in the fuel cycle, are economic drawbacks, but were not included in the Con Ed comparison.

So CEP performed other, more useful, comparisons calculating the generating costs from hypothetical oil plants that could have been sited at Indian Point in place of the nuclear plants. The analysis found that the fuel costs of the hypothetical plants (called "Oil Point") for 1974 would have been $95 million, to generate the same amount of electricity as Indian Point generated. Indian Point's 1974 uranium fuel costs were $23 million, which admittedly gave Indian Point a $72 million advantage in fuel costs; but the comparison did not end there.

Nuclear plants cost much more to build than fossil-fuel power plants, forcing a utility to borrow more money with more interest to be paid throughout the life of the plant. Interest and such items as depreciation, taxes, and property insurance make up the annual "fixed charges" on a power plant. For 1974, the fixed charges on Indian Point were $27 million greater than those calculated for Oil Point.

Moreover, Indian Point was a rather unreliable plant. The same complexity of nuclear plants which makes them more expensive than fossil plants makes them more temperamental. Indian Point operated with combined "capacity factors" of 45.7 percent in 1974. That is, the Indian Point plants produced only 45.7 percent of the electricity they would have produced had they operated at full power every day of the year. CEP concluded from statistics gathered by the Federal Power Commission and the Edison Electric Institute—a national trade association representing the private utilities—that a more likely capacity for Oil Point would have been 70 percent.[2]

Unreliable power plants are doubly expensive for electric utilities. They cost more to repair than reliable plants, and the utility must buy replacement power to serve its customers while its own plant is out of service. This must be bought from other utilities, usually at prices higher than their own costs. The extra repairs at Indian Point made its maintenance costs $9 million higher than for Oil Point, and the replacement power cost for Indian Point over Oil Point was $6 million for 1974. Totaling the above costs shows that Indian Point was $30 million ($72 million less [$27+9+6] million) less expensive to operate in 1974 than Oil Point would have been.[3] Although this comparison shows nuclear power to be more favorable than oil power, the savings are considerably less than Con Ed loudly proclaimed. CEP's analysis, moreover, continued.

CEP calculated the costs to Con Edison of not having built Indian Point at all, by assuming that all the electricity generated at Indian Point would have been bought from other utilities or generated by other Con Ed plants operating at higher capacity factors. The calculated costs for electricity that would have been supplied in this manner in 1974 were $106 million. The total costs—fuel, maintenance, and fixed charges—of owning and operating Indian Point in 1974 were $102 million. This means that if Con Ed had not built Indian Point at all, its customers would have paid only $4 million more in 1974 than they paid for Indian Point.[4]

But CEP developed still another comparison using hypothetical coal-powered plants ("Coal Point") that could have been built at the same location as Indian Point. Again the fuel costs of the coal plant are more than the costs for uranium—but by only $30 million per year. These excess costs are more than offset by the lower fixed costs, lower maintenance costs, and lack of costs for replacement power for the coal plant. The Council on Economic Priorities showed that Coal Point would have been $26.8 million less expensive to operate than the Indian Point nuclear plants in 1974.[5]

In spite of Con Ed's public relations campaign, it appears that Indian Point saved Con Ed's customers little or no money in 1974. Moreover, it should be pointed out that 1974 was a relatively good year for Indian Point. The reliability of the Indian Point nuclear plants had been much lower in earlier years. Furthermore, for the immediate future Indian Point 1 probably will not produce any power. On October 31, 1974, Indian Point 1 was ordered to shut down to install an improved Emergency Core Cooling System (ECCS). The installation will require at least two and a half years.[6] This situation will make the phantom savings from Indian Point even more illusory.

David Dinsmore Comey is an environmental gadfly who has investigated nuclear power for some time. In August 1968, he was head of the Research Institute on Soviet Science, in Ithaca, New York, and knew little about atomic reactors. But that month, he heard three Cornell biologists discuss the thermal pollution from a nuclear station on Cayuga Lake, proposed by the New York State Electric and Gas Corporation. Comey agreed to become executive director of the Citizens Committee to Save Cayuga Lake (CSCL), and the battle was on. Through Comey's ability to mobilize public opposition, NY State E&G eventually abandoned its plans, and CSCL became one of the few groups in the country to have actually stopped a nuclear plant.

In May 1970, Comey moved to Chicago to become director of environmental research for Business and Professional People for the Public Interest (BPI), a law firm dealing with urban and environmental problems. At BPI, Comey's intervention against the Palisades nuclear plant, proposed by Consumers Power of Michigan, led to an unprecedented agreement by the utility. Consumers promised to install cooling towers and an improved radioactive effluent treatment system, which had not been included in the utility's original plans.

In 1974, Comey received the First Annual Environmental Quality Award from the U.S. Environmental Pretection Agency "for services that have immeasurably improved the design and safety review of nuclear reactors." In November of the same year, Comey contributed to the joint Sierra Club–Union of Concerned Scientists critique of the AEC's Reactor Safety Study. In 1976, Comey left BPI to become executive director of Citizens for a Better Environment, a Chicago organization supported by a broad base of citizen contributors.

Comey also did ground-breaking research on nuclear power's economic problems. In September 1974, his report to the Federal Energy Administra-

tion on reliability problems at nuclear plants drew attention as one of the first detailed analyses of the fiscal difficulties of the atomic industry. Comey also found that his local utility, Commonwealth Edison, was having its own problems with unreliable nuclear plants.

Consolidated Edison is not the only utility which hides the questionable economics of nuclear power with public relations statements. In April 1975, the Atomic Industrial Forum, the industry's trade group, made the claim that Commonwealth Edison's nuclear plants saved its customers $100 million in 1974.[7] With seven operating nuclear plants, Commonwealth Edison, the major utility serving the Chicago area, is also the utility with the greatest nuclear capacity in the country.

In May 1975 Comey debated George Travers, a ranking executive with Commonwealth Edison, on a Chicago radio program. Comey claimed that in 1974, Commonwealth's six largest nuclear plants delivered electricity at a price about 27 percent higher than the company's five largest coal-fired plants—which meant that instead of saving consumers money, Commonwealth Edison's nuclear plants cost their customers money. Travers disputed Comey's exact figures, although he was forced to admit that 1974 was a relatively bad year for Commonwealth's nuclear plants, and he claimed that the utility's costs at its nuclear and coal plants were essentially the same, for 1974.[8]

In the light of Traver's admission the question must be asked: On what basis was the $100 million "savings" claimed? As with Con Ed's $95 million savings, Commonwealth's $100 million was no less phantom. The supposed savings were apparently based on the *fuel* costs of burning western coal to produce the same amount of electricity that was produced by Commonwealth's nuclear plants in 1974.[9]

The Flaws of Nuclear Economics

The conventional wisdom of nuclear power economics is that although nuclear power plants cost more than fossil fuel plants to construct, the fuel costs of uranium are much smaller than those of fossil fuels. Over the life of the plant, nuclear power's total costs supposedly are less than lifetime costs for conventional plants.

But these hypothetical considerations have been seriously upset by practicalities. The utilities expected nuclear plants to operate with high reliability when in fact they are out of service more than fossil plants—much more than had been expected. The first few nuclear plants sold to utilities were also sold at reduced costs by the reactor manufacturing companies to encourage the utilities to order more plants.[10]

In essence, the utilities became "hooked" on nuclear power before they realized its true costs. Construction costs of nuclear plants have skyrocketed and show no signs of leveling off. The one advantage that nuclear power plants supposedly maintained, fuel costs, is rapidly becoming less favorable. Uranium prices are also soaring—they have quadrupled over the past two years—and threaten to rise to levels comparable with coal and oil. With all of these economic uncertainties, and with the disappointing performance of nuclear plants to date, it is understandable that the utility industry's ardor for nuclear power has cooled. A definitive statement on nuclear economics is the number of plants that have been cancelled or deferred. By November 1975, 130,000 Megawatts-electric of nuclear capacity had been cancelled or deferred, representing over two-thirds of all cancellations or deferrals of power plants within the utility industry.[11]

One major reason for the deferral of so many plants is that electrical consumption in 1974 and 1975 grew at much lower rates than the utilities had predicted. But there are other problems and uncertainties which account for the overwhelming number of the nuclear plant cancellations. Some of these uncertainties are reviewed below.

CAPITAL COSTS

The largest portion of the costs of nuclear power are "fixed charge" or "capital" costs—costs of construction and interest charges on the initial investment. These capital charges represented about 80 percent of the lifetime costs of a nuclear plant—until uranium prices started their dramatic rise—so it is crucial that capital costs be accurately estimated and held down during construction. The industry's performance in accurately estimating capital costs, however, has been dismal.

In early 1975 a group of economists from Harvard and the Massachusetts Institute of Technology (MIT) published an article in *Technology Review* on nuclear capital costs. The economists found that the gap between estimated and actual costs was substantial and has been growing. Reactors in 1965 were thought to have construction costs of $130 per kilowatt-electric (kwe). By the time they were built, actual costs were $300 per kwe.[12] Plants estimated in 1968 to cost $180 per kwe were actually built for $430 per kwe.[13] Not only had nuclear capital costs been underestimated in the past, but they showed every indication of continuing their upward spiral. The *Technology Review* article showed that costs for nuclear plants had risen much more steeply than inflation or costs of other facilities such as fossil fuel power plants and oil refineries.[14]

Design problems are one of the reasons for the climbing nuclear costs because these problems are frequently found during construction, rather than before the plants are built. This situation, which requires costly retrofitting, exists because the AEC and the reactor vendors have been more interested in promoting nuclear power than in solving safety problems.

The General Accounting Office (GAO) performed a study of cost overruns at the Sequoyah nuclear power plant project now underway in Hamilton County, Tennessee, for the Tennessee Valley Authority (TVA). The GAO identified twenty-three design changes requiring that a structure or component be torn out, rebuilt, or added. Eleven of these changes had a "major impact" on the construction cost, which the GAO defined as a cost increase of $2 million or more. One of the major changes involved the Westinghouse ice condenser—a new containment concept. The concept "had not been tested," according to the GAO, and extensive delays occurred while the condenser was being redesigned. TVA officials believe that problems with the ice condenser are among the main reasons for the delay of the plant, which was to have begun operation in 1973, but which is not expected to start up until 1977.[15] Meanwhile, TVA continues to pay the interest costs, passing them on to consumers.

The Indian Point shutdown may represent the ultimate retrofit; unit 1 will be "down" for an estimated two and a half years for installation of an ECCS. A more recent problem occurred when a workman with the four-inch candle ignited a fire at the TVA's Browns Ferry, Alabama, nuclear station. The shutdown for repairs and relicensing lasted nearly eighteen months, and cost the TVA's consumers $18 million per month in replacement power costs.[16] The ramifications for the atomic industry could be even greater, because the fire could spur the Nuclear Regulatory Commission to require improvements in fire safety provisions on an industry-wide basis.[17] The fact that it took the Browns Ferry accident to make the industry consider a subject so basic as fire protection is an indication that design problems will continue to crop up, and plant costs will continue to climb.

Nuclear plant costs have also been evaluated by the Investor Responsibility Research Center (IRRC), a firm in Washington, D.C., which reviews public issues for investors. The IRRC study contained these findings on capital costs:

> An official of the Atomic Industrial Forum recently noted: "Estimating capital costs for power plants is like shooting at a moving target."[18]
>
> Charles Pierce, president of Long Island Lighting Co., noted that the Shoreham nuclear plant [in Suffolk County, New York], which in 1969 was expected to cost $278 per kilowatt of capacity to build, now is expected to cost $848 per kilowatt.[19]
>
> As was pointed out earlier, the AEC's 1974 estimates of capital costs appear to give undue weight to the desirability of nuclear power plants. A reasonable alternative set of projections would be that a 1,000-mw nuclear plant could be constructed at a cost of $811.1 million, and a comparable coal-fired plant would cost $638.4 million (as compared with the AEC's figures of $660 million and $550 million, respectively).[20]

The *Technology Review* article echoed the IRRC study, concluding:

> We just have little firm idea of what the actual cost in deflated dollars of reactors ordered subsequent to 1969 will turn out to be.[21]
>
> The capital costs of large light water reactors show no signs of stabilizing and, indeed, are apparently still climbing at alarming rates.[22]

However, Daniel Ford, of the Union of Concerned Scientists, has put the capital cost issue more bluntly: "By emphasizing fuel cost advantages, the AEC and nuclear industry looked at nuclear plant economics in the wrong perspective. They solved the less important part of the puzzle, and badly handled the more crucial capital cost question."[23]

RELIABILITY

Another problem with nuclear power plants is their unreliable operation. This is particularly important because the large fixed charges of nuclear plants must be paid even if the plant generates no power at all. The more electricity a power plant can produce, the less will be the costs of that electricity to consumers. Unreliable nuclear plants mean that consumers pay for repair costs, for replacement power, and for the capital costs of the plant.

In 1975 David Comey, then with Business and Professional People for the Public Interest, found major problems with nuclear power plant reliability. Comey analyzed nuclear plants larger than 100 Megawatts-electric (MWe) which were in commercial operation in 1973 and 1974, the only full years for which nuclear capacity factor data were available. His figures showed that the average capacity factor for the atomic industry as a whole was 55.2 percent.*[24] This is a striking disappointment compared with the 80 percent capacity factors which were the basis for utility cost-benefit analyses of nuclear power and which utilities apparently expected to obtain from nuclear plants.[25]

Some members of the atomic industry have argued that Comey's comparison is unfair—that the reliability of nuclear plants should be compared not with an 80 percent capacity factor but with the capacity factors of fossil-fuel power plants which have also been disappointing. Comey responded that the data available show yearly average capacity factors for fossil-fuel plants between 1964 and 1973 to be 68.9 percent. The average capacity factors for fossil plants larger than 390 MWe were 63.2 percent.[26]

Fossil plants are more reliable than nuclear plants on an absolute scale, but the difference is even more striking when one considers the way power plants are operated. The construction costs of nuclear plants are much higher than those of fossil plants, while the fuel costs to the utility are lower. An electric utility with a nuclear plant would want to run it as often as possible at a power level as high as possible—a condition referred to as "base-loaded"—so that the utility could make up its initial investment as rapidly as possible. A nuclear plant is therefore more likely to be base-loaded than a fossil plant. This makes the disparity between fossil and nuclear plant capacity factors even greater.

Comey took the capacity factors for coal and nuclear plants and applied them to the Investor Responsibility Research Center capital cost estimates for 1000 MWe power plants whose construction would start in 1975. IRRC had projected that the coal plant would cost $638 million and the nuclear plant $811 million. Comey then added operation, maintenance, and fuel costs that were represen-

* Comey updated his work to include figures for 1975. On an industrywide basis, nuclear plants in 1975 achieved a 54.9 percent capacity factor (David D. Comey, "Capacity Factors Stay Constant in 1975," in *Not Man Apart* [Friends of the Earth, March 1976], pp. 10–11).

tative of Commonwealth Edison's large coal and nuclear plants. At historical capacity factors, Comey projected busbar* costs of 32.9 mills per kilowatt hour for nuclear plants and 30.0 mills per kilowatt hour for coal plants. (One mill is one-tenth of a cent.) Even with the arbitrary assumption that nuclear capacity factors would increase by 20 percent—to 66.2 percent—while coal capacity factors would increase by 10 percent—to 69.5 percent—coal would still be more economical. At these capacity factors, coal costs would be 27.8 mills per kilowatt hour vs. nuclear costs of 28.2 mills per kilowatt hour.[27] Only if coal and nuclear plants were both to operate at 80 percent capacity—an assumption much more unreasonable for nuclear than for coal plants—would nuclear power have an advantage, of only 1.3 mills. Costs would be 24.3 mill/kwh for nuclear and 25.6 mill/kwh for coal.[28] Figure 1 shows Comey's estimates of busbar costs vs. capacity factor.

Comey's analysis gave every benefit to nuclear power. His use of IRRC's figures still may understate nuclear capital costs. The IRRC figures, for one, are less than the costs of the Shoreham plant, which is nearly completed. It should also be recognized that nuclear capital costs are rising faster than coal plant costs. According to a 1975 *New York Times* article, expected construction costs for nuclear reactors operating in 1985 are $1135 per kwe[29] Uranium costs have also risen far beyond the 1975 AEC estimates that Comey used in his report.

Comey also investigated nuclear plant reliability as it may be affected by age. The data are seriously limited, since there are only three plants which have been in service for twelve to fifteen years, and their designs are different from those of newer plants. However, Comey found as a general trend that these older plants require repairs more frequently and are more unreliable than newer nuclear plants.[30]

Nuclear reactors, just like any other product of imperfect technology, wear out and break down as they get older. In fact, a 1973 article in the *Wall Street Journal* referred to nuclear plants as "atomic lemons" and stated that "their unreliability is becoming one of their most dependable features."[31] In a 1974 article, David Comey explained in more detail the plants' problems as they age: ". . . corrosion problems set in, leaking fuel becomes a problem, and system

* "Busbar cost" is the cost of generating electricity at a power plant. It does not include transmission, distribution, or administrative costs.

Figure 1. *1982 Busbar Costs vs. Capacity Factor*

David Dinsmore Comey, "Nuclear Power Plant Reliability" (Chicago: Business and Professional People for the Public Interest, February 1975)

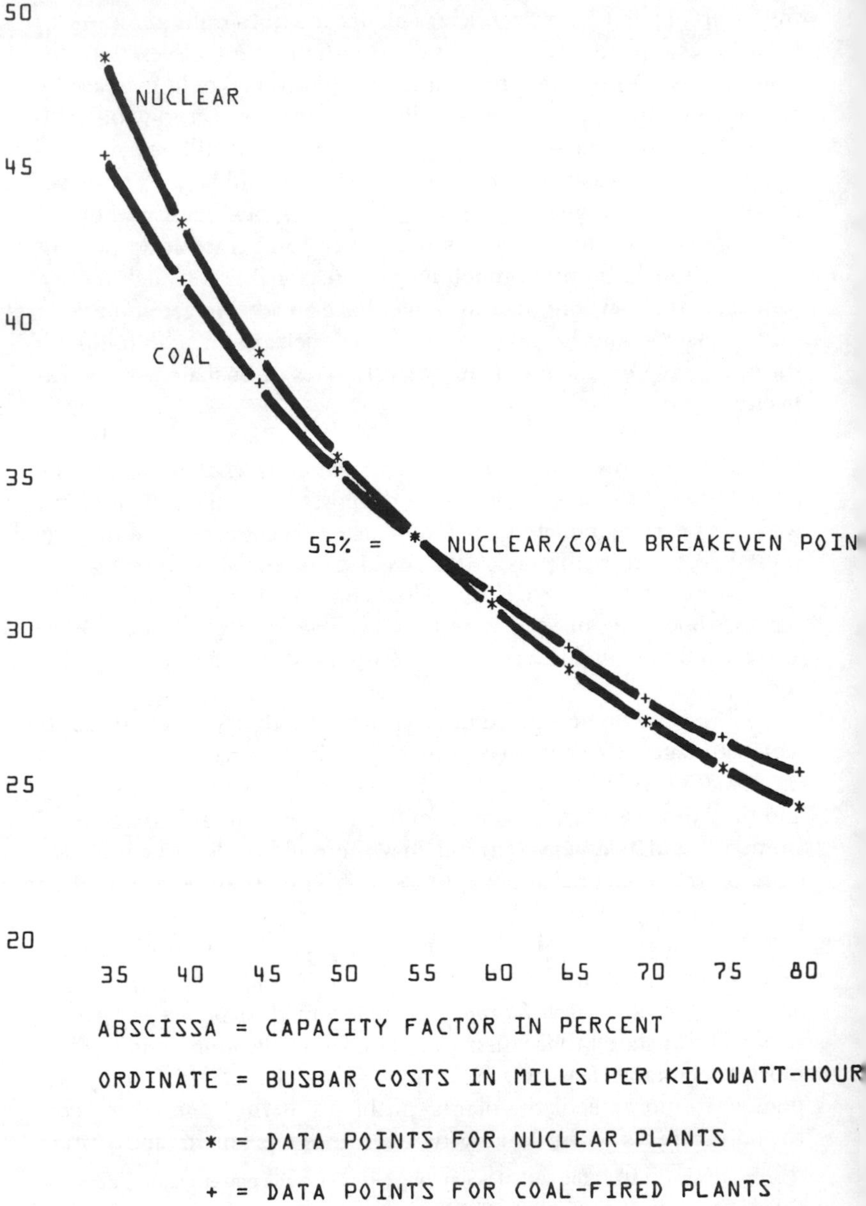

components break down due to fatigue and other wear-related problems. An additional hazard is the accumulation of highly radioactive crud in the primary system, which means that any repair work on this system will consume enormous amounts of time and personnel in order to avoid excessive radiation exposure."[32] As an example of the last problem mentioned, Comey later pointed to a prolonged outage at Commonwealth Edison's Dresden station that required 350 men to make repairs that twelve men could have performed quickly if the same problem occurred in a fossil-fueled plant.[33]

Not even the AEC expected that nuclear power plant reliability would improve in the future. One of its December 1974 documents, obtained by the Union of Concerned Scientists, suggested that the reliability problems resulted from the basically poor management and poor design of nuclear plants. The utilities, according to the AEC memos, have "marginal management capability" and are not "sophisticated" enough to include reliability considerations in nuclear plant design. The utilities, in minimizing initial capital investments, fail to insist on plants that can be maintained well enough to achieve high reliability. Similarly, the architect-engineer firms which construct the plants fail to build reliability in their designs.[34]

In a January 1976 report, David G. Snow, financial analyst for the New York investment firm of Mitchell, Hutchins, Inc., reached similarly critical conclusions over the utilities' inability to manage nuclear technology. Snow reviewed a series of technical problems faced by the industry—poor fuel performance, poor construction practices, General Electric's fuel-reprocessing white elephant, and the Browns Ferry fire. Snow's comment: "As one might expect, poor management lies behind this series of technological failures. The managerial shortcomings are on two levels: competency and honesty—neither of which is consistent with the trust the public now places in this industry for 8 percent of the electricity produced in the nation and the bulk of mid-term energy supply growth."[35]

Problems with plant reliability have led to legal infighting among members of the nuclear industry. Consumers Power Company of Michigan is suing Bechtel Corporation and Combustion Engineering for $300 million for negligence in construction and design of the Palisades nuclear plant, which has experienced extensive shutdowns due to equipment malfunctions. The Nebraska Public Power District has similarly initiated a $50 million suit against several nuclear industry companies—including General Electric and Westinghouse—for defects at the Cooper nuclear station.[36]

The poor plant reliability that results from this technical mismanagement harms consumers in several ways, since utility costs ultimately are passed on to them. A broken-down plant requires extra maintenance and repair costs. Replacement power must be purchased at high costs while the plant is shut down. Then the capital costs are passed on to the consumer, whether the plant produces any electricity at all. It is the combination of exorbitant capital costs and poor reliability which make nuclear power such a questionable economic enterprise.

FUEL

The one area in which nuclear power supposedly would clearly be superior was fuel costs. But that advantage is rapidly being eroded, as has been shown by David Snow's report for Mitchell, Hutchins, Inc. Snow found that from 1973 to 1976 uranium prices rose from $6 per pound to $50 per pound, and could be expected to rise to $100 per pound by about 1978. The total costs of a nuclear plant paying $100 per pound, Snow estimated, would just break even with an oil plant paying $12 per barrel for oil. Snow concluded that domestic uranium mining capacity could not be expanded rapidly enough to avert a supply shortage through 1980 and beyond, and that the importation of foreign uranium would become necessary.[37]

In discussing uranium prices, the structure of the energy industry is an important factor. The world market price for oil is set by a cartel of the major oil-exporting nations called the Organization of Petroleum-Exporting Countries (OPEC). Since late 1973, OPEC has sharply escalated the world market price of oil. Although OPEC-priced imported oil supplies less than 20 percent of the energy consumed annually in the United States, it is influencing domestic energy prices. In a competitive energy market, one might expect domestic producers of the other 80 percent of the nation's energy supply to compete for a larger market share by underpricing imported oil. However, the American energy industry is dominated by a few giant oil companies which also have major investments in natural gas, coal, and uranium.[38] Hence, rather than using lower prices to compete for expansion of market shares, these "horizontally integrated" energy companies prefer to raise the prices of their products to the equivalent of the OPEC cartel's monopoly price for oil. By using the artificially high OPEC monopoly price as a price leader, they expect to reap monopoly profits without violating anti-monopoly laws. Hence the 40 percent of domestic oil production which was free from price controls during 1974 and 1975 rose to the OPEC price before Congress imposed price ceilings in early 1976. Also, the prices of unregulated intrastate natural gas, coal, and uranium have risen rapidly toward the equivalent of the OPEC oil price.

Indeed, in 1974 and 1975 the price of uranium rose at a higher percentage rate than the price of oil.[39] The future prospect for the price of uranium, in short, is that the giant energy corporations will insure that it rises in tandem with other energy prices. Hence it is incorrect to expect that the rising prices for fossil fuels will improve the economics of atomic power by increasing the cost advantage of uranium.

Snow's report also pointed out that uranium reserves outside the United States are concentrated in a handful of countries: Canada, South Africa, France and French Africa, and Australia. There is thus a serious danger that these nations will form UPEC—a Uranium-Producers' Export Cartel that will match or exceed the power of OPEC.[40] Promoters who push atomic power as necessary for "energy independence" may be wrong on two counts. First, the U.S. may have to become dependent on foreign uranium producers. Second, even if there is no requirement to import uranium, the alternative will be for the country to become dependent on the pricing whims of oligopolistic domestic energy corporations.

Rising uranium prices, along with overcommitments on the part of Westinghouse Electric Corporation, led to that company's decision in September 1975 to break uranium supply contracts with over twenty utility companies. In addition to manufacturing reactor equipment, Westinghouse also fabricates uranium fuel rods for utilities and had made agreements with some utilities to supply fuel rods, in some cases for twenty years. The contracts to supply uranium, at prices prevailing when the contracts were signed, were from $6 to $10 per pound of uranium oxide.

But Westinghouse did not have sufficient reserves and had to purchase additional uranium to meet its commitments to utilities. The company found that by January 1976 it would have obligations to supply about 81 million pounds of uranium oxide, but that it had only 15 million pounds in inventory or on order. Westinghouse was faced with the prospect of buying uranium at 1976 prices—$40 per pound or more—and selling it at $10 per pound. This meant the loss of a billion dollars or more if it fulfilled its contracts with the utilities. Rather than do so, Westinghouse announced in September 1975 that it was breaking its supply contracts and demanding that the utilities renegotiate the contracts under new prices. Twenty-seven utilities subsequently sued Westinghouse to require it to uphold the terms of the original contracts. Westinghouse, in turn, responded in October 1976 by suing twenty-nine foreign and U.S. uranium producers, charging them with illegally raising and fixing uranium prices and with refusing to sell Westinghouse the uranium it needed.[41]

Besides creating internal legal warfare among members of the atomic industry, the Westinghouse decision made the economics of nuclear power even more questionable. Union Electric Company, near St. Louis, one of the Westinghouse customers, decided to reorder from another supplier at $40 per pound—at least four times the price on which the company based its nuclear economic analyses.[42] Such economic surprises, along with poor reliability and monstrous construction costs, are causing even the utilities to reevaluate their traditional misconceptions of atomic economics.

Hidden Costs and Economic Distortions

Beyond the economic analyses of utilities operating nuclear plants, there are the distortions of hidden costs that favor nuclear power throughout the fuel cycle. Even if electricity from nuclear plants could be sold to users at low costs, it would only be through major subsidies, direct and indirect, throughout the fuel cycle.

In tallying nuclear power's hidden subsidies, one should keep in mind David Comey's conclusion that nuclear power's advantage under the best of circumstances is very tenuous. If economic distortions in favor of nuclear power worth just a few mills per kilowatt-hour were removed, nuclear's supposed economic advantage would vanish.

One example of an economic distortion is uranium enrichment, provided to the utilities at reduced cost by government plants. As of August 1976, ERDA was charging $61.30 per Separative Work Unit (SWU) of enrichment. In 1975, ERDA estimated that increasing its prices by $23 per SWU would increase the cost of nuclear electricity by about 0.5 mill/kwh.[43] Industry sources have estimated that a private, commercial enrichment industry, if it existed, would charge about $110 per SWU, and a European consortium building an enrichment plant has already announced that it will charge $100 per SWU.[44] ERDA's lower prices thus artificially reduce the busbar costs of atomic power. The economic distortion from ERDA's enrichment services alone could be worth $50 per SWU, or about 1.0 mill/kwh, to the industry.

The Price-Anderson Act represents another hidden subsidy for companies operating nuclear plants. By setting a limit on liability, Price-Anderson reduces the insurance premium that an operator must pay. Herbert S. Denenberg, former insurance commissioner of the state of Pennsylvania, has estimated the indirect subsidy of Price-Anderson. Commissioner Denenberg used the results of the WASH-740 update to conclude that the damages from a nuclear plant accident could be $40.5 billion, a figure which includes the damages for health effects as well as property damage. Denenberg used a representative annual insurance premium of $580 per million dollars of liability insurance, and concluded that the premium necessary to cover a nuclear plant for this $40.5 billion accident would be $23.5 million per year.[45]

Denenberg then noted that the annual operating costs for a nuclear plant—including fuel and maintenance costs—are about $23 million per year, by comparison. The actual liability insurance premiums paid by nuclear plant owners are $300,000 to $400,000, which means that Price-Anderson reduces their premiums by about $23 million per year.[46] On a per-kilowatt-hour basis, the distortion provided by Price-Anderson could be as large as 3.8 mills/kwh in favor of nuclear.[47]

Even greater economic distortions are found in the accounting tricks that the federal government allows electric utilities to use. Two accounting dodges in particular—the accelerated depreciation allowance and the investment tax credit—are particularly effective in reducing utility tax payments to minimal amounts. The electric utility industry pays so little in taxes that in 1975, Senator Lee Metcalf (D., Mont.) introduced a bill to exempt utilities from federal income taxes. Senator Metcalf noted that in 1974, the utility industry's federal taxes represented 1.3 percent of revenues. The senator said that his bill would "accomplish openly what has already occurred through hidden tax loopholes."[48] Under present practices, utilities charge their customers for tax payments under one accounting system, then reduce the taxes actually paid under another accounting system—and the utilities keep the difference. Senator Metcalf's bill would halt this practice of collecting unpaid taxes from consumers.

The practical effect of these tax loopholes is to subsidize power plant construction. A group of researchers at the Lawrence Berkeley (California) Laboratory, in a 1976 study performed under contract with ERDA, concluded that the tax loopholes were so large that: "Effectively the federal government pays about 20 percent of the cost of each new power plant that comes on line (10 percent is tax credit and about 10 percent is accelerated depreciation allowance)."[49] This study estimated that between 1975 and 1985, federal tax dodges would provide about $37 billion worth of utility construction—almost $4 billion per year.[50] The lost tax revenues from the utilities will have to come from the pockets of individual taxpayers. In fact, there are even loopholes within loopholes. For example, an eighteen-hour test at Portland General Electric's Trojan plant in Oregon qualified the plant for a $7 million tax break for 1975. By operating at 10 percent capacity during the test, in December 1975, Trojan fulfilled requirements for the depreciation tax allowance, for the entire year of 1975.[51]

Although federal tax subsidies apply to construction of all power plants, nuclear tends to be favored, both because a nuclear power plant has higher construction costs (and therefore the magnitude of subsidy is greater) than for a fossil plant, and because the utility industry plans to build more nuclear plants than other types of power plants.

In addition to these hidden subsidies, there is the more overt subsidy of research, development, and demonstration projects funded by the government on behalf of the nuclear industry. The Investor Responsibility Research Center estimated in January 1975 that about $5 billion had been provided by the federal government for the development of civilian nuclear power.[52] This figure apparently did not include funds spent on military programs, such as the naval nuclear propulsion program, which enabled General Electric and Westinghouse to develop the nuclear technology they sell to utilities. If military programs are included, government expenditures would be far greater than $5 billion.

But even using the $5 billion figure, government expenditures are substantial. Considering the total power produced by commercial nuclear up to January 1, 1975, this means that direct support from government research at that time had been equivalent to 13.9 mill/kwh.[53] This is a significant factor to take into account when considering nuclear's total economic picture.

Also to be included are future costs of nuclear power that can only be estimated, as well as economic uncertainties throughout the nuclear fuel cycle. There are the costs for decommissioning, for which the industry admits it has only vague estimates. There are the costs for radioactive waste management which can only be estimated—and such wastes might require perpetual storage at government facilities. There are the costs—in dollars and civil liberties—of the private and federal security system that will be necessary if the atomic industry is allowed to recycle plutonium.

There are also major uncertainties in the reprocessing step of the fuel cycle. The first uncertainty is when reprocessing will be available, and how long power plants can operate before their storage pools fill up with spent fuel. The second uncertainty is the cost of reprocessing, which seems to be escalating as rapidly as AEC estimates of power plant capital costs. A 1974 AEC publication estimated reprocessing costs to the utilities at $35 per kilogram (kg) of treated fuel. By 1975, the AEC had revised its estimate to $100 per kg, equivalent to 0.39 mill/kwh.[54] Marvin Resnikoff, physicist at the State University of New York, has estimated that reprocessing costs would be $145 per kg.[55] But even this estimate might be too low. Nuclear Fuel Services, before it announced it was withdrawing from the reprocessing business, told one of its prospective customers that NFS would charge $1,062/kg to reprocess spent reactor fuel.[56]

With the uncertainties in the nuclear fuel cycle, the economic problems of nuclear power plants, and the legal infighting and suits of companies within the atomic industry itself, the industry is coming to Uncle Sam to bail it out with still more government funds. The next round of nuclear subsidies promises to be much larger and more overt:

Until January 1977, when he died of a heart attack, Joe Cury was a maverick. When the price of sugar shot up in the summer of 1974, the Jacksonville grocer continued to stock his shelves. But he also added a sign saying, "The price of this sugar is outrageous, and I advise you not to buy it," until prices went down.

In January 1974, Cury began fighting the Jacksonville Electric Auth-

ority (JEA) when the electric bill for his independent store, the Mandarin Supermarket, suddenly doubled. Soon Cury was pressing the Federal Energy Administration to investigate JEA's oil supply contract with a company named Venn-Fuel, Inc. That local investigation led to congressional hearings in Washington and national publicity on the Venn-Fuel case.

Encouraged by this success, Cury organized a consumer group called POWER (People Outraged with Electric Rates) and began questioning the wisdom of JEA's plans to buy floating nuclear power plants from a subsidiary of Westinghouse, Offshore Power Systems (OPS). Cury and POWER argued that the floating power plants, an untried and untested concept (none have ever been built), were not needed to supply Jacksonville's forseeable electric power demand. POWER charged that a multi-billion-dollar investment, which would only create excess generating capacity, would be an economic disaster for Jacksonville ratepayers, rather than the promised bonanza of cheap electricity and new jobs. The opposition of Cury and POWER was a major factor in the Jacksonville City Council's decision to deny approval of JEA's planned purchase of the two floating plants.

JEA reacted by ordering investigations into the private lives of Cury and Hartley Lord, another POWER activist. This only led to more trouble for JEA, as the FBI and the Justice Department began an investigation of JEA for invasion of privacy. Their investigation in turn led to a grand jury investigation of certain JEA and OPS officials.

Today the Jacksonville City's nuclear development plans are obsolete and the Westinghouse subsidiary, OPS, is in trouble. But large companies don't go bankrupt anymore, they go to Washington. Westinghouse asked the Federal Energy Administration to buy four of its floating nuclear plants, at about $1 billion apiece, and lease them back to electric utilities. The Ford administration, happy to provide a shoulder for corporate management, took the proposal under consideration.[57]

Westinghouse is not the only nuclear company asking for direct bailouts. Allied Chemical Corporation and Gulf Oil Corporation have asked ERDA to subsidize their nuclear reprocessing plant in Barnwell, South Carolina. These companies have spent about $250 million on the plant, but they may have to ask for federal bailouts totaling $500 million if the reprocessing plant is ever to operate.[58] ERDA, in addition, developed the proposal to provide $8 billion in federal guarantees to develop a private uranium enrichment industry.

The Ford administration proposal to create an Energy Independence Authority (EIA) would have squandered even more of the public's funds on the atomic industry. The EIA, inspired by Vice-President Nelson Rockefeller, would have been empowered to provide $100 billion in loans and loan guarantees to energy companies. Since nuclear power is experiencing the most severe squeeze on capital markets, a large portion of the EIA's budget would go to nuclear. As government funds are gobbled up by capital-intensive nuclear plants, whether onshore or offshore, or by energy "parks," there will be that much less money left over to develop non-nuclear technologies.

The atomic industry's public relations claims of "savings" from nuclear power become even more illusory if the "savings" to the industry come at the expense of higher taxes. The economic burden placed on the citizen is still onerous in the light of the limited choice: Decisions are made by large energy corporations and monopoly utility companies, whose lobbyists push federal and state subsidies, along with government policies to further the companies' own special interests. In addition to those utility and energy costs that are passed directly to the ratepayer, the energy industry's reduced tax burden must be recovered from individual taxpayers. If atomic power grows, citizens will clearly pay for it in two ways—once as utility ratepayers, and once as taxpayers.*

The financial troubles of the nuclear industry were summarized in early 1976 by a letter from Maurice Van Nostrand, then chairman of the Iowa Commerce Commission, to Federal Energy Administrator Frank G. Zarb. Concerned about uncertainties in nuclear construction costs, uranium prices, the costs of reprocessing, and the storage of radioactive waste, Van Nostrand told Zarb that there were "monumental unanswered questions in the economics of nuclear generation."[59] The arguments against the health and safety problems of nuclear power certainly are compelling on their own merits. But nuclear power's economic problems leave the industry

* One subsidy which has not even been mentioned is nuclear power's "energy subsidy." Some researchers have concluded that because of the energy input necessary to support nuclear plant construction, a rapidly growing nuclear program will require more energy than it produces. For studies on nuclear's "net energy" problems, see: Peter Chapman, "The Ins and Outs of Nuclear Power," *New Scientist*, December 19, 1974, p. 866; John Price, *Dynamic Energy Analysis and Nuclear Power*, Friends of the Earth, Ltd, London, December 1974; and P. N. Lem, H. T. Odum, and W. E. Bolch, "Some Considerations That Affect the Net Yield from Nuclear Power," paper delivered at Health Physics Society Nineteenth Annual Meeting, Houston, Texas, July 1974.

with no justification for itself. Even if rising citizen awareness is unable to stop the atomic industry, the industry, without the infusion of massive government subsidies, will crumble under the weight of its own economic burdens.

XIV

Alternatives to Nuclear

MANY NUCLEAR POWER PROMOTERS will concede some or all of the problems of nuclear power which have been discussed in previous chapters. But these promoters will argue that with all its faults, nuclear power must still be developed rapidly because there are no other alternatives. They dress their visions of the future in scenarios in which energy consumption can only grow at compound rates. The only energy sources available, they argue, are imported oil, dirty coal, and nuclear power. Indeed, there is no question that if decision-makers lock themselves into such scenarios every energy source must be rigorously developed, including nuclear power. However, the underlying assumptions are grossly flawed and can be challenged decisively.

Conservation

One alternative that is only beginning to creep into the vocabularies of nuclear promoters is energy conservation. Energy conservation means greater efficiency in the use of energy so that the nation's energy supplies can be extended. Conservation means doing better, not doing without.

One can consider several different ways of conserving energy, some of which overlap. There is conservation induced by higher energy prices, something the country is already experiencing. There is conservation from routine thrift measures, such as turning off energy-using equipment when not needed. There is conservation from technological improvements, which make energy-using systems more efficient. Finally, there is conservation from changes in life-styles—changes as uncomplicated as occasionally shifting from automobiles to mass transit, or even from 5,000-pound cars to 3,000-pound cars. But energy conservation resulting from the use of energy-efficient devices could reduce the country's per capita energy consumption by half, with an improvement in consumer well-being.[1]

Conservation caused by higher energy prices has already been observed. Electrical use in the United States, over the twenty years previous to 1974, grew at about 6 percent per year; total energy use grew at about 3.5 percent per year.[2] But this compound growth occurred during a time when energy prices were actually dropping in "real" (as opposed to inflated) dollars.[3] Since late 1973, higher energy prices have dampened this growth rate. Total United States energy use in 1975 was nearly 5 percent lower than in 1973, this country's record year for energy consumption.[4] Electricity use in 1975 was only 2 percent higher than in 1973 (an average of 1 percent per year), according to the Federal Energy Administration (FEA).[5]* So those prognosticators who base their visions of a nuclear future on historical energy growth ignore the likelihood of slow growth in energy usage, due to higher prices alone.

Part of the decrease in energy growth has come from such routine thrift measures as turning off lights and appliances when not in use, reducing unnecessary driving, and being more conscious of energy usage. Roger W. Sant, assistant FEA administrator for conservation until his resignation in 1976, has stated that businesses can cut their energy costs 15 to 25 percent "without investing a thing."[6] The federal government itself has substantiated Sant's statement with its own conservation program. Twenty-six federal agencies were able to cut their use of energy in the third quarter of fiscal year 1975 by 28 percent from their use in the third quarter of fiscal year 1973.[7]

The city of Los Angeles provided another dramatic illustration of the potential savings from thrift measures. During the 1973 oil embargo, the city instituted a strict conservation program and reduced its electricity consumption an average of 18 percent, including

* It is not surprising that electricity consumption grew faster in 1976 and early 1977. The economy was coming out of a recession, causing increased industrial and commercial activity, and the nation faced its coldest winter in 177 years. Still, the rise in consumption did not come close to the industry's overblown projections. Although the Edison Electric Institute predicted that 1976 summer peak demand would be 9 percent higher than in 1975, the actual rise was 3 percent—which still left the utility industry with 35 percent more generating capacity than it needed.

a 28 percent reduction in the commercial sector (office and commercial buildings). This reduction occurred with few significant economic problems. A Federal Energy Administration report, reviewing the Los Angeles experience, stated: "Almost without exception, the establishments in the survey reported that the ordinance placed no economic hardship on their business. Even the service establishments that modified their hours of operation—dry cleaners and laundromats, for example—did the same business in a shorter period of time."[8]

Technical innovations produce even more dramatic energy savings. Consider the experience of Ohio State University. The University implemented a program of retrofitting six buildings to improve the efficiency of their ventilation and heating systems, remove excess lights, and adjust the controls on their cooling systems. After several months of monitoring the energy use at these six buildings, the University found average reductions of 36 percent in electricity consumption and 61 percent in natural gas consumption. The retrofit investment of $209,000 was paid off in 7.4 months by reduced energy costs, and the University will continue to save $338,000 per year in energy costs at these buildings in the future. Encouraged by the results, the University is proceeding to modify six more buildings and is studying sixteen others for later modification.[9]

An estimate of the potential savings from technical improvements came in a report to the Ford Foundation Energy Policy Project, which examined the potential energy savings in six energy-intensive industries: iron and steel, petroleum refining, paper and paperboard, aluminum, copper, and cement. On a per-product basis, the report concluded that energy requirements for these industries could be cut by one-third over the remainder of the decade by implementing technologies already available. If technologies on the threshhold of development are considered, the potential for additional energy efficiency is even more striking.[10]

Lee Schipper, physicist and Information Specialist with the Energy and Resources Group, University of California at Berkeley, estimates that "proper conservation measures today would reduce the energy requirements of autos by 50 percent, reduce the energy requirements of buildings by 50 percent, and reduce industrial needs by 25 to 33 percent."[11] In another study, the American Institute of Architects estimated that a commitment to developing energy-efficient buildings by 1990 could *alone* save more energy than nuclear power is projected to supply even at historical growth rates; and the pay-back time for the capital investment for more energy-efficient buildings would be much shorter than for nuclear plants.[12]

The last example demonstrates that, with its ability to readily dwarf nuclear power's energy output, conservation should be viewed as the country's most viable energy source. Furthermore, technologies that increase energy efficiency can be implemented with less expense and in shorter time periods than technologies to increase energy supplies.[13] In addition, the benefits of an aggressive conservation program would be shared proportionally by large and small users alike, with a minimum burden.

Hysterical atomic power promoters have claimed that conservation could lead to economic chaos. But one should remember that Sweden, Denmark, and Switzerland consume about one-half of the per capita energy consumed by the United States, yet in 1974 each had a higher per capita gross national product than the U.S.[14] Common sense will show that conservation improves the economy by curbing waste, reducing pollution, and curtailing inflation.

Conservation also benefits employment, since the most energy-intensive industries are also the least labor-intensive. That is, products which take much energy to produce usually require little manpower. Moreover, a dollar invested in constructing and operating an electrical power plant will produce fewer jobs than a dollar invested in almost any other industry. According to the Edison Electric Institute, the trade association which represents private utilities, in 1971 it took $173,000 to produce one job in the electric utility industry. For all manufacturing industries on the average, $22,000 would create one job.[15]

Thus energy efficiency has the potential both to replace nuclear power and to stimulate employment. More efficient energy use frees money that otherwise would have been spent on energy waste. This makes more money available for employment, and conservation technologies also induce a higher demand for labor.[16] The Ford Foundation Energy Policy Project concluded that energy conservation could have major beneficial economic effects:

> Substantial economies are possible in U.S. energy input with the present structure of the economy, without sacrificing the continued growth of real incomes. . . .
>
> Our adaptation to a less energy-intensive economy would not reduce

employment; in fact, it would result in a slight increase in demand for labor. . . .

Other Project-sponsored studies also support the conclusion that we can safely uncouple energy and economic growth rates.[17]

An understanding of energy pricing trends can further put the economic effects of energy conservation into perspective. Because major oil companies also have substantial interests in natural gas, oil, coal, and uranium, unregulated domestic energy prices will tend to float upward to the highest-priced energy sources—at present, imported oil. If the energy use of the country continues its present trend, the U.S. economy will be at the mercy of the OPEC cartel on the one hand and the Exxon-led cartel on the other hand. Money spent on energy waste will continue to line the pockets of large energy corporations and divert funds from other segments of the economy, including jobs. A strong case can be made, therefore, that the country's economic problems are in substantial part *caused* by its inability to use energy more efficiently. Reducing energy waste will clearly improve the economy and its competitiveness abroad, as well as render nuclear power unnecessary.

Domestic Oil and Gas

On the energy supply side, one should not ignore oil and natural gas as alternatives. Much publicity has been given to the pessimism of recent reports on oil and gas reserves; little attention has been paid to the assumptions that accompanied the pessimism. When carefully read, these reports lead to the conclusion that domestic resources are much greater than the energy industry would admit.

As an example, the latest survey by the U.S. Geological Survey (USGS) concluded that remaining recoverable reserves of domestic oil are 112 to 189 billion barrels. This represents a thirty-seven- to sixty-two-year production life at the 1974 domestic production level of 3.04 billion barrels. By comparison, total U.S. oil consumption in 1974 was about 6 billion barrels. Recoverable reserves of domestic natural gas, according to USGS, are equivalent to thirty-six to fifty-one years of production life.[18]

But the USGS report admitted that its analysis was done for pre-1974 oil prices, and that no estimates had been made for the additional oil that might be obtained from present prices.[19] The USGS

Survey also noted: "Excluded from consideration were oil shales, tar sands, and heavy hydrocarbons and tight gas sands not currently productive. Also excluded was offshore potential beyond 200 meters of water depth. All of these resources or areas for resource development have significant future potential measured in tens or hundreds of billions of barrels."[20] Finally, the USGS, in acknowledging that earlier surveys (1965, 1972, and even 1974) had been much more optimistic in its estimation of domestic oil and gas reserves, did not offer any explanation as to why these earlier surveys might have been so inaccurate.

Another review widely regarded as pessimistic on oil reserves was made by a committee of industry and academic researchers for the National Academy of Sciences (NAS), an organization established to advise the federal government on scientific and technical questions. The NAS report estimated that undiscovered, recoverable resources of domestic oil were 113 billion barrels and that undiscovered, recoverable resources of domestic natural gas were 530 trillion cubic feet.[21] When measured and inferred reserves are added to these figures, they are roughly equivalent to the higher of the USGS estimates.

One important qualification of the NAS study was that oil yet to be recovered from known U.S. fields represents at least an additional 200 billion barrels.[22] Other sources estimate the figure to be 300 billion barrels or more.[23] If only half of that estimated 300 billion barrels could be recovered, that single source *alone* could provide *all* of U.S. oil consumption at present levels for twenty-five years.

But the most significant aspect of the NAS report is that it reached conclusions radically different from those of the nuclear promoters. In view of its own estimates of oil reserves, the NAS study group recommended conservation. In fact, the primary recommendation of the entire NAS study was that the federal government should strongly encourage conservation of all natural resources. The NAS study also took a dim view of energy-demand forecasts projecting traditional growth patterns, on which a rapidly expanding nuclear industry is crucially dependent. The study group warned that such projections, although unrealistic because of likely higher prices, can be "self-fulfilling."[24]

The NAS also commented that reliable data on mineral resources are difficult to obtain, in part because the industry considers such information a trade secret.[25] The federal government does not generate its own fossil fuel reserve data and is forced to rely on industry figures. But the industry's measured reserves, supposedly the most credible data, may be understated. For example, the American Petroleum Institute (API), a trade association representing the oil companies, reported a decline in the measured reserves of oil and natural gas at the beginning of 1975. The announcement met with considerable skepticism, because API had ignored a basic fact of life: as oil prices rise, as they had in 1974, so will economically recoverable reserves.[26]

Even in early 1974, Hendrik Houthakker, professor of economics at Harvard, stated: "Recent [oil] price changes have been so large they invalidate all projections [of reserves], including those of the National Petroleum Council [another oil industry trade association], that ignore the effect of prices on supply and demand."[27] Houthakker knew that higher prices make secondary and tertiary techniques—such as injecting water or gas into an oil deposit to push the remaining oil to the surface—more economical. The use of such techniques could recover the large amounts of oil still in known U.S. fields.

Moreover, there are likely to be large resources that remain to be discovered. Enormous quantities of natural gas, for example, are believed to exist in onshore deposits on the Gulf Coast, at depths of one and a half miles or more. Estimates of the natural gas in these deposits range upward to 37,000 trillion cubic feet—more than one thousand times the nation's present annual use of natural gas. Myron Dorfman, of the University of Texas, acknowledges there will be many problems in exploiting this gas, "but it must be remembered that the difficulties are minimal in comparison to those associated with other technologies, such as nuclear energy."[28]

It should also be recalled that the oil industry has been warning of "shortages" for over eighty years.[29] So there are many reasons to have a healthy skepticism of reserve figures produced by an industry attempting to justify setting domestic oil prices equivalent to oil import prices. Certainly, this skepticism should not be taken as an excuse to waste fossil fuel reserves. Rather, it should serve as a reminder that a headlong rush into nuclear power is unnecessary, and unjustified by unsupported warnings of oil and gas shortages.

Persons who remain convinced that domestic reserves are scanty can refer to experienced geologist M. King Hubbert. His estimates of reserves are even more pessimistic than the latest USGS or NAS estimates.[30] But Hubbert, along with over 2,000 other scientists, recently signed a statement noting that the dangers of nuclear power were "altogether too great" and urging a "drastic reduction" in nuclear plant construction.[31] Hubbert on another occasion admitted that nuclear power "scares the hell out of me." He believes that the proper course for the future is development of solar power and strong conservation programs to "phase out nuclear as fast as we can."[32]

Solar Energy

Just as energy conservation is the option that can best render nuclear power unnecessary over the next thirty years, solar energy is the option that will render nuclear power unnecessary over the long term. Solar energy is a "renewable" resource. Although the sun will one day die out, it will continue to radiate heat and light for billions of years, making solar energy a practically perpetual source that will be available long after fossil and mineral fuels are depleted. Challenges to the development of solar energy include its diffuse nature (it is spread out over a large area) and the requirement of storage systems for times when the sun does not shine. Different technologies are being developed to tap solar power in all its direct and indirect forms, including solar heating and cooling, wind power, bioconversion, ocean thermal power, solar thermal conversion, and photovoltaic cells.

SOLAR HEAT

The technology of heating and cooling is closest to mass commercial use. In 1975 there were about 30 homes in the United States presently using solar heat which demonstrate feasibility.[33] Houses planned or under construction raised that number to about 200.[34]

Comparing the prices of solar and electric home heating systems verifies that solar is a better investment in all areas of the country—with the exception of the Pacific Northwest, which has relatively cheap electricity from hydroelectric power.* Solar heat is

* According to Arthur D. Little, a consulting and research firm, a solar heating and cooling system for an average all-electric home would cost $2,000 to $2,500 more than a conventional system. However, the solar system would reduce electrical use in comparison with conventional electric systems by about 6000 kilowatt-hours per

also close to becoming as economical as oil heat, a situation which is striking because mass-production has not yet begun with solar devices.[35]

Solar heating systems presently consist of long, shallow flat-plate "collectors" which hold a "fluid" such as air or water to absorb the sun's heat. The collector is mounted on a roof or, in some cases, on the ground. The fluid is pumped or blown through the collector, where it absorbs the heat of the sun's rays, and is then transported through the home, where it releases or transfers its heat. This system will also include a heat storage apparatus, such as a large insulated tank, to retain heat on days when the sun does not shine brightly. Figure 1 is a schematic of a typical solar home system.

Air conditioning with solar energy is also possible. One process uses an ammonia and water mixture passed through a solar collector which turns the ammonia to vapor, separating it from the water at a higher pressure. The ammonia is passed through a condenser, becoming a liquid-gas mixture which is passed through a heat exchanger. Here the heat from the air is absorbed to cool the home. As it absorbs heat, the ammonia mixture expands and is reabsorbed by water returning from the solar collector. The cycle can then be repeated. In the early 1960s, an ammonia-water cooling system was built and tested (to demonstrate technical feasibility) at the University of Florida.[36]

A variation of heating systems is the solar hot water heater. These units are much smaller than home systems and are already in wide-scale use. A study by the A. D. Little Corporation found that 160,000 solar hot-water units were sold in Japan in 1974; up to 8,000 such units are operating in Florida and the Southwest.[37] Hot water heating is one of the major energy uses in the home, accounting for 15 percent of residential energy consumption and 3 percent of total energy use.[38] This is the same amount of energy supplied by nuclear power in 1975.

Solar heating systems obviously are most effective in the Southwest, where the sun shines more than in other areas of the country. But they can be used efficiently in other areas of the country as well. Robert F. Schmitt, a professional builder and engineer, has con-

---

year, or by $420 to $500 per unit. Because of this saving, the solar system would pay back the original investment in four to seven years (Ron Lanoue, *Nuclear Plants: The More They Build, the More You Pay*, [Center for Study of Responsive Law, Washington, D.C., 1976], p. 27).

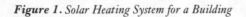

***Figure 1.** Solar Heating System for a Building*

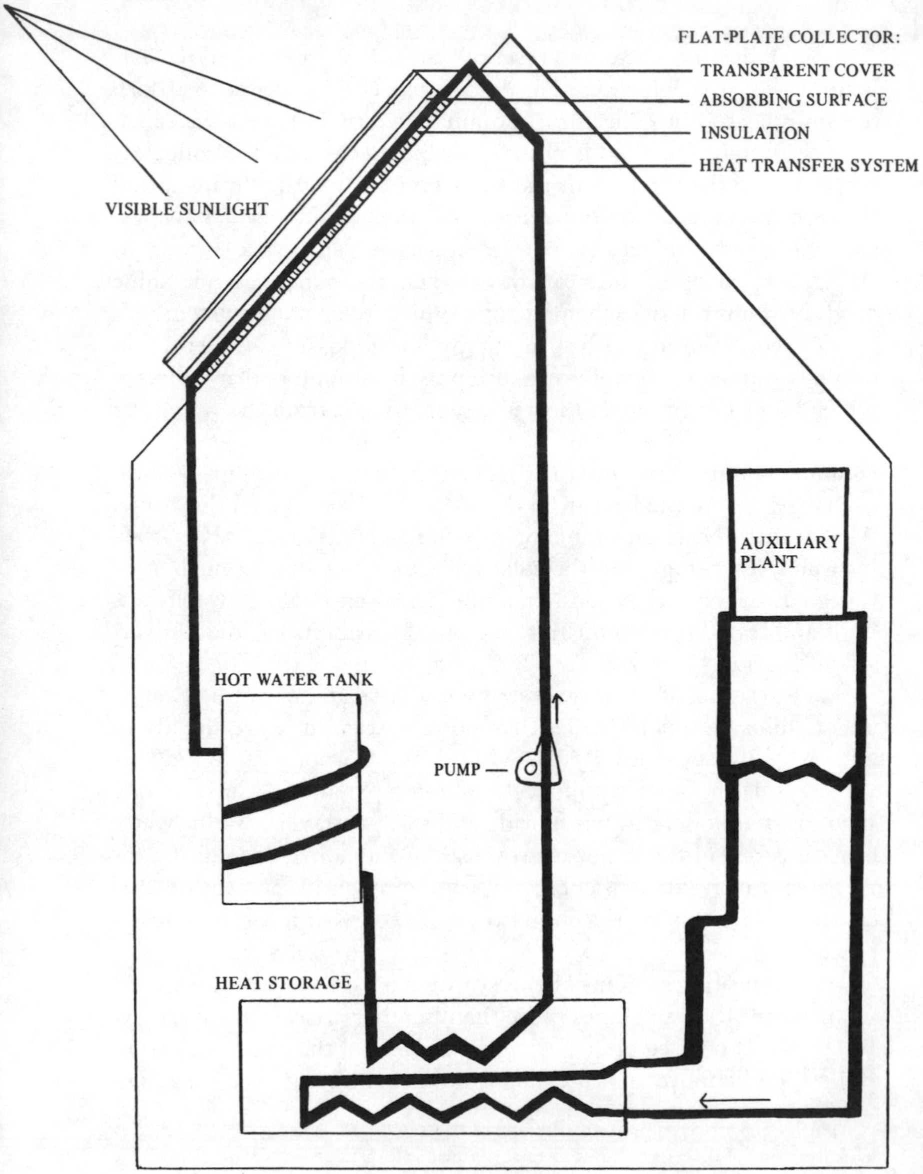

structed a solar home near Cleveland, Ohio, that doesn't even use solar collectors. Instead, the house, which is well insulated, absorbs solar heat, which is distributed throughout the building, from four large skylights above an atrium. A fireplace in the center of the atrium burns wood or coal for the only supplemental heat. The annual energy costs of the Schmitt home are 40 to 60 percent less than those of a comparable "conventional" home.[39] Schmitt has thus demonstrated that innovative design could make solar energy practical in all parts of the country.

The Timonium Elementary School, near Baltimore, Maryland, utilizes a solar heating system and a solar-driven air conditioning system.
(J. E. Westcott, U.S. Energy Research and Development Administration)

WIND

A second solar technology which has been demonstrated is wind power. Energy from the wind is considered solar in nature because the sun heats air masses and the earth. Temperature differences create winds by causing the air masses to move. The wind is converted to electrical energy by the vanes of a windmill that is attached to an electric generator. A windmill can also be used to pump water, as it did on most farms in the U.S. until the 1930s.

Windmills have already been used in the U.S. to supply electricity in areas as different as Massachusetts[40] and New Mexico.[41] U.S. Representative Henry S. Reuss (D., Wis.) operates a windmill on his property that produces about 70 percent of the average monthly household use.[42] Consolidated Edison Company is opposing a group of environmentalists who constructed a windmill to supply electricity to a New York City tenement building. Con Ed has warned that if the windmill operators feed their surplus electricity into its power grid, the utility will disconnect service to parts of the building.[43]

Larger-scale wind energy systems would also be feasible. A 1.25 megawatt wind generator operated at "Grandpa's Knob," near Rutland, Vermont, and fed power into the Vermont power grid from 1941 to 1945. The unit was not operated after 1945, when one of its blades broke. But its performance record has been interpreted as demonstrating that a cluster of similar machines could generate a total of 9 megawatts for a cost of $400 per installed kilowatt, in 1974 dollars.[44] This figure is half or less the projected capital costs of nuclear power plants.

There are even more ambitious wind projects on the drawing board. William Heronemus, professor of civil and electrical engineering at the University of Massachusetts, has proposed offshore wind power systems. One proposal would place floating wind stations, each with a capacity of about 6 megawatts, across the prevailing winds off the New England coast. Heronemus estimates the total potential capacity at 82,000 megawatts—more than enough to supply the six New England states with the amount of electricity they now use.[45]

PHOTOVOLTAICS

The conversion of sunlight into electrical energy by the use of photovoltaic devices such as the silicon cell is an established technique. Silicon is a semiconductor—it is both an electrical insulator and a conductor. When silicon discs are coated with materials such as boron, a positive electrical layer is created on top of an underlying negatively charged silicon layer.

If units of sunlight, called "photons," strike the cell they are converted into free electrons. The positive (boron) layer accepts the electrons and the negative (silicon) layer rejects them. Thus, a flow of direct-current (DC) electricity is set into motion between the two layers. The current is diverted by an electrical conductor imbedded in the surface of the cell. Typical solar cells are able to convert 10 percent of the incident sunlight into energy—which is a significant amount, considering that at an energy conversion efficiency of 10 percent less than 2 percent of the U.S. continental land mass could supply the nation's total current energy needs.[46]

Other materials and cell designs will improve both the efficiency and manufacturing techniques of solar cells. In the last several years, there have been important breakthroughs which suggest that the way is open for further advancements.

Joseph Lindmayer was puzzled by the stagnation of the silicon solar cell industry in the 1960s and 1970s, especially since important advances were being made in the development of semiconductor technologies such as the use of silicon chips in miniature calculators. The solution, he decided, was to replace laboratory techniques for the production of solar cells with manufacturing processes developed in other, related, industries.

In May 1972, Lindmayer, working with a Communications Satellite Corporation (COMSAT) research team in a Washington, D.C., suburb, produced a remarkable new cell using the modernized techniques of the semiconductor industry. The "violet cell," so called because of its ability to convert more sunlight from the violet/ultraviolet range of the light spectrum than previous cells, had a 15 percent efficiency and reduced by one-half the silicon needed in a single cell.

Shortly afterward Lindmayer left COMSAT and founded Solarex Corporation in Rockville, Maryland, where he pioneered a new technology that led to a dramatic reduction in the cost of silicon cells while maintaining their high efficiency. Solar cells from Solarex are presently used in over 1,000 locations around the world. Power from the cells is used in such applications as short-wave and police radios, the Israeli army's communications systems in the desert, and landing beacons at an Anchorage, Alaska, airport.

If the potential of photovoltaic cells appears too remote, it should be remembered that through the efforts of Lindmayer and others the cost of delivered solar-cell electricity was reduced from 1,000 times present utility costs to only 50 times the cost within a two-year period. Further breakthroughs will make the cells even more competitive.[47]

In 1974, Paul Rappaport, research scientist for the RCA Corporation, reported that a panel of twenty-one industry experts concluded that solar cells could produce power at costs of five cents per kilowatt hour by 1985 and one cent per kilowatt hour by the year 2000. This is a favorable comparison, for residential consumers presently pay two to nine cents per kilowatt hour for their electricity. Estimated costs of the program to develop photovoltaic electricity at five cents per kilowatt hour by 1985 were $250 million.[48] Photovoltaics could be used in large central generating stations or in rooftop systems for individual homes.

In fact, engineers at the Massachusetts Institute of Technology (MIT) have already developed a solar energy home unit that produces both heat and electricity using solar cells. A series of mirrors concentrates sunlight, reducing the number of photovoltaic cells required for the solar "array." The increased heat from the concentrated sunlight can damage cell efficiency, but the MIT researchers reduced the losses by cooling the cells with water. The heated water can then be used for home heat or hot water, just as a "conventional" solar heating system.[49]

BIOCONVERSION

An indirect form of solar energy is provided by "bioconversion"—the process of changing natural, or organic, materials such as plants into methane (a form of natural gas), oil, hydrogen, or alcohol. This is considered a solar energy source because the plants that are converted to fuel receive their original energy from the sun. A demonstration plant in Pittsburgh, operated by the U.S. Bureau of Mines, is processing organic wastes—such as sawdust, animal manure, sewage sludge, and paper—to produce a heavy oil which can be used as fuel, via a chemical reaction with carbon monoxide and water.[50]

The potential for bioconversion energy sources was explored by Farno L. Green, a General Motors Corporation engineer, in 1975. Green examined the energy content of "farm residues," or non-edible parts of plants (such as cornstalks and husks, wheat straw, grain sorghum, soybean, cotton, and other stalks remaining after the principal food is removed.) Assuming that two-thirds of the U.S. farm residues in 1972 had been removed from fields, leaving one-third to prevent wind and water erosion and to improve water infiltration and soil quality, Green estimated that a heat value equivalent to 30 percent of the coal mined in the U.S. during 1972 could have been produced.[51]

Professor Fred Benson, dean of the Texas A&M College of Engineering, has suggested that crops could be developed for the sole purpose of serving as energy sources. The crops could be converted to ethanol or methanol, alcohols that can be used as automobile fuel. Brazil already produces alcohol, used in industrial processes and as auto fuel, from molasses. In 1974, the Brazilians extended their supply of gasoline by 2 percent with alcohol, and the government intends to raise the alcohol content of its gasoline to 10 percent by 1980.[52] Dean Benson of Texas A&M has estimated that enough liquid supplement for gasoline could be produced from energy crops to eliminate the need for importing oil into the U.S.[53]

Benson's proposal is supported by the experiments of Thomas Reed, research chemist at the Massachusetts Institute of Technology. Reed has been running two cars on methanol-gasoline mixtures over the last few years, with no alterations made to the cars. He has found that the mixtures can result in improved fuel efficiency, increased acceleration, lower exhaust temperatures, and less pollution.[54]

The U.S. Navy Undersea Center, at San Diego, California, has been investigating a bioconversion option which uses kelp (seaweed) as its crop. The kelp can be grown and harvested at large ocean "farms," then used as feed for fish or animals, or converted into food for humans. Material not used as food can be converted to synthetic natural gas. The Undersea Center estimates that a kelp farm covering a square of ocean approximately 470 miles on a side—roughly the distance from San Francisco to San Diego[55]—could produce the equivalent of the total food energy and natural gas energy presently being consumed by the United States each year.

OCEAN THERMAL

Other solar technologies are aimed at producing electricity from central generating stations. One proposal would use the surface water of the ocean to heat a "working fluid" such as propane. As it expanded, the propane would force a turbine, housed in a partially submerged plant, to drive a generator. The operation of the plant would be analogous to the steam cycle of a conventional power plant, with propane replacing steam as the working fluid. After leaving the turbine, the working fluid would be condensed by cold water from the deeper part of the ocean. This process, using the different temperatures of the ocean, is called "ocean thermal power" (OTP).

Professor Clarence Zener, physicist at Carnegie-Mellon University in Pittsburgh, has studied the ocean thermal concept and concludes that a plant 300 feet in diameter would be able to produce about 100 megawatts of electricity. As an indication of ocean thermal's potential, he states: "Such plants would have to be spaced about 10 miles apart in a limited region so surface waters would not be cooled off too much. To give an idea of the resource base, such a network over the entire tropical ocean could supply the world's population in the year 2000 with the energy per capita now consumed in the United States.[56] Professor Zener recognizes that if an OTP plant were built today, it might require twice the capital costs of a nuclear plant. But he also believes that cost reductions are possible through advanced technology, appropriate manufacturing methods, including mass production, and selection of favorable ocean sites. He concludes that these improvements could make the capital costs of an OTP plant less than those for a nuclear plant—and, of course, fuel costs would always be less—for the OTP plant.[57] Figure 2 is a diagram of an OTP plant and a schematic of its operation.

THERMAL CONVERSION

You won't find the energy corporations beating a path to Professor Otto J. M. Smith's door, but he believes that he has designed a way to harness the sun economically to generate electricity. A hard-headed electrical engineer at the University of California at Berkeley, Smith calls his mirror-tower plant "a practical solar-thermal-electrical power plant that can be built at a reasonable cost with available materials and conventional engineering design techniques."

What he wants to see built is a pilot solar plant of a size (100 megawatts) that could supply the residential needs of a community of 100,000 people. On about 700 acres of land, the plant would be supplied from 1,100 towers 100 feet high. A hexagonal field of 312 mirrors would illuminate each tower receptor. "The absorbed heat from the hot receptors", according to Smith, "would be carried by heat exchange fluids through pipes to a central station power plant containing heat exchangers to preheat and boil water and to superheat steam for a conventional turbine and electrical generators."

This proposed solar-electric plant was not a sudden idea of Professor Smith. It was a meticulous process covering the creative thought, design, and costing-out of components; outside evaluation; and continual refinements. Smith said when the solar-electric plant design was completed, two engineering reviews, one at the Sandia Laboratories in Livermore, California, and

Figure 2. *Conceptual Drawing of Ocean Thermal Power Plant*

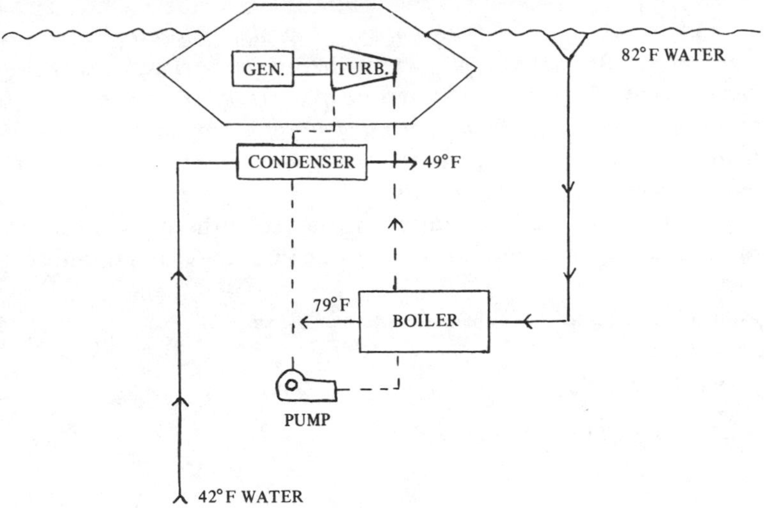

SKELETON MODEL OF OCEAN THERMAL POWER PLANT

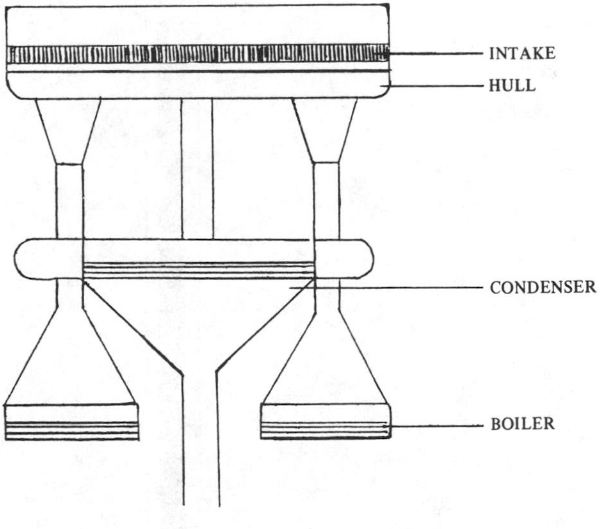

the other at the Naval Weapons Center in China Lake, California, were favorable.

Smith claims that his calculations for capital, operating costs, and maintenance costs make the proposal competitive now with nuclear energy and oil. He believes, however, that with mass production economies for the mirrors, the cost will go down. The cost of atomic power plants, on the other hand, is expected only to go up—fast. Smith believes that his proposal is "the present alternative to nuclear energy."[58]

Figure 3 is a diagram illustrating the "solar thermal conversion" plant, a variation of Professor Smith's concept. A large number of

Figure 3. *Example of Solar Tower Generation Concept*

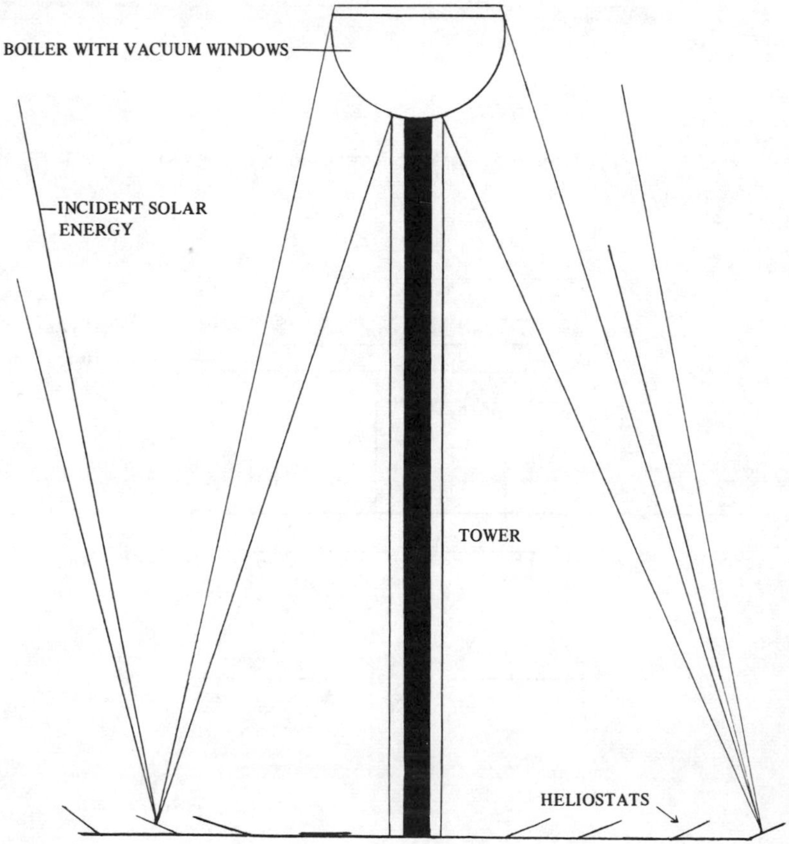

flat mirrors would focus sunlight on a boiler at the top of a high tower. The boiler would produce high-temperature steam to drive a turbine.

Another solar thermal proposal would use a trough-like concentrator to focus sunlight onto a central pipe. Heat collected in a fluid flowing through the pipe could be stored, or the fluid could drive a turbine directly.[59]

STORAGE

Storage systems also must be developed to provide energy at times when the sun is not shining. There are several promising possibilities for this. Solar electricity could either be stored in high-efficiency electrical batteries or used to break down water into its components, oxygen and hydrogen. The hydrogen could be stored, or transported to other areas and burned as fuel.

Storage via the mechanical energy in rotating bodies is also possible. Excess electricity could operate a motor-generator as a motor and set a flywheel in rotation to very high speeds. When the immediate supply of solar electricity declined, the inertia of the flywheel would drive the motor-generator as an electrical generator.

Direct solar heat can also be stored. In home heating solar systems which use water as the fluid circulating through collectors, heat is stored in insulated water tanks. In heating systems that use air as the fluid, heat can be stored in clusters of rocks or gravel. Heat can also be stored in molten salt mixtures at high temperatures, for use at a later time.

Some of the solar technologies require more fundamental research. But for the other technologies most of the research has been completed and no technical breakthroughs are needed. This is particularly true for the heating and cooling option, which needs only demonstration projects to improve design, and mass production economies to reduce costs. It should also be recognized that developing solar technologies is chiefly a matter of developing sophisticated plumbing systems. No great scientific breakthroughs are required.

But a viable solar program requires the determination of the federal government to develop solar energy. That determination is now absent and is unlikely to appear because of the private vested interests and bureaucratic momentum to expand nuclear power.

Development of solar energy will not be without technical

problems, but a solar world avoids the serious *institutional* problems of nuclear power. Moreover, solar energy is a renewable resource which will be available long after other energy sources have been depleted. Solar energy's environmental effects should also be small; even the earth's heat balance will undergo little net change, since most of the energy tapped would radiate upon the earth in any case. Nor is solar energy hampered by problems such as limited liability or the need for emergency evacuation plans, which afflict atomic energy.

Even the pro-nuclear Federal Energy Administration, which was established under President Nixon as the federal agency responsible for ensuring adequate energy supplies, estimated during the end of the Ford Administration that by the year 2000 solar energy could provide 7 quadrillion BTUs of energy.[60] This estimate, which may well understate solar's potential, is seven times FEA's prediction for the contribution from the breeder—with which solar must compete for research funds—and about ten percent of the nation's current energy use. Solar's ultimate potential, of course, is much greater. Solar technologies are the options which over the long term will make nuclear power completely unnecessary. Solar energy is therefore too valuable a resource to allow its development to be delayed by massive nuclear power subsidies.

Other Energy Sources

POWER FROM WASTE

There are other non-nuclear energy sources in addition to solar energy. One such source is solid waste. Where bioconversion changes plant material to burnable form, solid waste utilization burns urban wastes for fuel. Garbage is shredded and metal and glass separated from combustible material. The combustible material is burned as low-sulfur fuel; the scrap metal and glass are reused.

A St. Louis utility is converting two of its coal-fired boilers to allow them to burn a mixture of 80 percent coal and 20 percent dried garbage. The state of Connecticut has plans to construct ten plants which can separate garbage into combustible and non-combustible material. Other countries are far ahead of the United States in the use of solid waste for energy. The Dutch have been using garbage to produce steam heat and electricity for over fifty years. Copenhagen and Paris have been using similar systems for nearly as long. About

25 percent of the population of West Germany uses electricity generated by garbage.[61]

In addition to the fact that it generates low-polluting energy *and* recovers materials, solid waste utilization represents a means of disposing of urban waste, a measure which is welcome because many cities are running short of landfill or dumping space. Use of solid waste for fuel could recover up to $1 billion worth of metals, could reduce the waste disposal problem, and could conserve the equivalent of 290 million barrels of oil per year.[62]

"Cogeneration" is an energy source which allows more efficient use of an electrical turbine's waste heat—which ordinarily would be dissipated to the atmosphere or a body of water. The cogeneration option is applicable to companies which use steam in their industrial processes. In one form of cogeneration the temperature and pressure of the industrial steam is raised first enough to power a steam turbine, which generates electricity. The steam exhausted from the turbine is then used for the normal, lower-temperature, process steam needs. The energy cost of generating electrical power in this way is the additional fuel required to increase the quality of the process steam enough to power the turbine, which makes industry generation of electricity twice as efficient as central power plant generation. Cogeneration of electricity thus requires only half the fuel and half the capital required by central power plants.

A study performed for the National Science Foundation by the Dow Chemical Company (and other authors) suggested that it would be profitable for industry to construct 71,000 MWe of electric generating capacity by 1985, which would be enough to supply all industrial electricity needs and still sell some electricity back to the utilities.[63] The industrial generating capacity would permit utilities to forego $38 billion of new investment but would require only $19 billion invested by industry due to cogeneration's greater efficiency.

GEOTHERMAL POWER

Another source of energy is geothermal power. Geothermal power is also a "renewable" resource, since its source is the heat from inside the earth. Geothermal energy can be utilized in several different categories. First, there is "dry steam"—steam with very little moisture—which comes to the earth's surface at only a few places on the earth, including the geysers in northern California. The dry steam can be used to drive a turbine directly to produce electricity. Pacific Gas & Electric power plants at the geysers produce enough electricity to supply over half the population of San Francisco.[64] Other countries, including Iceland, Italy, and New Zealand, use geothermal steam for electrical production or direct heat.[65] Total worldwide geothermal electrical capacity was estimated at 1085 MWe in 1973.[66]

Other forms of geothermal heat are hot water, "wet steam" (steam with water droplets), or brine. These are all heat sources which could not be used directly on a turbine, because they either are not hot enough or contain water or brine which could corrode a turbine. These heat sources would have to be passed through a secondary heat exchanger to generate steam or boil some other working fluid to drive a turbine.

The third type of geothermal power would be the "hot rock" option. This would require boring holes through the earth down to some heat source. Water would then be pumped through the holes, would absorb the heat from the earth, and would flow back to the surface. At the surface the water would give up its heat and then be forced back to the heat source. Exploiting the "hot rock" resource would require development of inexpensive and efficient deep-drilling techniques, but the hot rock potential is enormous. The Cornell University Workshop on Energy and the Environment in 1972 concluded that geothermal energy alone would be capable of meeting all U.S. power requirements for "several centuries" if hot dry rocks could be made a practical energy source.[67] An optimistic 1973 report endorsed by then Secretary of the Interior Walter J. Hickel predicted that by the year 2000, with an aggressive research and development program, geothermal power could supply 395,000 MWe of electrical capacity.[68] This figure represents nearly the present total capacity of the nation's electric utility industry.[69]

COAL SEAM METHANE

Another energy source is methane, a form of natural gas found in underground coal deposits. Methane is an explosion hazard when confined, so it is normally forced out of the mines by ventilation. But the U.S. Bureau of Mines (BuMines) has conducted pilot projects to use the methane as an energy source, instead of just wasting it. Boreholes are drilled into mining deposits to extract the methane before mining begins. The methane can then be distributed through commercial gas pipelines. A BuMines project in West Virginia has been operating for three years, distributing methane from coal seams via pipelines. The methane drained each day from the West Virginia project could meet the daily cooking needs of 18,000 average households.[70]

In addition to the West Virginia project, BuMines plants in Alabama and Pennsylvania have also demonstrated the practicability of tapping methane from coal seams.[71] Robert Stefanko, professor of mining engineering at Pennsylvania State University, has estimated that the potential for methane extraction from mineable coal beds in the United States is 260 trillion cubic feet, equivalent to this country's total natural gas consumption for a dozen years. It also represents more energy than nuclear power is projected to supply from now until the year 2000, even under the most optimistic development estimates.[72] In addition to its potential as an energy source, the extraction of methane from underground coal mines will remove an occupational hazard before mining begins.

COAL

Then there is coal itself as an energy source. The 1973 Cornell Workshops on Energy Research and Development estimated that recoverable coal reserves could supply 33,700 quadrillion BTUs* of energy.[73] This means that coal alone could supply the nation's energy needs, at present levels of consumption, for 440 years.

Coal's environmental and occupational problems must be recognized, and use of coal as an energy source cannot be completely acceptable until these are corrected, although coal will still be widely used, at least until the year 2000, even with a rapidly expanding nuclear program. On the question of coal mining safety, it should be recognized that U.S. mines cover a range of safety records. The safest mines in this country have one-tenth or less the injury rate of the worst mines. Even on a national basis, the fatal injury rate for underground coal mining in the U.S. is two to four times the rate in Great Britain's coal mines. This fact has led the Ford Foundation Energy Policy Project to conclude: "Wide variation among U.S. companies and the much better record of British coal mining

* BTU is an abbreviation for British Thermal Unit—a measure of heat energy. One BTU is the energy necessary to raise one pound of water 1°F in temperature. It takes about 150 million BTUs to heat the average house for a year. For the year 1975, the total United States energy consumption was about 72 quadrillion BTUs. (A quadrillion is 1,000,000,000,000,000.)

strongly suggests the high human cost being suffered by U.S. coal miners is not necessary. . . . There is no inherent reason why the underground mining of coal cannot be made a reasonably safe occupation."[74]

Coal's environmental problems stem from the presence of impurities—such as traces of mercury and radioactive thorium; but the worst pollutant by far is sulfur, which is found in all grades of coal. When coal is burned, sulfur oxides are produced and are emitted from a coal plant's smokestack with the exhaust gases from coal combustion. The sulfur oxides can damage the lungs and aggravate breathing problems. There are, however, promising solutions for removing sulfur from coal before it is burned, or removing sulfur oxides from exhaust gases after coal is burned.

Coal power plants operating today can install stack gas desulfurization systems, or "scrubbers," to remove the sulfur oxides. Utility companies, particularly American Electric Power, Inc.—the nation's largest private power system, which services the Ohio-Kentucky area—have claimed that scrubbers are unreliable and unworkable. But a September 1974 statement by the Environmental Protection Agency should have put the scrubber controversy to rest:

> While a few utilities continue to argue that Flue Gas Desulfurization systems [scrubbers] have not yet demonstrated reliable operation, actual experience over the past few months clearly refutes these claims. Experiences at Louisville Gas and Electric, Arizona Public Service, and Southern California Edison are illustrative. LG&E's Paddy's Run unit has been available to the boiler for a total of five months at well over 90 percent reliability. APS's Cholla unit has operated continuously for eight months with a reliability of 90 percent or better, and SCE's Mohave unit has operated continuously from January to September 1974 at 84 percent reliability.[75]

"Fluidized bed combustion" is a process which has promise for the future, and which has proven effective in the removal of more than 90 percent of sulfur dioxide formed by coal combustion. In a fluidized bed combustor, an upward stream of air suspends small particles of limestone and coal. The coal burns, generating heat to produce steam. The limestone particles react with sulfur dioxide to form calcium sulfate, a chemically inert substance. Methods are being developed to recover the sulfur (for resale) from the limestone, which would allow recycling of the limestone for later use. Systems incorporating the fluidized bed combustor are expected to be less expensive, more compact, and more efficient than conventional power plants burning high-sulfur coal with scrubbers.[76]

Sulfur can also be removed from coal before it is burned, which would seem the most logical manner to deal with the sulfur problem. Battelle Memorial Institute, a research center in Columbus, Ohio, has announced the development of its "hydrothermal" process, which removes sulfur from the coal via a chemical reaction. The sulfur can be recovered for resale, the chemical solution used to react with the coal can be regenerated for reuse, and the product of the process is a solid form of coal which is low in sulfur. The process is simple, it is environmentally attractive because it does not produce large volumes of waste, and it promises to be relatively inexpensive. Battelle estimates that commercial plants treating coal with the hydrothermal process could be developed by 1982.[77]

The hazards of the coal fuel cycle should certainly not be played down. Nor should the hazards of other alternative energy sources be overlooked: Geothermal power can cause land subsidence and can release its own pollutants, including sulfur-based gases. The manufacture of solar cells or even metal panels for solar heating will require the use of raw materials and will undoubtedly have some environmental effects. For these reasons the nation's immediate energy option should be conservation. Energy efficiency has only beneficial effects to the environment, to jobs, and to the economy.

If conservation is developed to its full potential, it is clear that the choices for the future are not limited to nuclear vs. coal, as the atomic industry often claims. But even if the choices were so limited, nuclear power would lose, for two major reasons. The first is that the problems of coal power are largely technical—safer mines, reclamation of land from strip mining, and pollution control measures. Similarly, geothermal and solar energy will present their own technical problems, but all of these technical problems are amenable to technical solutions. Moreover, the atomic industry's warnings of the effects of alternative technologies are rather incongruous when one realizes that these promoters have employed nuclear power for over twenty years without solving such monumental problems as nuclear waste disposal.

The problems of nuclear power, in contrast to other energy technologies, are institutionally serious as well as technically unresolved. Nuclear power tempts saboteurs, terrorists, and hostile nations. The export of nuclear power plants also exports the means for weapons proliferation. Nuclear power will strain civil liberties; it will require massive, government-financed corporate socialism; and the government handouts necessary to perpetuate nuclear technology could forestall the development of non-nuclear options. Nuclear power will burden future generations and will require incredible stability in individuals and human institutions if its present inventory and future nuclear garbage are to be controlled.

The second reason nuclear power must be rejected is its interrelated nature as a technology. Even proponents of nuclear power, such as Senator John Pastore, acknowledge that to be a viable energy alternative it must be free of catastrophic accident. Were a catastrophe to occur, at any plant at any step in the nuclear fuel cycle, what would be the reaction of a citizenry that had been told for years that "it will never happen" by government officials and an industry grown smug from its own propaganda? There would be such an outcry that nuclear power plants all over the country might be shut down—forever. The nation would then be faced with simultaneous radioactivity and energy crises. Clearly this country cannot let such an unstable technology become a major energy source, now or in the future.

3. The Institutional Setting

XV

The Industry Profiled

WITH ALL the faults of atomic power, and all the abundant alternatives to make it unnecessary, why then does it continue to be promoted? Why do corporations and government institutions continue to foist a nuclear economy on the nation and the world? In fact, decisions on nuclear power are made or implemented by the very organizations which have brought atomic energy to its costly, hazardous, and unreliable position. Their attitude, investment, and overall stake in the atom leave them no "options for revision."

The next three chapters focus on the promoters of atomic power. This chapter deals with the structure of the atomic industry, pointing out which companies benefit from nuclear power. The following chapter examines the federal structure—the congressional bodies and federal agencies which have spent most of their tenure developing atomic power policies and at the same time developing a bureaucratic inertia in favor of atomic power, to the detriment of alternatives to it. The third chapter discusses the international nuclear problem—that export of reactors also exports the means of producing atomic weapons. With reactor exports, corporate and government promoters work in combination to spread the problems of atomic power and atomic weapons proliferation throughout the rest of the world.

The atomic industry is represented by the Atomic Industrial Forum (AIF), an industry trade association based in Washington, D.C. Members of the AIF include most of the companies which manufacture reactors or reactor equipment, the utilities that own and operate nuclear power plants, uranium mining companies, and other firms that operate plants as part of the nuclear fuel cycle. Obviously, the companies that profit from atomic power in concert wield a great deal of economic and political power. The segments of the atomic industry are outlined below.

Electric Utilities

About 200 private, investor-owned utilities (IOUs) produce about 78 percent of the electricity in the United States. The remaining 22 percent is generated by government-owned systems and rural electric cooperatives. Although municipal government systems and rural cooperatives often own percentages of nuclear power plants (they will pay a certain small percentage of plant costs in order to insure that the IOU will supply them from the nuclear plant), in most cases a nuclear plant will cost so much that only the IOU can raise the necessary capital to order the plant.

Each IOU operates as a monopoly, since it is presumed inefficient to have two or more electric companies serve the same area, each with its own set of power stations and transmission lines. In return for agreeing to supply all of the customers in its service area, a utility is allowed the right to earn a guaranteed rate of return, which is set by the state utility commission regulating electric utilities.

The utility industry's commitment to nuclear power began in 1955, when the Atomic Energy Commission established the Cooperative Power Reactor Program. This Program provided almost $260 million in direct assistance through 1974 for sixteen nuclear reactors. The total costs to the utility participants was $800 million, but the federal assistance represented a significant subsidy.[1] As earlier chapters have noted, the government provides other services to encourage utilities to build nuclear plants, including research and development support, government indemnity and limited accident liability from the Price-Anderson Act, and uranium fuel enrichment services.

Another inducement for nuclear power exists because of the rates of return guaranteed utilities. Because a nuclear power plant costs approximately 25 percent more than a fossil-fuel power plant, the nuclear plant gives the utility a larger rate base, which in turn means that the utility's earnings, in absolute dollars, will be greater. The utility will thus have a greater cash flow to compensate existing investors and to attract new capital for the construction of new power plants. Utilities are interested in building nuclear plants because their higher construction costs enlarge the rate base.

In spite of these economic incentives to build nuclear plants, only the largest IOUs are able to generate or attract the enormous amounts of capital needed to finance a nuclear plant. As of July 1976, only thirty-seven utilities were the major owners of the nation's fifty-eight operating nuclear power plants.[2] (As recognized above, several utilities might each own a small percentage of a nuclear plant, but by and large most of the investment comes from the IOU which is the principal owner.)

The ten largest nuclear utilities, and the state locations of their operating plants, are given below. These ten utilities accounted for over half of the 39,000 Megawatts represented by the fifty-eight nuclear plants.[3] It is thus clear that most of the "power" of atomic power is concentrated with a small number of large utilities.

Utility	Nuclear Capacity (Megawatts Electric)
Commonwealth Edison *Illinois*	5,204
Duke Power *South Carolina*	2,613
Florida Power and Light *Florida*	2,142
Tennessee Valley Authority *Alabama*	2,134
Philadelphia Electric *Pennsylvania*	2,130
Consolidated Edison *New York*	1,838
Northern States Power *Minnesota*	1,615
Virginia Electric and Power *Virginia*	1,576
Carolina Power and Light *North and South Carolina*	1,521
Northeast Nuclear *Connecticut*	1,480

Nuclear Vendors and Suppliers

In 1946 General Electric (GE) was awarded the contract to make plutonium for the Army at the Hanford Reservation—a remote 575-square-mile area in southeastern Washington State. During the early 1960s the newest of the nine production reactors built at Hanford was modified to provide electricity for the Washington Public Power Supply System. It has continued to produce both plutonium and electricity through 1974. By 1967 GE had transferred control of the Hanford works to McDonnell Douglas, United Nuclear, and Atlantic Richfield. GE still operates a weapons production plant in Pinellas, Florida.

Since 1950 GE has been building submarine reactors at the Knolls Atomic Power Laboratory in New York State. GE also operates ERDA-owned facilities for submarine reactors at Windsor, Connecticut, where Combustion Engineering sets up its Reactor Development Division. GE developed sodium-cooled reactors, one on land and one in the Seawolf submarine, which was later converted to a Westinghouse Pressurized Water Reactor. Although GE does not sell PWRs to utilities, it has designed them for submarines.

In 1956, following the launching of the AEC's Cooperative Power Reactor Program, GE began to design commercial reactors, using technology developed for the AEC by the University of Chicago.[4]

The world of reactor manufacturers, or "vendors," as they are called, is even more concentrated than the world of nuclear utilities. There are four vendors in the United States. Westinghouse Electric Corporation and General Electric Company each supply about 36 percent of the nation's reactors. Babcock & Wilcox Company and Combustion Engineering, Inc., split the rest of the market, each accounting for about 14 percent.[5] Westinghouse, Babcock & Wilcox, and Combustion Engineering all manufacture Pressurized-Water Reactors. General Electric is the only company manufacturing Boiling-Water Reactors. General Electric and Westinghouse also manufacture turbine-generators, which means that they can supply an entire power plant—both the non-nuclear and nuclear parts.

The above material on General Electric indicates that federal government assistance in the weapons and nuclear submarine programs gave GE and Westinghouse their start in nuclear technology. The AEC's Cooperative Power Reactor Program, in addition to inducing the utilities to build nuclear plants and opening a market for the vendors, also opened the way for architect-engineering firms to become involved with nuclear power. These firms draw up the plans for a power plant and, in many cases, supervise the plant's actual construction. Major components are ordered from the four reactor vendors, and other pieces of equipment (individual pumps, valves, etc.) are ordered from smaller companies. The nine major nuclear power architect-engineers in this country are Ebasco Services, Gibbs and Hill, Bechtel Corporation, Gilbert Associates, Burns and Roe, Stone and Webster, Sargent and Lundy, Pioneer Service and Engineers, and United Engineers & Constructors.[6] Some utilities—for example, Duke Power—provide their own architect-engineering services.

Even with the Cooperative Power Reactor Program, however, utilities were still reluctant to enter the nuclear field. So, to stimulate sales, the reactor vendors began selling plants to the utilities for prices less than the cost of producing them. GE and Westinghouse took losses as large as $100 million per plant on these "loss-leaders." The vendors hoped that these early deficits would be offset by future sales, if only they could get the utilities interested in nuclear power.[7] The strategy worked, and by 1975 the overall investment in nuclear power and its fuel cycle was over $100 billion.[8]

One disturbing anti-competitive aspect of the nuclear industry is that the reactor vendors are involved in multiple stages of the nuclear business, and are expanding their involvement. For example, although Westinghouse does not manufacture reactor pressure vessels, both Babcock & Wilcox and Combustion Engineering do. General Electric and Chicago Bridge & Iron form a joint venture[9] called CBI Nuclear, which also manufactures pressure vessels. Moreover, all four of the reactor vendors manufacture uranium fuel rods and assemblies, which they supply to the utilities.[10] General Electric, in its 1972 Annual Report, noted that this arrangement contributes to the company's enthusiasm about nuclear power: "Our potential revenue base in a nuclear plant, for example, is some six times that of a fossil fuel plant because we can supply the reactor, the fuel, and fuel re-loads as well as turbine generators and their auxiliary equipment."[11]

The reactor vendors are also developing the capability to fabricate fuel rods with plutonium, in the event that the NRC approves plutonium recycle. General Electric, Babcock & Wilcox, and Westinghouse all presently own plants with the capability of fabricating plutonium fuel rods. Westinghouse has applied to the NRC for a construction permit to build a commercial fabrication plant in Anderson, South Carolina—in the same region as the Barnwell, South Carolina, reprocessing plant which would extract plutonium from spent reactor fuel.

The most alarming development in the expansion of the reactor vendors, however, was the December 1976 merger between the General Electric and Utah International, a large mining company with extensive reserves of coal, uranium, and other natural resources. Utah International, which had 1975 sales of $686 million, owns both uranium mines and mills.[12] Acquisition of Utah International gave General Electric most of the front-end of the fuel cycle (except enrichment and conversion). GE can thus provide a utility with the reactor, the turbine generator, fuel rods and uranium, and, if GE's Midwest Fuel Recovery plant begins operation, reprocessing services. Even with a coal-fired plant, GE would be able to offer coal from Utah International, as well as the components of the plant. Although General Electric, to satisfy the Justice Department's Antitrust Division, was forced to place management of Utah International's uranium operations with independent trustees until the year 2000, profits from uranium operations will flow to GE immediately.

With the vendors' past investment in nuclear power and their plans to control more of the fuel cycle in the future, it is easy to understand why the reactor vendors are committed to a nuclear future. Reginald H. Jones, president of General Electric, explained why GE is determined to stay with nuclear energy: "We think we can make a great contribution, and here we're not entirely altruists. Certainly there is a need for this [nuclear power], not just in our nation, but in the world. But, beyond that, there is a great opportunity also for General Electric share owners."[13] John W. Simpson, former president of the Power System Division of Westinghouse, in 1973 estimated exactly how much nuclear power could mean to his company: "Between now and the year 2000, the potential return to Westinghouse, just assuming it maintains its present share of the nuclear reactor market, could be $300 billion."[14] With stakes such as these, the atomic industry will be extremely resistant to any plans for a non-nuclear future.

Energy Corporations

Energy corporations used to be called oil companies, until they diversified into natural gas, coal, uranium, and other sources of energy. The Ford Foundation Energy Policy Project explained the damaging effects that this diversification can have on free enterprise:

> The extension of oil companies into all branches of the energy industry has implications for interfuel competition. If an electric utility, for example, can choose among three kinds of fuels, the degree of concentration in the industry supplying each fuel may not be of much relevance. Even if one company has 100 percent control of one of the fuels, it would still have to face the competition of other fuel supplies in setting prices. If one company dominated supplies of all three sources of fuel, however, there would be no competitive safeguard and that company could set the prices it wished. The spread of oil and other companies into the production and reserve holdings of other forms of energy, becoming in effect energy companies, could in this way eventually diminish interfuel competition.[15]

The major oil companies have already been successful in raising unregulated domestic oil prices to levels comparable to OPEC oil prices, even though production costs have not risen enough to begin to justify such dramatic price increases. But with oil companies also deeply involved in other energy sources, coal and uranium prices will also be raised until all prices approach the highest common denominator—at present, OPEC oil.

In fact, even geothermal power prices will rise, if the energy industry has its way. A May 1970 contract between Union Oil Company, which "owns" the geothermal steam at the geysers in California, and Pacific Gas & Electric, which operates geothermal power plants there, states in effect that the price of geothermal steam shall be tied to the weighted price of nuclear and fossil fuel–generated electric power. As Senator James Abourezk (D., S.D.) has recognized, the consequence of this agreement is that "as the weighted average price of nuclear and fossil fuels increases, geothermal prices will increase by a proportional amount, *irrespective of the cost of producing geothermal power*" (emphasis added).[16]

The exact extent of the "oil" companies' holdings is alarming, as the Energy Policy Project found:

> Of the 14 largest petroleum companies (ranked by 1969 assets), seven (including the four biggest) had diversified into all other branches of the

energy industry—gas, oil shale, coal, uranium, and tar sands. The other seven companies produced oil, natural gas, and at least one other form of energy. . . .[17]

Partly due to the recent entry of oil companies into coal production via the merger process, the eight largest fossil fuel producers consist of seven oil companies, plus Peabody Coal. . . .[18]

Oil companies own over 50 percent of the uranium and tens of billions of tons of coal. Exxon is not among the top twenty coal producers, but its vast coal reserves assure it a leading position in coal production in the future.[19]

Concentration of power in the uranium mining industry is even greater than in the oil and gas industry. The eight largest oil companies own 64 percent of the oil and gas reserves.[20] But the eight largest uranium mining companies control over 80 percent of uranium reserves—and five of those eight are oil companies—Continental Oil, Exxon, Gulf, Getty, and Kerr-McGee. In fact, two companies, Kerr-McGee and Gulf, control over half the reserves. The remainder of the eight major uranium miners are Phelps Dodge-Western Nuclear, United Nuclear, and Utah International.[21]

Moreover, oil companies are involved not only in uranium mining but in other fuel cycle steps. Continental Oil mines and mills uranium; Exxon mines, mills, fabricates fuel rods, and is interested in building a reprocessing plant; Getty Oil mines, mills, and owns the Nuclear Fuel Services reprocessing plant. Kerr-McGee, which also has coal, oil and natural gas investments, mines and mills uranium, operates a plant to convert uranium oxide to UF_6, and owns uranium and plutonium fuel fabrication facilities; General Atomic, a subsidiary of Gulf Oil, is involved in mining, milling, fuel fabrication, nuclear reactors, and reprocessing; Atlantic Richfield Corporation operates ERDA's waste storage facilities in Hanford, Washington, and in 1977 acquired Anaconda Company, which ranks ninth in uranium reserves.[22]

The growth of the atomic industry has thus been accompanied by ominous trends in concentration. There are only four major reactor manufacturers, and they are active in expanding their operations into parts of the fuel cycle. The uranium supply industry is dominated by large mining companies, including large oil companies—and the oil companies are also moving into other steps in the fuel cycle. Should these trends continue, competition in the nuclear industry will be further reduced, and the questionable nature of nuclear power's economics will be more fully exposed as uranium and other energy sources rise to the same artificially high prices.

The Energy Production System

"Dr. John Teem, who is directing the solar effort for the new Energy Research and Development Administration (ERDA), announced today that construction has begun in New Mexico on a joint government-industry shallow solar pond project designed to heat industrial processing water with the sun's rays.

The water, normally heated by oil-fired boilers, will be used for processing uranium in a milling operation at Grants, New Mexico. . . .

"ERDA's Lawrence Livermore Laboratory in California is designing the basic system and providing technical assistance. The project is being carried out as a cooperative effort between the Sohio Petroleum Company of Cleveland, Ohio, and ERDA."

—ERDA Press Release 75–23, February 26, 1975

This brief excerpt from the ERDA press release says volumes about the energy establishment. Sohio is a subsidiary of the Standard Oil Company of Ohio. The solar energy will be used to process uranium which will go to a nuclear reactor which will produce electricity to heat homes. It would seem that several middlemen could be eliminated if solar energy heated homes directly. If solar energy is good enough for an oil company's uranium mill, it should be good enough for individual citizens. But there is a great deal of resistance to the "solar economy."

The various segments of the nuclear economy—the utilities, the reactor vendors, and the energy corporations—all benefit from an energy system which is highly capital-intensive and centralized. As more and more of the energy system becomes dependent on fewer and fewer suppliers, the political and economic power of the energy and electrical corporations increases.

Conservation and solar energy challenge this centralized system. None of the major "actors" in the nuclear industry are interested in conservation, which, because it means less energy waste, also means lower sales for utilities, reduced profit for oil and other energy companies, and fewer plants to be purchased from the reactor vendors. By the same token, solar energy also means less profit for those who control energy supplies. Mineral rights to the sun cannot be purchased, there is no need to build a transmission system from the sun, and solar technology is simple enough that a handful of manufacturers cannot control it. But, it is more convenient for ERDA to deal with centralized technologies. For example, it is easier to administer a uranium mill project or the breeder reactor in the sense of dealing with one or a few corporations, than to administer a nationwide solar or energy efficiency program directly involving much larger numbers of manufacturers and individual consumers.

The Ohio State experience, with an eight-month payback time for retrofitting buildings to make them more energy-efficient, is a dramatic demonstration that it is more economical to save energy than to build plants to supply energy.[23] In addition, energy conservation provides more jobs than nuclear power. A solar energy industry is also very likely to be less capital intensive and more job intensive than nuclear power.[24] Thus, the decentralized options of energy efficiency and solar power will be good for the economy and good for employment, and, in addition, can reduce inflation and pollution. These options are also beneficial for consumers, since they reduce energy and electrical bills and can make consumers less dependent on large energy corporations.

But decentralized energy systems do not benefit those who have the most to gain from energy consumption and energy waste: the utilities, reactor vendors, and energy corporations. Even as nuclear power crumbles under its own weight, the industry will lobby for temporary solutions and federal financing to keep itself alive. The political crunch will come when the federal government must decide whether to bail out the atomic industry or to encourage the development of a decentralized energy system. The federal agencies that will be involved in such a decision are reviewed in the next chapter.

XVI

Atomic Promotion: The Federal Push

In 1967, Anthony Mazzocchi, legislative director of the Oil, Chemical, and Atomic Workers' International Union, testified before the Joint Committee on Atomic Energy (JCAE) during hearings on lung cancer hazards to uranium miners. When Mazzocchi claimed that with regard to atomic occupational hazards "there has been much talk but no action" by the committee, the members erupted. Representative Chet Holifield (D., Calif.) rose to defend the JCAE and to badger the witness. As he was concluding, the congressman met unexpected resistance from the audience:

REPRESENTATIVE HOLIFIELD: *"There has never been a time when a letter from a representative of a labor union to this committee asking to be heard or asking for conferences with the staff or with the members has been turned down.*

"So, my friend, if you have—"

MR. LEO GOODMAN (*from the audience*): *"You yourself turned me down several times, sir."*

REPRESENTATIVE HOLIFIELD: *"Sit down or leave the room!"*[1]

In 1969, Holifield, as JCAE chairman, initiated the following exchange on nuclear power plants:

CHAIRMAN HOLIFIELD: *"Will you give us a little bit of a report? I know you have participated in a number of public hearings in Vermont, New Hampshire, and Minnesota. Would you give us an analysis of what these meetings have been like?*

"I know some of them have been kind of rough and maybe some of them not quite so rough."

MR. RAMEY: *"I could give you a few, shall we say, impressions. . . ."*

"At each of these meetings that I have gone through—the Vermont and Minnesota ones—there has been a convergence of certain factors. I will be fairly candid, if

I may, on this. One of these factors is that there are some professional 'stirrer-uppers' involved in each one of the meetings."

CHAIRMAN HOLIFIELD: "*That is a good name, 'stirrer-uppers.'. . .*"

MR. RAMEY: "*Second, there is a group of younger scientists, some of whom might be a little bit on the extremist side, who seem to always be talking on matters beyond their own professional competence. They discuss and comment on areas in which they have not performed their scientific work.*"

CHAIRMAN HOLIFIELD: "*Well, we have a certain number of book writers, too, of sensational books.*"

MR. RAMEY: "*They are usually journalists and public relations men.*"

CHAIRMAN HOLIFIELD: "*That is right; with no scientific background or competence. . . .*"[2]

CHET HOLIFIELD AND JAMES T. RAMEY represented the classic characteristics of the old "Atomic Establishment." Chester Earl Holifield, who preferred to be called "Chet," was a Democratic congressman from California. In 1946 he was a member of the first congressional Joint Committee on Atomic Energy (JCAE), and he served on the committee for 28 years. From 1961 to 1970, Holifield was either chairman or vice-chairman (the chairmanship rotates every two years between House and Senate, since the JCAE's membership is taken from both bodies). In 1970 he vacated the chair, while remaining on the JCAE, to accept the chairmanship of the House Committee on Government Operations.

As the ranking Democratic representative on the JCAE, Chet Holifield became a defender of the faith, using his position to bully, harangue, and intimidate critics of nuclear power who dared come before the JCAE. Leo Goodman's particular sin, in the eyes of Holifield, was that in 1959, before the JCAE, he recommended that the committee take stronger action to prevent radiation exposure to uranium miners. Throughout the 1960s and into the 1970s, Holifield refused to allow Goodman, who was an atomic power consultant to the United Auto Workers and other groups, to testify before the committee. Leo Goodman had the last work, however: in 1975, after Holifield retired, Goodman testified again before the JCAE.

James T. Ramey represented the inbreeding and insularity of the atomic establishment. In 1962, Ramey was serving as executive director of the JCAE's staff. When two vacancies arose among the five commissioners of the Atomic Energy Commission (AEC), Holifield and the JCAE demanded that Ramey be appointed commissioner. From inside the AEC, Commissioner Ramey was a direct link to the JCAE, a relationship which increased the coziness between the JCAE and the AEC.

Craig Hosmer, a representative from California and Holifield's Republican counterpart on the JCAE, was equally ardent in his support of the AEC's promotion of nuclear power. By 1963 Hosmer was the ranking Republican representative on the committee, and he remained so until his retirement in 1974. Holifield and Hosmer considered themselves experts on nuclear energy because they had been involved with the Joint Committee for so long. Their objectivity, however, was curious. Government witnesses, such as Ramey, and others who testified before the JCAE in favor of nuclear power, were treated with courtesy. The atomic advocates on the committee were always pleased to hear praise for nuclear power.

But if persons critical or skeptical of nuclear power appeared, the self-proclaimed "experts," such as Holifield and Hosmer, were the first to belittle the scientific ability of the critics. It was through the efforts of men such as Holifield, Hosmer, and Ramey that the JCAE, established as the watchdog of the AEC, underwent a transformation from a healthy adversary to the AEC's leading apologist, protector, and partner.[3]

Holifield, Ramey, and Hosmer are no longer in the U.S. government. Holifield retired from Congress at the end of 1974, as did Hosmer. Hosmer, however, is still an active atomic promoter: he directs the American Nuclear Energy Council, an organization in Washington, D.C., which lobbies Congress on behalf of the atomic industry. Ramey served as AEC commissioner until 1973, when he returned to the JCAE staff. In 1974, Ramey left to consult for Stone & Webster, an engineering firm based in Boston which designs and builds nuclear power plants. With the departure of these men, the dominating influence of nuclear apologists on the Joint Committee began to erode, however slowly.

The AEC and the JCAE

Following the explosion of the first atomic bomb, Congress, by the Atomic Energy Act of 1946, established the Atomic Energy Commission, which, under the direction of five civilian commissioners, was to develop atomic energy for peaceful purposes. The act also established the congressional Joint Committee on Atomic Energy, composed of nine senators and nine representatives. The JCAE's members, as with other committees, were chosen by the congressional leadership—the ranking members of each party in the House and Senate.

The JCAE was given unusually extensive powers because control of nuclear material and nuclear technology belonged exclusively to the government.[4] The Atomic Energy Act of 1954, a revision of the 1946 act, authorized the dissemination of nuclear information to private industry and "a program to encourage widespread [industry] participation in the development and utilization of atomic energy for peaceful purposes."[5] The 1954 act established procedures by which the AEC would grant licenses for possession of nuclear material and for operation of nuclear power plants. Thus the AEC was given the conflicting mandate to both promote and regulate a nuclear industry.

From its inception, the AEC was a glamor agency. Its task was highly technical and scientific; among its employees were some very reputable technologists. Despite its image the agency suffered a major embarrassment, perhaps its first, in the 1950s. The AEC had been pushing a major propaganda campaign to convince the American public that fallout from nuclear weapons testing was innocuous. As more and more information leaked to the public, the dangers of atomic fallout became more and more apparent. The AEC's credibility was thus dealt a serious blow. In 1957 Chet Holifield, who had not yet reached the position of power he would later attain, found it necessary to criticize the AEC:

> I believe from our hearings that the AEC approach to the hazards from bomb test fallout seems to add up to a party line—"play it down." As custodian of official information, the AEC has an urgent responsibility to communicate the facts to the public. Yet time after time there has been a long delay in issuance of the facts, and oftentimes the facts have to be dragged out of the agency by the Congress. Certainly it took our investigation to enable some of the Commission's own experts to break through the party line on fallout.[6]

The fallout "incident" would be followed by others, resulting in further damage to the AEC's credibility. Many of the AEC's culpable actions (or its inactions) have been described in earlier chapters: the exposure of miners to lung cancer; the discovery of radioactive mill tailings in streams, in drinking water, and under homes; the harassment of John Gofman and Arthur Tamplin when they attempted to release scientific evidence developed in the AEC's own laboratory; the events in Lyons, Kansas, where the Kansas Geological Suvey had to do the homework neglected by the AEC; the Emergency Core-Cooling System hearings, in which the AEC chose to ignore the damaging evidence presented even by its own scientists; and the suppression of information on the WASH-740 update and other subjects. Rather than deal with these issues the AEC preferred to play a role of nuclear apologist, apparently inspired by former Chairman Glenn T. Seaborg's vision of a plutonium future. Nuclear energy was seen as the nation's energy salvation and the plutonium breeder reactor as the nation's number one energy research program.

The AEC's abuse of its regulatory powers might have been checked had the JCAE been an effective watchdog. However, the committee also found it easier to accept the nuclear genie on faith rather than dig into the myriad questions surrounding the technology. It therefore came to view the AEC as a partner in atomic promotion. Because of the mystique of atomic energy, and because of the JCAE's excessive powers, other members of Congress by and large allowed nuclear energy to become the private fiefdom of the JCAE.

There was no other committee in Congress quite like the JCAE. It was established by law as a result of the Atomic Energy Act of 1954. Until 1977, the JCAE was the only joint committee which could sponsor legislation. For issues other than nuclear power, there are companion committees in both the House of Representatives and the Senate from which legislation is introduced. During floor debate in each house, amendments will be added to a particular bill, so differences in the two versions are likely. The congressional leadership then selects a conference committee from both houses, the conference committee resolves the differences in the two versions of the bill, and submits a "conference bill" to the House and Senate for approval. If both houses approve the bill, it goes to the president for his signature or veto.

For legislation dealing with atomic energy, however, the process was different—the JCAE acted as both legislative committee (that is, it introduced the bill) *and* conference committee. If bills were unfavorably amended in the House and Senate during floor debate, the JCAE could drop the amendment in conference. Invariably, the conference bill submitted to both houses for a final vote would be identical to the original bill submitted to the Congress by the JCAE.

One example of the Joint Committee's hegemony came during the controversy over the proposal of former President Nixon to sell reactors to countries in the Middle East. Both houses of Congress introduced legislation requiring congressional approval or disapproval of international agreements for cooperation on nuclear technology. The Joint Committee bill, however, allowed the president the initiative: unless Congress disapproved an agreement, it would automatically go into effect. In the house, Representative Clarence Long (D., Md.) successfully introduced an amendment requiring affirmative action by Congress before nuclear agreements could take effect. If the Congress took no action on a particular agreement for cooperation, the amendment would not go into effect. The Joint Committee, predictably, dropped the Long amendment from its conference bill. An attempt to recommit the conference bill to the Joint Committee with orders to add the Long amendment was narrowly defeated,[7] and the Joint Committee's bill was signed into law by President Ford on October 26, 1974. This was merely one example of the exceptional powers wielded, and abused, by the Joint Committee on Atomic Energy.

Energy Reorganization: ERDA and NRC

The history of the AEC and the JCAE can soberly be described as outrageous, and as the mounting abuses of the AEC became more apparent, Congress was compelled to abolish the agency and its schizophrenic mandate to promote as well as regulate nuclear power. Significantly the legislation to abolish the AEC came not from the JCAE but from the Senate Government Operations Committee.

The Energy Reorganization Act of 1974 split the AEC into two new agencies. The AEC's research and development side, including the Division of Military Application, became part of the new Energy Research and Development Administration (ERDA). Also transferred to ERDA were the Office of Coal Research (from the Department of the Interior), and those sections of the National Science Foundation which dealt with solar and geothermal energy. ERDA's legislative mandate is to conduct a broad-based energy research program, investigating not just nuclear energy but also fossil, solar, geothermal, and conservation sources. In March 1977, President Carter announced that ERDA would become part of his proposed new Energy Department.

Under the Energy Reorganization Act, the regulatory branch of the AEC became the Nuclear Regulatory Commission (NRC). The major divisions of the NRC include the Office of Nuclear Reactor Regulation, the Office of Nuclear Material Safety and Safeguards, and the Office of Nuclear Regulatory Research. The first two offices are intended, respectively, to regulate nuclear reactors and nuclear fuel cycle facilities. The Office of Regulatory Research exists to provide the other offices with an independent research capability characterized as "confirmatory assessment" of the safety of commercial reactors and other facilities subject to regulation.[8]

"Common Cause's analysis of the data reveals that:

"52.3 percent (or 73) of the 139 top ERDA employees came from private enterprises involved in energy activities;

"75 percent (or 55) of these employees came from private enterprises that were recipients of ERDA contracts. This represents 40 percent of the 139 top employees of ERDA.

"71.5 percent (or 307) of the 429 NRC senior personnel have been employed by private enterprises active in the energy field;

"90 percent (or 279) of these 307 employees came from private enterprises holding licenses, permits, or contracts from NRC. This represents 65 percent of NRC's top personnel."[9]

Employees of industries and the government agencies that supposedly regulate them often seem interchangeable and indistinguishable. In addition to the large numbers of former industry personnel working for the government, the employees and advisors of the nuclear industry include several former government officials. Craig Hosmer, now president of the American Nuclear Energy Council, and James Ramey, now with Stone & Webster, have already been mentioned. Robert E. Hollingsworth—who, as AEC general manager from 1964 to 1974, was responsible for the functioning of the AEC's promotional branch—is now with the Bechtel Corporation. Bechtel, which lobbied ERDA for government loan guarantees for construction of uranium enrichment plants, also hired two former cabinet members. George P. Schulz, former secretary of labor and secretary of the treasury, became Bechtel's president. Caspar W. Weinberger, former secretary of health, education, and welfare, became special counsel to Bechtel. Richard W. Roberts, ERDA's assistant administrator for nuclear energy, left the agency in Febru-

ary 1977 to return to General Electric, for whom Roberts had worked for thirteen years before joining the government.

The Washington, D.C., firm of LeBoeuf, Lamb, Leiby & MacRae, which represents utilities against intervenor groups, hired two of the atomic industry's "regulators": William O. Doub, former AEC commissioner, whose prime responsibility had been to improve the regulatory branch of the agency, and L. Manning Muntzing, who had been the AEC's director of regulation. Another example is Howard Larson, who moved from Allied General Nuclear Services, which was applying for a reprocessing license for its Barnwell facility. In 1974, Larson became director of the NRC's division of materials and fuel cycle facility licensing. In 1975, Larson left the NRC to become vice president at the Atomic Industrial Forum. With such tidy relationships between the atomic industry, the industry's advocates, and government agencies, it is not surprising that both ERDA and NRC are all too often sympathetic to the industry's position. The NRC, for example, accepts as its own guidelines, standards which are written by industry committees.

ERDA

When ERDA was created, it began its existence with an institutional bias to nuclear power. The personnel transferred to ERDA included 5,988 from the AEC, 1,106 from the Department of the Interior, and 13 from the National Science Foundation. The relative percentages for the transfer were: from the AEC (chiefly nuclear power), 84.3 percent; from Interior (chiefly fossil fuels) 15.5 percent; and from the NSF (solar) 0.2 percent. Nor did the figure of AEC employees moving to ERDA reflect the approximately 85,000 individuals who worked for the AEC as employees of outside contractors that operate AEC-owned facilities.[10]

The first budget submitted by ERDA was heavily imbalanced toward nuclear energy and the breeder reactor. Even after a revised budget request which was supposed to reduce nuclear funding was submitted in June 1975, ERDA's budget still had a very heavy emphasis on nuclear power. Hopes that ERDA would move expeditiously towards more balanced budgets in the future were dashed by the agency's submission of its budget request for fiscal year 1977 (FY 77). The FY 77 budget request emphasized nuclear power even *more* than the FY 1976 request. Such a situation could only indicate ERDA's blind faith in nuclear power and determined discrimination against solar, conservation, and other energy alternatives. The charts below indicate ERDA's budget authority requests for "Direct

Energy" programs—which include research and development for energy sources and exclude military applications such as weapons production.[11]

	FY 76 Request (Millions of $)	Percent	FY 77 Request (Millions of $)	Percent
Fossil	417	24.8	542	20.8
Solar	89	5.3	160	6.1
Geothermal	32	1.9	50	1.9
Conservation	73	4.3	120	4.6
Other Energy	44	2.6	96	3.7
Non-Nuclear Total	*655*	*38.9*	*968*	*37.1*
Fission (Breeder)	763 (430)	45.4 (25.4)	1250 (655)	47.8 (25.0)
Fusion	264	15.7	392	15.1
Nuclear Total	*1027*	*61.1*	*1642*	*62.9*
Direct Energy Total	*1682*	*100.0*	*2610*	*100.0*

Moreover, the "Direct Energy" expenditures do not reflect other portions of the ERDA budget which had a heavy emphasis on nuclear power. There is also uranium enrichment, which exists in large part to support the atomic industry, and an ERDA category called "Physical Research" is heavily biased towards nuclear physics research.

ERDA also inherited the AEC's Division of Military Affairs, which not only adds to the nuclear bias but also detracts from the agency's work on energy. In fiscal year 1976, for example, over $1.5 billion, or about one-third of the ERDA budget, was earmarked for weapons development.[12] Because of their priority and complexity, the weapons programs distract ERDA from non-military research and development tasks.

There is serious question about ERDA's ability to develop decentralized, "low-technology" energy options such as solar energy or conservation—because of ERDA's major bias towards centralized, high-technology, and, particularly, nuclear, options. John M. Teem, ERDA assistant administrator for solar, geothermal, and advanced energy systems, resigned in January 1976 and later charged that the Ford administration was not giving solar energy the priority or support it deserved.[13] During 1976, ERDA worked surreptitiously through its San Francisco office to distribute information, speakers, and literature against the California Nuclear Initiative, which if passed would have restricted nuclear power development in that state.[14] In addition to its support for the breeder reactor program and the subsidized development of private uranium enrichment plants, ERDA held a series of meetings with nuclear industry companies to receive their advice on government support for private plutonium processing and reprocessing plants.[15] There has been much discussion of ERDA funding the Barnwell, South Carolina, reprocessing plant as a "demonstration" unit, at a cost to taxpayers of $250 million or more.

ERDA has also been insensitive to the diminution of fuel competition in the energy industry—represented by large oil companies which are gaining substantial interests in coal, natural gas, and uranium. Gulf Oil Company is one of the co-owners of the Barnwell, South Carolina, reprocessing plant; and has sought ERDA's help in funding its High-Temperature Gas Reactor.[16] Companies meeting with ERDA on uranium enrichment and reprocessing projects have included the Atlantic Richfield Oil Company and Exxon Nuclear Company, a unit of the oil company.[17] ERDA has also offered to sell government-owned land at Oak Ridge, Tennessee, to Exxon Nuclear for the site of a projected nuclear reprocessing plant.[18]

NRC

The performance of the Nuclear Regulatory Commission during its brief lifetime has been similarly disappointing. The major issue with which the NRC has been grappling is the use of plutonium as Light-Water Reactor fuel. In May 1975, the NRC announced a provisional decision on "plutonium recycle" that generally was applauded by consumer and environmental groups. In November 1975, however, the NRC caved in to pressure from the atomic industry and completely reversed its earlier announcement.

The plutonium controversy began in August 1974, when the AEC, shortly before disbanding, announced that it intended to approve the use of plutonium as LWR fuel.[19] The announcement drew criticism from environmental and consumer groups, and eventually from the Council on Environmental Quality (CEQ), a federal body established to advise the president on environmental policy matters. The major criticism was that the AEC had not performed an adequate analysis of the measures necessary to safeguard plutonium from theft, nor of the possible civil liberties ramifications of such measures. Even the AEC itself had admitted that current theft-prevention measures were not adequate for plutonium recycle.[20]

The Council on Environmental Quality agreed with the public criticism and recommended that the NRC complete a more detailed safeguards analysis before making a decision on approval or disapproval of plutonium recycle.[21] In its response to CEQ, NRC indicated that licenses for facilities to support a plutonium recycle industry would not be approved until a final environmental statement, including an analysis of alternative safeguards, had been completed.[22]

The NRC in May 1975 announced a provisional decision on the procedures to be followed before it would approve the use of plutonium as a fuel. This decision would have committed the NRC to complete the safeguards analysis. The decision also included the view that the NRC would grant licenses to individual facilities to process plutonium on an "experimental" basis in order to demonstrate technical feasibility. The NRC announced that it did not expect final approval or disapproval of the use of plutonium before late 1977 or early 1978.[23]

The NRC announcement was not, to be sure, a ban on plutonium recycle. It merely addressed the procedures that would be followed by the agency in deciding whether plutonium would be used. But the decision did seem to represent a prudent approach, and was generally applauded by environmental and consumer groups.[24]

The Atomic Industrial Forum (AIF), the trade group which represents the nuclear industry, called the provisional decision "deplorable," and immediately launched a lobbying campaign against the NRC.[25] The industry argued that it had always assumed that plutonium would be used as reactor fuel, and that the reprocessing step in the fuel cycle depended on plutonium recycle for its economic viability. Delaying plutonium recycle would delay reprocessing, the industry maintained; this in turn would mean that reactor storage pools would continue to fill, and spent fuel could not be solidified for disposal as waste.[26]

These industry arguments are, of course, irrelevant to the NRC's duty to protect the health and safety of the public. To argue that the industry always expected plutonium recycle ignores the safeguards problems of doing so, and it ignores the requirements of the National Environmental Policy Act (NEPA) that all the environmental effects of a proposed policy be examined in detail. The lack of reprocessing and spent fuel storage, on the other hand, are problems which have been caused by the industry itself. The nuclear industry has expanded too rapidly to handle its own waste. Industry negligence has resulted in the absence of operating reprocessing plants, in the shortage of fuel storage space, and in the lack of an ultimate solution for nuclear waste disposal. These problems should dictate extreme caution, instead of a headlong rush to the use of plutonium. The logical approach for a prudent nuclear industry would be to stop or slow growth of atomic power until solutions are found to its self-generated problems.

Nevertheless, to advance its own cause, the Atomic Industrial Forum recommended that the NRC adopt procedures that would speed the decision on plutonium recycle, including: (a) immediate publication of a final environmental statement on all issues except safeguards; (b) publication of a draft statement on safeguards; (c) holding public hearings immediately following the final statement on non-safeguards items (further, the industry recommended that hearings be legislative in nature—a form which restricts public participation by denying members of the public the power to cross-examine and subpoena witnesses); (d) holding legislative hearings on safeguards, immediately following publication of the final environmental statement on safeguards; (e) issuance of the NRC decision to approve or disapprove plutonium recycle by the end of 1976.[27]

Responding to industry pressure, the NRC in November 1975 announced a decision which, in essence, adopted all the AIF recommendations. The only minor difference was that the earliest date for a decision on approval of plutonium recycle was estimated as early 1977.[28] Moreover, the NRC adopted the recommendation of other segments of the atomic industry that it grant "interim" licenses to facilities to reprocess, fabricate, or use plutonium fuel.[29] The fact that individual facilities would be licensed to use plutonium, before the NRC had made a decision on the wide-scale use of plutonium, would generate tremendous pressure on the NRC to approve wide-scale use.

The NRC's decision, in short, demonstrated that the agency had little stomach for regulating the atomic industry. Citizen groups responded to the plutonium announcement with a lawsuit. On December 19, 1975, the Natural Resources Defense Council (NRDC) and five other public interest groups representing over 150,000 members filed suit against the NRC. The suit was brought because the commission's decision was "legally deficient" and because plutonium is "uniquely threatening."[30]

The NRDC suit was upheld by the Second Circuit Court of Appeals, which ruled that the "interim licensing" provision of the NRC's procedures was illegal.[31] The NRC's hearings on plutonium recycle began in November 1976 (there was a slippage of several months of the agency's original timetable), but the pressure that would have been created by interim licenses has been removed.

Plutonium recycle may be the most important issue on which the NRC will rule for the remainder of the decade. The NRC's rush to aid the atomic industry in handling plutonium is, therefore, indefensible. In one sense, the NRC's November 1975 decision was beneficial, for it dispelled any illusions the public might have had about the willingness of the agency to be a regulator. As for the effect on the commission's credibility, the NRC staff in an earlier internal memo had, itself, predicted the justifiable response from the public when the NRC backed down from its May 1975 provisional decision: ". . . it will expose the Commission to charges of promoting Pu recycle before the matter of safeguards is completely resolved and will not enhance the image of the Commission as an impartial regulatory body operating in a careful deliberative manner."[32]

Legislative Issues—Ninety-fourth Congress

Two major nuclear power issues before the Ninety-fourth Congress in 1975–76 were funding for the breeder reactor and renewal of the Price-Anderson Act. While amendments offered by critics of the atomic energy program were defeated, votes on the issues indicated that in response to growing citizen concern congressional skepticism over the nuclear option is also growing.

PLUTONIUM BREEDER

The original plutonium breeder funding request for fiscal year 1976 (FY 76) was $490 million. This would have been one-third of ERDA's total civilian energy research and development budget, and more than the combined allocations for fossil fuels ($331 million), solar energy ($57 million), geothermal energy ($30 million), advanced energy systems ($21 million), and energy conservation ($38 million).

In spite of the breeder's safety and economic problems, the Joint Committee approved the entire ERDA request for $490 million after receiving testimony from only government witnesses. No consumer or environmental critics were invited. The JCAE decided not to wait for impending reports on the breeder program from ERDA, from the General Accounting Office, or even from the JCAE's own ad hoc Subcommittee on the Liquid-Metal Fast-Breeder Reactor.[33]

In June 1975 the ERDA budget went to the House of Representatives, where nuclear opponents won partial victories. Representative Lawrence Coughlin (R., Penn.) submitted an amendment to cut $94 million from the breeder budget. The budget cut would have delayed the order of the expensive major components for the Clinch River Breeder Reactor (CRBR), which is ERDA's demonstration plant project, scheduled for operation by 1983. Although the Coughlin amendment was defeated by a vote of 227 to 136, the level of skepticism in the House was high enough so that nuclear promoters, led by Congressman Mike McCormack, were forced to make two major concessions. First, the House adopted the section of the Coughlin amendment requiring annual review and authorization of the CRBR. Industry forces had sought an open-ended authorization that would not have been reviewed. In the second concession, in order to undercut the Coughlin amendment, McCormack and the Ford administration were forced to recommend a $60 million cut from the breeder program for FY 76. Another defeat for the atomic establishment came when Congressman Fred Richmond (D., N.Y.) moved to increase the authorization for solar power to $194 million. The Richmond amendment was passed by voice vote.[34]

The congressional debate on the breeder then moved to the Senate, where Senator John Tunney (D., Calif.) submitted an amendment similar to Coughlin's in the House. The Tunney amendment, which would have cut money for the expensive major components at Clinch River, and would have required a one-year reevaluation of the breeder program, was defeated by a vote of 66–30. Despite the loss, the vote was an indication that the legislative consensus enjoyed by nuclear energy programs was crumbling in the face of growing opposition and the breeder's own economic woes.[35]

In 1976, when the House again considered the future of the plutonium breeder program, Congressman Coughlin offered another amendment to control the costs of the Clinch River Breeder Reactor. In 1976, ERDA informed Congress that the CRBR's cost, estimated a year earlier at $1.7 billion, had risen to an estimated $1.95 billion.

The 1976 Coughlin amendment would have required the private participants in the CRBR to pay 50 percent of any cost overruns beyond $2 billion. Since private participants were in fact contributing only about 13 percent of the CRBR's costs, the Coughlin amendment would have provided a strong incentive to bring the CRBR costs under control and protect the funding for non-nuclear research projects from being consumed by breeder cost overruns.

The 1976 amendment was also defeated, but Congressman Coughlin picked up thirty-seven more votes in the House than a year before. In the Senate, a similar amendment offered by Senator Floyd Haskell (D., Colo.) was defeated, with no increase in voting strength over the 1975 Tunney amendment.[36]

PRICE-ANDERSON

The other legislative issue for 1975 was renewal of the Price-Anderson Act, which provides for limited-liability nuclear insurance. The act requires each utility operating a nuclear plant to obtain the maximum available private insurance, which in 1975 was $125 million. As of 1975, the federal government provided another $435 million in nuclear liability insurance. Were a nuclear accident to occur, liability of the reactor operator and the federal government would end at $560 million. This figure is less than one-twentieth the property damage in the worst accident predicted by the NRC's Reactor Safety Study (WASH-1400). Health effects would add to damages significantly, and more serious accidents than the WASH-1400 accident are possible.

The Price-Anderson Act therefore provides insufficient compensation for nuclear industry accidents—by any measure. If the chances of a reactor accident are really as remote as the nuclear industry claims, it would be no hardship for the industry to assume full financial liability for accidents it claims, in essence, will never happen. If, on the other hand, reactor accidents are more likely than the claims of the industry, then the public ought to be protected by unlimited industry liability.

The Ford administration and the JCAE submitted to the Ninety-fourth Congress legislation that would have extended, with some minor changes, the Price-Anderson Act. The pro-nuclear forces wanted to get the extension through Congress in 1975, because the act would otherwise have expired in 1977. Price-Anderson originally had been passed, in 1957 for ten years, because the utility industry would not have entered the nuclear field without the limit on liability. In 1965, the atomic industry was successful in extending the act for another ten years, to 1977.

The basic difference between the 1975 administration-JCAE legislation and the Price-Anderson Act was that the 1975 bill established a "retrospective premium" plan empowering the NRC to set a premium of between $2 million and $5 million for each atomic power plant. These retrospective premiums would be deferred until a nuclear accident actually occurred, and would require only a promise by the utilities to pay when an accident occurred. Such a retrospective premium program is not, of course, available to individual citizens who insure their homes, autos, or lives. Neither did the administration legislation require the utility to set aside a fund to hold these retrospective premiums in escrow. This means that the chances would be poor that the utility industry would have sufficient cash available to honor retrospective premiums. A serious nuclear accident would put major strains on capital markets as all utilities attempted to raise their retrospective premiums to provide compensation for accident victims.

If the maximum premium were assessed and collected, the total retrospective premium would be only $2.1 billion by 1990. (This assumes 420 nuclear plants in the United States by that year.) This amount is still far below the $14 billion in property damage alone caused by the worst accident analyzed by the Reactor Safety Study.

In summary, the administration's legislation did not change any of the basic objectionable elements of the Price-Anderson Act; it did not remove the artificial subsidy to atomic power from a limit on liability, nor did it insure the availability of adequate funds to fully compensate the victims of a nuclear plant accident.[37]

In November 1975 the JCAE sent the Price-Anderson extension to the House immediately after the NRC's release of the final Reactor Safety Study. The JCAE hurried a vote on the bill before any sort of detailed review of the Safety Study could be completed. The reason for the JCAE's unseemly haste was simple: the committee could publicize selected portions of the Safety Study (which supposedly showed how safe the reactor industry was) and avoid any scientific review of the defects of the final study before a quick vote on the Price-Anderson extension. The severe defects in the draft Safety Study were a strong indication that the final version, which reached similar conclusions, could not stand up to detailed scrutiny.

The Joint Committee therefore wanted to stampede Congress into a quick vote to avoid serious review of the atomic industry's shaky basis for its claims of safety.[38]

Nuclear power skeptics introduced in each house of Congress amendments which would end the Price-Anderson limit on liability. These amendments, sponsored by Representative Jonathan Bingham (D., N.Y.) in the House and Senator Mike Gravel (D., Alaska) in the Senate, would have made the nuclear industry fully liable for any damages that the industry would cause with an accident. The amendments were defeated in both houses, but there was concern among nuclear advocates for the first time ever that Price-Anderson might be changed.

The Gravel amendment in the Senate originally had a very good chance of passing because Senator Hugh Scott (R., Penn.), agreed to be co-sponsor. But intensive lobbying by the atomic industry, including Westinghouse—which considers Scott one of its Senators because the company has its headquarters in Pittsburgh—caused him to withdraw his support. This seriously undercut the Gravel amendment, and it was defeated, 62–34. In the House, the Bingham amendment was defeated 217–176. Even so, the Bingham amendment gained forty more votes than the Coughlin breeder amendment had received earlier in the year, indicating significant growth in congressional doubt on nuclear power over only a few-month period. The continued existence of Price-Anderson, moreover, remains an albatross around the neck of the atomic industry. The basic fact that the industry insists on limited liability is an indication that the industry itself does not believe its own reactor safety propaganda.[39]

The JCAE's Demise

In spite of the Joint Committee's success in beating back the opposition to the breeder and Price-Anderson, the committee's days were numbered. The first attempt to loosen the JCAE's stranglehold had actually taken place in 1974, when the House of Representatives reorganized some of its committee jurisdictions. Convinced of the logic of splitting up the AEC, the House tried to remove some of the JCAE's powers. A proposal by the Select Committee on Committees, chaired by Richard Bolling (D. Mo.), would have removed all non-military matters from the Joint Committee's jurisdiction. The ranking JCAE members had too much power, however, and the House adopted a more modest proposal providing supplemental oversight by other House committees. The Interior Committee was directed to exercise oversight of the nuclear regulatory functions. The Science and Technology Committee was directed to periodically review non-military nuclear research and development.[40]

Even with the retirements of committee members Holifield and Hosmer, the Joint Committee in 1975 was still led by unabashed nuclear advocates. Members such as Senators John Pastore (D., R.I.) and Howard Baker (R., Tenn), and Congressmen Melvin Price (D., Ill.), John Anderson (R., Ill.), and Mike McCormack (D., Wash), made the JCAE a bastion of atomic promotion amidst increasing congressional skepticism.

Still, it was not until the closing days of the Ninety-fourth Congress that the committee experienced any setbacks. Toward the end of 1976, the JCAE was defeated when it tried to rush through three major policies: on uranium enrichment, on nuclear proliferation, and a nomination for NRC commissioner.

The Nuclear Fuel Assurance Act, President Ford's legislation to establish a private uranium enrichment industry, is a vivid illustration of the hand-in-glove relationship between major U.S. corporations and federal agencies. On May 30, 1975, Bechtel Corporation, a giant West Coast construction firm, submitted an unsolicited proposal to the Energy Research and Development Administration (ERDA) for the construction of a large uranium enrichment facility. The Bechtel proposal included requests for extensive government assurances and guarantees. On June 26, 1975, less than one month later, President Ford sent Congress the Nuclear Fuel Assurance Act (NFAA), remarkably similar to the Bechtel proposal.

The act had two major points. First, it turned over to private industry government-owned uranium enrichment technology. Enrichment plants had been run by the AEC or ERDA for over thirty years, for security reasons. Second, the act empowered ERDA to make promises and up to $8 billion in guarantees to private companies, such as Bechtel, which decided to enter the uranium enrichment business. ERDA could promise to acquire the assets and liabilities of the private ventures if they failed. ERDA could buy enrichment services from the private companies. If no market existed when the facility was finished, the U.S. government would buy the unneeded material. In addition, ERDA was empowered to guarantee that the technology it provided would work during the period of commercial operation.

The Ford administration proposal did not get a friendly reception, even from the Joint Committee, which wanted to build a government facility at Portsmouth, Ohio, instead. When the JCAE asked for a General Accounting Office (GAO) analysis of the Nuclear Fuel Assurance Act, the GAO reported that the administration proposal was "not acceptable" and recommended that the Portsmouth facility be built instead.[41] The White House remained adamant in its support of the Fuel Assurance Act, and the Joint Committee decided not to oppose the president. The JCAE compromised by reporting both the NFAA and the authorization for the Portsmouth facility.

On the floor of the House, the legislation was attacked from all sides. Some members objected to the obvious accomodation of the Bechtel Corporation's proposal. Others argued that since the NFAA authorized two large gaseous diffusion facilities, when only one was requested by the White House, the NFAA would create an enrichment glut. Many representatives were concerned that passing the NFAA would lead to nuclear proliferation through the export of enrichment facilities. Turning enrichment technology over to private industry might inevitably lead to its export. Lastly, a number of conservative representatives opposed the legislation because of its extensive government guarantees.

Opposition coalesced in the House of Representatives around an amendment offered by Congressman Bingham. The Bingham amendment deleted all of the privatization and guarantee elements of the legislation, leaving only the authorization of the Portsmouth facility. During a dramatic vote on July 30, 1976, the Bingham amendment passed by two votes, 170–168. The Joint Committee, realizing that its bill had been gutted, pulled the NFAA off the floor.[42]

Opponents of the Bingham amendment, through a parliamentary maneuver, managed to force a second vote on August 4. This time, the House reversed itself and defeated the amendment, 193–192. Although Democrats voted 174–76 in favor of the amendment, its defeat hung on the votes of the three top Democratic leaders. Majority Leader Thomas P. O'Neill (D., Mass.) and Majority Whip John J. McFall (D., Calif.) voted no to create a tie, throwing the deciding ballot to House Speaker Carl Albert (D., Okla.).

The speaker voted no, in support of the Ford administration. With the Bingham proposal beaten, the full House passed the Joint Committee's bill and sent it to the Senate.[43]

Senator James B. Allen (D., Ala.), with the support of JCAE Senators Baker and Pastore, introduced the NFAA on September 29, 1976. This last-minute introduction of the NFAA was done in the hope that it could be rushed through the Senate in the few days before Congress was to adjourn for election campaigns. But the NFAA's opposition, led by Senators Abourezk, Durkin, and Stevenson (D., Ill.), was successful in a motion that the NFAA be tabled, which effectively killed the bill.[44] The motion to table, which passed 33–30, was perhaps the first major legislative defeat for the JCAE in the Senate.

The Joint Committee was also embarrassed by nuclear nonproliferation legislation drafted by the House International Relations Committee (IRC). The IRC bill was aimed at preventing the U.S. export of reprocessing facilities, the reprocessing of U.S. supplied fuel, and the reprocessing of non-U.S. fuel used in exported U.S. reactors. The IRC targeted reprocessing, which can extract the reactor-grade plutonium that can be used in weapons. The International Relations Committee recognized, however, that if it submitted its proposals as a bill the JCAE would gain jurisdiction and could refuse to act on the bill. So the IRC decided to enact its proposals as amendments to the Export Administration Act during floor debate.

The Joint Committee quickly drafted its own proliferation bill, a watered-down document which did little more than codify existing administration policy. The full House rejected the JCAE bill and passed the International Relations amendments in September 1976. Pressure from the White House and the need for Congress to adjourn, however, caused the amendments to be dropped in the House-Senate conference. Nevertheless, House passage of the International Relations Committee amendments was another unprecedented slap in the face for the JCAE.[45]

The JCAE's last move was reminiscent of the appointment of James Ramey to the AEC. In 1962 the chief actors had been Holifield, Hosmer, and Ramey. In 1976, the main characters were Pastore, Baker, and Murphy.

Senator Pastore, the JCAE chairman, had announced his retirement at the end of the 1976 congressional session. But before he left the chairman wanted to appoint George F. Murphy, the JCAE executive director, to a vacant seat on the NRC. Congress had held up action on several of then President Ford's nominations, since there was a good chance that after November the president would be Jimmy Carter, who could make his own nominations. Murphy, however, stood a much better chance of Senate confirmation. He was a Democrat and would have the backing of Pastore, a member of the Senate "establishment" who could be an aggressive retaliator when opposed. As a member of the JCAE staff, Murphy had the backing of Senator Howard Baker (R., Tenn.), the JCAE's ranking Republican senator, who had recommended the nomination to President Ford. Although Senator Baker claimed he made the recommendation in May, it was not until September 20, 1976, just a few weeks before Congress was due to recess for the election campaign, that the president made the nomination.

Some senators believed, however, that the last-minute nomination did not give them time to adequately review Murphy's qualifications. There were also fears that confirmation of Murphy would undermine the NRC's ability to operate as a regulatory agency, and undermine the constitutionally prescribed separation of Congress and the executive branch. As a member of the JCAE staff for eighteen years, Murphy had become indoctrinated as a promoter of atomic power, which could make him less willing to strictly regulate the atomic industry.

Faced with senatorial opposition, Pastore lost control. On the floor of the Senate, he attacked his colleagues as "sneakeroos" and railed against people "from downtown" (presumably public interest organizations) who opposed the Murphy nomination.[46] But his rhetoric was to no avail. When Senators Proxmire (D., Wis.), Durkin (D., N.H.), and Abourezk (D., S.D.) indicated their willingness to filibuster, the Senate leadership withdrew Murphy's name.[47] Pastore had been defeated in his desire to appoint his staff director to the NRC. With Senator Pastore's retirement at the end of 1976, the Senate lost its most ardent supporter of atomic power, setting the stage for further erosion of the Committee's authority.

In December 1976, the newly elected Democratic majority members of the Ninety-fifth Congress voted in caucus to strip the Joint Committee of its legislative powers, transferring them to other committees. On January 4, 1977, the full House approved the division of the JCAE's legislative jurisdiction among five committees: Armed Services, Interior, International Relations, Commerce, and

Science and Technology.[48]

The JCAE remains in existence, but it is a shell of its former self. It has lost its unique and tyrannical distinction of being the only congressional body to serve as both legislative and conference committee. In the end, it was the Joint Committee's unwillingness to change which was more effective than any other factor in causing its downfall.

Senator Mike Gravel, first elected in 1968 to represent Alaska, has been the leading congressional opponent of atomic power since shortly after assuming office. The senator's concern began in 1969, when he found that the Atomic Energy Commission was not telling him the whole truth about an underground nuclear bomb test at Amchitka Island, in the Aleutian chain, off the Alaska coast.[49] *As a result, the senator became more interested in atomic power and began to probe the AEC's reassuring statements about reactors. The more he probed, the more he found nuclear power unacceptable.*

Egan O'Connor was the senator's staff member who did most of the research on nuclear matters. In 1969, O'Connor had developed her own questions about nuclear energy when she investigated a claim by Howard Hughes that the effects of a scheduled Atomic Energy Commission underground nuclear test in Nevada could damage the hotels, casinos, and mining stakes he controlled in Las Vegas. In reviewing the AEC's own environmental report on the proposed test, O'Connor discerned the AEC's credibility gap. Her experience in examining the Nevada test made her ably suited to researching the Amchitka blast for Senator Gravel, and she had soon expanded her investigations into the entire nuclear power question. In 1973 O'Connor drafted a bill, introduced by Gravel, which called for a moratorium on the use of nuclear power to generate electrical power.

Egan O'Connor now works with the Committee for Nuclear Responsibility, an association of concerned scientists and public figures dedicated to a more rational national energy policy and chaired by John Gofman—the same scientist who dared to challenge the AEC on its radiation standards. Before she left Washington, O'Connor organized the Task Force Against Nuclear Pollution, which is conducting a national campaign to collect signatures on a petition for a nuclear moratorium.[50]

Mike Gravel was reelected in 1974, which means that he should continue to oppose nuclear power until at least 1980. The list of members of Congress who share the senator's skepticism on nuclear power gets longer every day. But Mike Gravel and Egan O'Connor led the way in encouraging Congress to ask questions about nuclear power that challenged the reassuring public statements of federal atomic agencies.

XVII

The International Spread of Atomic Power

In 1965 Clifford Beck had supervised the WASH-740 update, but in 1972 he was reviewing foreign reactors for the Atomic Energy Commission. In December 1972, Beck visited India's Tarapur nuclear reactors, built by two United States firms, General Electric and the Bechtel Corporation. He reported that Tarapur had been "beset with a variety of problems from the beginning."

Atomic fuel shipped from the U.S. to Tarapur had been leaking substantial amounts of radioactivity—which meant that the plant's liquid and gaseous effluents contained abnormally high levels of radioactive material. Radioactivity had been found along the shoreline of the Arabian Sea, the waters of which were used to cool the reactor, and in the bodies of the local fish-eating population. Inside the plant, radiation levels were so high that 1,300 workers had received their maximum radiation doses. To top it all off, the plant's radioactive waste system—the system which processed liquid waste from the plant before it was released to the environment–was being operated by Indian workers perched in the rafters of the Tarapur plant, using long bamboo poles.

Clifford Beck returned to Washington to tell his colleagues of his trip. Stephen H. Hanauer, special advisor for the AEC, later told a Bechtel executive that, based on his conversations with Beck, Tarapur was a "prime candidate" for "a major nuclear disaster."[1]

On May 18, 1974, the Indian government exploded a nuclear bomb in the Indian desert. The plutonium used to make the bomb is widely believed to have been produced from an Indian research reactor built by Canada, which utilized U.S.-supplied heavy water. Plutonium was extracted from the reactor, perhaps over several years, until enough to make a bomb was available.[2]

On March 2, 1976, three public interest groups—the Natural Resources Defense Council, the Sierra Club, and the Union of Concerned Scientists—filed a legal intervention to block the proposed U.S. export of nuclear fuel to India for the Tarapur reactors. In their intervention, the groups identified the health and safety problems of Tarapur—which had never been considered by the Nuclear Regulatory Commission in reviewing applications to export nuclear material or equipment. The groups also pointed out that India had refused to renounce the development of nuclear bombs, refused to sign the nuclear weapons Non-Proliferation Treaty, and refused to place all its nuclear facilities under international inspection.

The groups charged that their intervention was necessary because "the U.S., by fostering nuclear power growth around the globe, is providing the basis for nuclear proliferation and setting the stage for a world catastrophe."[3]

THE INDIAN EXPERIENCE demonstrates the dangers of nuclear reactor export. There is the danger that plutonium diversion (i.e., theft for weapons production) can be carried out by a nation as well as by a terrorist group. Indeed, the Indian explosion indicates that, at present, diversion by a nation could be the more likely threat. Export of nuclear reactors can also mean export of nuclear weapons.

In addition, there is the danger of reactor accidents and accidents in the nuclear fuel cycle. Nuclear power is not nearly safe enough for industrialized nations, but accidents are much more likely to occur in countries which do not have the technical and regulatory infrastructure necessary to manage this advanced technology. The export of reactors to tension-ridden areas such as Pakistan, Argentina, South Korea, or Egypt and Israel (all of which either have reactors or have plans to import reactors) means the export of sabotage targets as well.

This chapter will first discuss the process by which export licenses are granted. Following this review are discussions on the procedural difficulties and the dangers of atomic export.

The Export Process

There are three main U.S. government agencies involved in nuclear exports: the State Department, the Energy Research and Development Administration (ERDA), and the Nuclear Regulatory

Commission (NRC). The International Atomic Energy Agency (IAEA), an autonomous, intergovernmental organization affiliated with the United Nations, also comes into the picture. The IAEA was founded in 1957 to promote and regulate the "peaceful" uses of atomic energy throughout the world.

There are two general types of licenses for export: one for utilization facilities (referring to reactors) and one for special nuclear material (SNM—plutonium or enriched uranium, including reactor fuel). Export licenses are issued by the NRC, but before a license is issued "Agreements for Cooperation concerning the Civil Uses of Atomic Energy" must have been signed between the United States and the recipient country or between the U.S. and an international organization, such as IAEA. The Agreement for Cooperation is negotiated by ERDA, in conjunction with the Department of State. The agreements are then submitted to the president for approval and are also subject to review by Congress.[4]

An Agreement for Cooperation specifies the scope of cooperation between countries or with an international organization—that is, whether information and materials exchanged will apply to research reactors (usually used at universities for physical studies) or power reactors, or both. A typical Agreement contains guarantees by the participating parties that: (1) safeguard measures, specified in the Agreement, to prevent the diversion of special nuclear material, will be maintained; (2) no material or equipment supplied by the U.S. under the Agreement will be used for nuclear weapons, for research on nuclear weapons, or for any military purposes.

The United States presently has Agreements in force with twenty-one individual countries and two international agencies, the European Atomic Energy Community (EurAtom) and the IAEA.[5] Under the EurAtom pact, the U.S. supplies materials and equipment to EurAtom which EurAtom in turn transfers to reactors in Belgium, France, Italy, the Netherlands, and West Germany.[6] Under Agreements with IAEA, the U.S. supplies nuclear fuel to that agency, and the IAEA in turn transfers the fuel to reactors in Mexico, Yugoslavia, and Pakistan.[7] The IAEA also promulgates safeguard standards for methods of accounting for SNM. All of the U.S. Agreements for Cooperation, except with EurAtom, stipulate that IAEA safeguards will be in force in the recipient country.

Following the entry into force of an Agreement, licenses can be issued for the export by private U.S. companies of nuclear power reactors and SNM to fuel such reactors, and ERDA may conclude contracts to supply enriched uranium for power reactors abroad. There is no formalized application form for export of a power reactor. There have been no applications to export a completed reactor, but there have been many applications for export of equipment and components for use by a recipient country with a reactor. Applications for export of such equipment are made by letter, which includes general information on the company and the equipment it is shipping.[8]

When the NRC receives the application, a notice is published in the *Federal Register* (a compilation of statements and proposed actions of federal agencies) advising interested parties that they have fifteen days to ask for a hearing on the application or to submit a petition to intervene. The NRC also forwards to the State Department a copy of the application for export. The executive branch, with the State Department as the "lead agency" (the agency with overall responsibility), considers whether the proposed export would be "inimical to or constitute an unreasonable risk to the common defense and security."[9] The State Department sends copies of the application to ERDA, the Arms Control and Disarmament Agency (ACDA), the Department of Defense, and the Department of Commerce, to complete the executive branch review.

The procedure for issuing licenses for export of special nuclear material is slightly different. Again, export is allowed only to an area in which an Agreement for Cooperation is in effect. A company desiring to export SNM submits to the NRC an application which includes: (1) the names and addresses of the applicant, ultimate recipient, and any intermediate recipients; (2) a description of shipping and packing procedures, and a description of the material being shipped; (3) a declaration of the ultimate use of the SNM to be shipped.[10]

The NRC reviews the application to determine whether the containers to be used in the shipment are properly designed and constructed. A copy of the application is sent to the State Department and undergoes the same inter-agency review as a reactor application. If the decision of the executive branch is favorable, the NRC can issue the license.

The Problems with the Process

This simple explanation of the export process above does not reveal the multitude of problems with the process itself. The Agreements for Cooperation, for example, are carried out with little public or independent input, and because the Agreements are not treaties they are not subject to the advice and consent of the Senate. The Agreements are, however, submitted to the Joint Committee on Atomic Energy (JCAE) for referral to the Congress. In 1974 legislation was passed giving Congress the power to reject an Agreement by joint resolution, within sixty days of the Agreement's referral from the Joint Committee. If no congressional action is taken within sixty days, the Agreement goes into effect. A more acceptable process would be to require positive congressional approval before an Agreement be considered in effect.[11]

Two other agencies involved in the review process, the State Department and ERDA, are unabashed promoters of nuclear export. ERDA, for example, negotiates contracts with foreign countries to provide enriched uranium fuel from ERDA's gaseous diffusion plants. ERDA also decided, during the preparation and review of its environmental statement on nuclear export, that it could see no reason for reducing or halting the export program, pending the completion of a Final Environmental Statement.[12] The State Department's promotion of the proliferation of atomic power was demonstrated by the highly questionable proposals engineered by Secretary of State Kissinger and other State Department officials, to provide reactors to the Middle East.[13]

In addition to ERDA and State, there is still another federal agency, the Export-Import Bank (Eximbank), that has taken upon itself the responsibility of promoting nuclear export. Eximbank was created in 1934 to encourage U.S. exports by providing loans, loan guarantees, and insurance to recipient countries for the purchase of U.S. manufactured goods. The agency lends money up to its authorized ceiling (at present, $25 billion),[14] and returns the interest from its loans to the Treasury. Because Eximbank is interested in promoting a large dollar-volume of exports, it is biased towards supporting large, high-technology, capital-intensive items. Steel mills, chemical plants, airplanes, and nuclear power plants are examples of the products it generally supports. Through December 1975, Eximbank had loaned nearly $3.0 billion and had guaranteed $1.3 billion for export of nuclear reactors to eleven countries.[15]

Eximbank repayment procedures also discriminate in favor of nuclear power plants. Repayment by a country receiving an Eximbank loan does not begin until the product is operational. For a nuclear plant, repayment, stretched over ten or fifteen years, will not begin until about ten years after the loan. This procedure makes the loan terms for nuclear power plants more favorable for an importing country than, for example, aircraft, tractors, or industrial plants which are operational much sooner, thus requiring earlier repayment. If the recipient country believes inflation will occur, it would be particularly interested in beginning its repayments at a later date. The fact that Eximbank lends money to encourage the export of U.S. reactors represents another of the many hidden subsidies for nuclear power.

There are thus three federal agencies promoting nuclear export—the State Department, ERDA, and Eximbank. Arrayed against these promoters is the Nuclear Regulatory Commission. As the NRC was established by law to balance ERDA's promotion of domestic nuclear energy, so should the NRC balance the institutionalized promotion of nuclear export. But there is little indication that the commission intends to follow an independent course on export licensing matters.

To its credit, the NRC has made some progress in developing procedures, which the AEC Regulatory Staff largely lacked, for reviewing export applications. The NRC also agreed to hold hearings on the export of reactor fuel to Tarapur, and the commission has even had two dissents—both by Commissioner Victor Gilinsky—from decisions allowing the export of a reactor to Spain and nuclear fuel to Tarapur. In each case, Commissioner Gilinsky was concerned that controls over the plutonium produced as a byproduct of reactor operation would not be adequate.[16]

But, by the same token, there are many factors which the NRC still does not consider in its export license review. The commission does not review health, environmental, and safety aspects of an exported plant; nor the ability of the recipient country to maintain and operate the plant adequately; nor the availability of alternatives and whether nuclear power would be appropriate for the recipient. Nor, for that matter, are these questions reviewed by any other U.S. agency involved with the export process.

Moreover, by the time the NRC considers the license, several federal agencies have already approved the export, and the commission may feel obligated to issue the license. In fact, William A. Anders, then NRC chairman, indicated during April 1975 testimony before the Senate Government Operations Committee that the NRC would feel promotional pressures from ERDA and the State Department to rubber-stamp the license: "So, by the time the NRC receives a specific request for a license it seems to me that the intent of the Government has been generally indicated."[17]

The Library of Congress Congressional Research Service voiced the same concern about pressures on the NRC which undermine its ability to function as an independent regulator, in a report to a congressional committee:

> The situation is further complicated because the Administration has chosen to promise U.S. nuclear aid and products to various countries as a carrot to gain their cooperation with U.S. foreign policy. For example, commitment by the President to sell nuclear power reactors to Egypt or to Iran or to Israel could produce White House pressures on the NRC to approve the required licenses without argument or delay. How well the NRC could stand up to such pressure is a matter of concern.[18]

An even more important problem has been that neither the NRC, nor any other executive branch agency, has acted on the problems of plutonium and nuclear fuel reprocessing as they relate to proliferation. Congressman Morris Udall (D., Ariz.), in commending Commissioner Gilinsky for his dissent, objected to the Spanish reactor approval by the remaining NRC commissioners as a "business as usual" approach and "an example of insufficient concern over production of plutonium which could be used for nuclear weaponry."[19]

As long as the plutonium by-product from reactors remains in spent fuel rods, its diversion is less practicable, and a relatively long time would be necessary before diverted fuel rods could be reprocessed, and plutonium in a purified form could be made available for conversion to nuclear weapons. Albert Wohlstetter, professor at the University of Chicago, and his colleagues have pointed out, in a report for the U.S. Arms Control and Disarmament Agency, that once plutonium is processed, its fabrication into a weapon becomes a much greater threat:

> At the other extreme is the plutonium that can be found at the output or [fuel cycle] "back" end of reprocessing plants and at the input or front end of

plants fabricating plutonium or "mixed oxide" fuel. Such plutonium in the form of plutonium dioxide or plutonium nitrate could be converted to plutonium metal using handbook methods and without remote handling equipment or extensive shielding and the like, but only a glove box. And as we have said, it should take no more than a week."[20]

Thus, any serious initiative towards controlling nuclear proliferation must involve prohibiting or curbing the construction and export of nuclear fuel reprocessing plants. As Wohlstetter pointed out, in one case the United States was successful in convincing Korea to cancel plans for a French reprocessing plant—so progress is definitely possible.[21] But stronger steps are necessary. If the United States, as the world's recognized leading nuclear power developer, were to prohibit domestic reprocessing, there would be strong pressure on other countries to eschew reprocessing, thus reducing tendencies toward proliferation. The United States, however, has taken no such strong stand. The NRC is continuing with its proceedings on plutonium fuel, which, if approved, would add reprocessing and raw plutonium to the fuel cycle. If other nations follow the lead of the U.S. and develop their own reprocessing capability, there will be tons of raw plutonium available for diversion.

There are indications that ERDA and the Ford administration were considering licensing the Allied-Gulf Nuclear Services reprocessing plant in Barnwell, South Carolina, as a "demonstration" facility, which would allow it to begin operation without an NRC license. The rationale for the proposal would have been that the Ford administration, supposedly to reduce proliferation, would prohibit commercial domestic fuel reprocessing, and encourage other nations to do the same, until the safety and security of reprocessing could be "demonstrated." However, the demonstration would be carried out by operating Barnwell as a government-funded plant.[22] The distinction between operating Barnwell as a "commercial" facility or as a government "demonstration" facility is trivial. It is doubtful that other nations would see any distinction—they would recognize only what the United States had begun reprocessing and would thus be encouraged to develop their own capability.

In a pre-election statement, President Ford backed off from specifically proposing a demonstration plant at Barnwell. However, his language was so vague that it did not preclude the demonstration facility. Equally disturbing, President Ford's statement recommended greater dependence on nuclear power, both in the United States and throughout the world.[23] This dependence will only make plutonium reactor fuel, and eventually the breeder reactor, inevitable, as uranium reserves are depleted throughout the world. Thus, until the NRC prohibits reprocessing and plutonium fuel, and the country abandons the breeder, the prospect of reducing nuclear proliferation will be very dim, because of the inevitable reliance on a worldwide "plutonium economy."

The International Atomic Energy Agency

"In a recent interview with The Chronicle, *Dr. Rudolf Rometsch, inspector-general of the International Atomic Energy Agency noted that his Vienna-based organization has 67 inspectors now visiting nuclear plants around the world to monitor the plutonium inventories at commercial reactors in 20 countries.*

"Their job is to see that tiny quantities of the deadly metal are never diverted clandestinely, but they must rely on the cooperation of the countries themselves, and on the hope that no nation is doctoring its books to cover up 'MUFS' [materials unaccounted for].

"India, France, China, Israel, Egypt and several other states with nuclear power or potential remain outside the system.

"Rometsch estimates that his inspectors maintain surveillance over about 2,000 kilograms (2.2 tons) of plutonium today, and will double that quantity next year.

"Within five years, he said, that quantity of potential bomb material in the hands of nuclear nations will rise to 400,000 pounds—and this doesn't count the plutonium in the secret stockpiles of the nuclear weapons nations.

"What really worries Rometsch is the international shipment of plutonium for here his U.N. agency has no jurisdiction at all, and not even the voluntary power to inspect."[24]

Advocates of nuclear power respond to the problems of nuclear proliferation by pointing out that exported U.S. reactors are regulated by the International Atomic Energy Agency (IAEA). Such assurances, however, are not at all comforting because major defects within the IAEA severely restrict its effectiveness.

To begin with, the IAEA is both a promotional and a regulatory agency for nuclear power. The Atomic Energy Commission was abolished primarily because it promoted, to the detriment of regulation—a situation still characteristic of the IAEA. Three observers, writing on behalf of the Center for Law and Social Policy, a public-interest law firm in Washington, D.C., have commented on the IAEA's conflicting role: "Its promotionalism is often unabashed. The IAEA *Bulletin* is literally filled with advertising by nuclear suppliers, and it has been wildly optimistic about future growth of nuclear power worldwide."[25] Clearly, the combination of promotion and regulation is no more appropriate for the IAEA than it was for the AEC.

Many other weaknesses of the IAEA were identified in a study, completed under contract for ERDA, by Richard J. Barber Associates, a Washington, D.C., consulting firm. ERDA did not release the Barber report until August 1975, when its existence was reported by David Burnham of the *New York Times*—six months after the report was completed. Nonetheless, in contrast to the findings of the Barber report, ERDA officials had expressed, in testimony before the Senate Government Operations Committee in April 1975, their confidence in IAEA procedures.[26]

A more recent report on the IAEA, conducted by the General Accounting Office (GAO), provided strong support for the Barber report's findings. The GAO concluded:

> Although the global expansion of nuclear energy makes effective international safeguards crucial to U.S. and world security, international organizations have no authority to require physical protection measures, no authority to supervise, control, or implement such measures, and no authority to pursue and recover diverted or stolen material. Their inspectors have neither unlimited access nor authority to seek out possible undeclared or clandestine facilities or stockpiles of nuclear material. In addition, technical, political, financial, and staffing obstacles hamper the effective implementation of international safeguards.[27]

In addition to recognizing the defects of international safeguards, the GAO commented that the U.S. government does not even have available to it information to assess the adequacy of the IAEA system.[28] The IAEA's inspection reports are not released: "IAEA is committed to restrict the dissemination of inspection information, so neither the United States nor any other nation has access to IAEA inspection results. According to one State Department official, a nuclear supplier, even one that has transferred its bilateral safeguards responsibilities to IAEA, can get few details of the precise status of material it supplied."[29]

The IAEA's major deficiencies include its small inspection staff and small budget. Although responsible for reactors in thirty countries, the agency has only sixty-nine inspectors.[30] IAEA's 1975 budget for safeguards programs, including inspections, was a tiny $5.7 million.[31] The Barber report estimated that by 1990 the IAEA might need $169 million to $1.17 billion for safeguards alone.[32]

Furthermore, the IAEA's safeguards inspections are little more than auditing functions. The agency reviews records to determine that no special nuclear material has been diverted from reactors. Due to the inherent problems of accounting methods, the IAEA has little assurance that diversion of nuclear material might not already have taken place.[33] David M. Rosenbaum, who directed a special safeguards study for the AEC, stated that IAEA safeguards were "grossly inadequate" to meet the problem of plutonium diversion.[34]

The IAEA safeguards, designed only to detect diversion of plutonium for weapons production by a nation, do not address physical security against sabotage or diversion by a terrorist group. Nor are there provisions to apprehend nuclear thieves.[35] The Barber report concluded that terrorist activity was a real threat in nuclear plants:

> The physical vulnerabilities of present-day nuclear power facilities are, of course, very serious because security of such facilities has been accorded relatively little attention.
>
> Terrorist groups are also acquiring very sophisticated weapons, e.g., the Palestinians captured near European airports with portable ground-to-air missiles. Precision guided weapons (PGW) are already in the hands of some LDC [less developed country] governments and their eventual appearance in terrorist organizations is assured. Perhaps as important, terrorist organizations are beginning to collaborate with each other, e.g., the Popular Front for the Liberation of Palestine and the United Red Army of Japan; the Irish Republican Army and ETS, the Basque separatist group in Spain; and four important urban guerilla groups in Latin America.[36]
>
> The generally weak international response to terrorism to date provides further ground for concern. In brief, the terrorist threat vis-à-vis nuclear power facilities is plausible. If a U.S.-provided nuclear power station in a foreign location is involved in an incident, the domestic and international repercussions will be severe.[37]

There is also the question of how the IAEA would actually deal with a suspected diversion, were it to occur. The Barber report addressed this question:

> It is unlikely that an international inspector will ever witness a diversion; all he will have to go on will be indirect evidence—discrepancies between the suspect nation's domestic accounting system and the audit of the

> international inspectorate, breakdown in remote observation equipment, or suspicious procedural roadblocks in the way of international inspection. The nation involved could argue that such discrepancies were due to human error, mechanical failure, or normal process—and it could well be correct.
>
> In the end, after all the debate had subsided, and the slow wheels of the international system had been put in motion, it is possible that if the nation involved actually had diverted nuclear material, it would be too late to prevent their successful completion of one or maybe several nuclear weapons.[38]

Even if the IAEA were to detect a diversion, sanctions against the guilty country would have to be approved by the United Nations Security Council. There is strong reason to believe that the Security Council might not even agree to take action, much less devise adequate sanctions.

The Indian example demonstrates how the IAEA can be circumvented. Some Indian nuclear facilities were subject to IAEA regulations, but India refused to place all facilities under the IAEA. Thus, India could develop an explosive through facilities not subject to international inspection.[39]

Recently, Senator Abraham Ribicoff (D., Conn.) released information indicating that U.S. exports outside the IAEA system may have contributed to India's bomb. Senator Ribicoff found that in 1956, before the IAEA was established, the United States exported twenty-one tons of heavy water for use in a Canadian-supplied research reactor. In 1963, when the U.S. and India were negotiating safeguards agreements for the export of two power reactors to Tarapur, the State Department neglected to demand that the twenty-one tons of heavy water also be placed under international safeguards. The State Department thus has no assurance that the U.S.-exported heavy water was not used to produce the plutonium for the Indian weapon.[40]

In summary, the IAEA and the procedures it uses have significant deficiencies. It is insufficient for atomic promoters to address the proliferation problem by promising that international safeguards will be in effect. Such a promise is an empty defense.

The Exporters and the Recipients

Some nuclear proponents use the example of the Indian weapon to support their promotion of nuclear export. The bomb, they point out, was made from plutonium from a Canadian reactor, not a United States reactor. It is thus necessary, they argue, for the U.S. to export its reactors under strong safeguards, to prevent reactors from being exported by other nations under weak safeguards. One comment on this strange logic has been: "The curious proposition that United States promotion of fission power development minimizes the risks of proliferation" is "worthy of the Mad Hatter in *Alice in Wonderland.*"[41]

This same kind of "if we don't export reactors someone else will" argument has been undercut by a decision by the Canadian government, following the explosion of the Indian bomb, to withdraw all nuclear aid from India. The United States has ignored an excellent chance to follow Canada's example and explore ways to limit the spread of nuclear technology. Instead of supporting the Canadian government, the United States continues to export uranium fuel to India's Tarapur reactor.[42] Even with India's development of a nuclear weapon, it is important to recognize that United States heavy water probably contributed to the bomb.

As the Barber report recognized, United States companies benefit from sales of CANDU* reactors: "A network of Canadian and U.S. companies actually supply almost all the equipment needed to build CANDU nuclear power plants. Many of the Canadian firms are in turn subsidiaries of U.S. companies (i.e., Canadian General Electric and Westinghouse of Canada). In fact, one government authority estimates that 20 to 30 percent of the return on CANDU sales flow to U.S. companies."[43] The Canadian government has also purchased from Canadian General Electric (92 percent owned by General Electric) a plant in Nova Scotia to manufacture heavy water for the CANDU.[44]

United States technology is linked to reactor exports from many other countries besides Canada. United States corporations are part owners of some foreign reactor vendors, or have licensing agreements with other vendors, under which the U.S. firms provide technical advice or equipment and receive royalties in return. These relationships were demonstrated by the May 1976 sale of the first nuclear power reactor to South Africa, a non-singer of the Non-Proliferation Treaty. South Africa originally intended to buy its reactor from General Electric, but reversed itself and ordered the reac-

*CANDU stands for Canadian-Deuterium (referring to heavy water)-Uranium (referring to natural uranium fuel).

tor from a French consortium when civic and congressional opposition to the sale developed in the United States. However, the "French" consortium included the two major reactor manufacturers in the country: Framatome, which is 15 percent owned by Westinghouse, and Alsthom, which has a licensing agreement with General Electric.[45]

U.S. reactor manufacturers also influence, through licensing agreements or subsidiaries, reactor vendor companies in West Germany, Japan, Sweden, Italy, and Belgium. United States companies are major participants in numerous other foreign enterprises which do not merely produce reactors but are active in all aspects of the nuclear fuel cycle.[46] The chief "competitors" for U.S. reactor exporters are foreign extensions of American multinational corporations, which profit whether they or their overseas offices export the technology.

In addition to their indirect influence, U.S. firms are also directly responsible for an overwhelming majority—nearly 70 percent—of the world's reactors. Based on domestic and international reactors in operation and on order, General Electric and Westinghouse each have about 26 percent of the world market. Babcock & Wilcox and Combustion Engineering, the other two U.S. reactor vendors, each have about 9 percent. All foreign producers combined have a 30 percent share of the world market.[47]

In addition to the danger of proliferation of nuclear weapons, there are also serious questions about whether nuclear reactors, which are not safe enough for industrialized nations, are appropriate energy sources for developing countries. Reactors pose significant safety problems in general. A developing nation might have difficulty in establishing an adequate regulatory framework. Moreover, the lack of a technical infrastructure intensifies problems of reactor maintenance and repair. Obviously, the inability to maintain or repair a reactor properly can significantly increase accident risks.[48]

There will also be problems with other phases of the nuclear fuel cycle necessary to support foreign reactors. Storage of radioactive wastes will be no less a problem abroad than it is in the U.S. International transportation of nuclear material poses hazards in safety and in diversion of potential weapons material. An October 1974 internal AEC memo called the problem of safeguarding international nuclear shipments from theft "almost insurmountable"—but the author of the memo did not explain how he justified the use of "almost."[49]

Moreover, there is probably no single product that developing nations need less than nuclear power. Such countries may have an energy problem, but not necessarily an electrical power problem.[50] Amory B. Lovins and John H. Price, representing Friends of the Earth of London, have expanded on this theme:

> What matters, though, is not aggregate or even unit energy production, but ability to meet the energy needs of people in particular circumstances. Indeed, the energy technologies that most people in the world need are those which perform basic end-uses such as heating, cooking, lighting, and pumping; and these can be done admirably by simple devices based on sun, wind, and organic conversion. These are not glamorous technologies and have no military applications, so people interested in developing them tend to receive every assistance short of actual help.
>
> Of the criteria mentioned above for practical energy systems—small-scale, simple, low-technology (which does not mean unsophisticated), decentralized, nonelectrical—the last is perhaps the most controversial and, to thoughtful analysts, the most obvious, because electricity is the costliest form of energy to make, store, or transport in bulk. Electrification of most end-uses in an industrial economy is simply too expensive for any major country outside the Persian Gulf.[51]

In addition, a large electrical generating station, nuclear or otherwise, would represent a significant segment of a developing nation's electricty. When the plant shuts down, a large part of the nation's energy supply would be inoperative. The unreliable nature of nuclear power plants aggravates this problem. Moreover, "if the plant or plants performs like Indian Point #1 (0% CF [capacity factor] in 1973) or Millstone #1 (32% CF in 1973) or Dresden #1 (33% CF in 1973), the economic and other consequences will be severe and the decision considered most unwise. *Indeed a government could fall over just such an issue.*"[52] (Emphasis added.)

Lastly, renewable resources, highly appropriate for developing nations, may actually be much more abundant in such countries than in the rest of the world. Amory Lovins, commenting on the logic of implementing small-scale solar technologies, writes:

> Centralized conversion to electricity on a very large scale must, however, be distinguished from diffuse conversion, e.g., in single-dwelling collectors that yield hot water and (via heat pumps) hot and cold air. The latter

> type of device is in use in many countries today, albeit in rudimentary form, and would probably be competitive with conventional methods in most temperate latitudes if supported by modest development efforts. Indeed, there is good reason to believe that diffuse solar technology may already be competitive with conventional methods anywhere in the USA or similar latitudes. There seems no fundamental reason why such [small-scale] technologies could not be widely proliferated within the next few decades, producing vast savings of fossil fuels.[53]

Nuclear power technology is thus highly inappropriate. A developing nation would want nuclear power largely for prestige—that flows from the development of such a high-technology option; and for possible weapons manufacture.

Clearly, the nuclear export problem requires an international solution. But just as clearly the United States could exert enormous pressure by taking unilateral action to halt its own exports. Since the U.S. is directly responsible for about 70 percent of the world's reactors, and at least until 1985 will supply most of the world's enriched uranium, the gap in supply caused by a U.S. decision to halt exports could not easily be filled by other countries. A U.S. halt to exports would also dampen demand for reactors in other countries; such a decision could encourage developing countries to examine their own energy needs in a logical manner and could encourage other exporters to halt or limit their promotion of reactors. A U.S. decision to halt development of nuclear power *domestically* could encourage reevaluation of other countries' nuclear programs, particularly where there are growing citizen movements against nuclear power, as in Japan and Western Europe.

David E. Lilienthal, the first chairman of the Atomic Energy Commission, has called for an embargo on exports of U.S. reactors and nuclear material. Lilienthal in January 1976 stated that the proliferation of nuclear weapons was reaching "terrifying proportions." He urged that the United States unilaterally impose a moratorium on nuclear export in order to put moral and economic pressure on other nations to check the proliferation of atomic reactors and weapons.[54] The facts on reactor technology and the fuel cycle point overwhelmingly toward the need for a nuclear power moratorium in this country. The facts just as overwhelmingly justify a moratorium on atomic exports.

4. *What Can a Citizen Do?*

THE NECESSITY FOR CITIZEN ACTION

UNTIL THE PAST FEW YEARS, most people viewed nuclear power as a miracle form of energy which was clean, safe, and limitless. For many citizens as well, atomic power was wrapped in a mystique of technical terms and exotic references. Former President Nixon, whose administration fostered rapid expansion of atomic power and spent over a billion dollars on the breeder reactor, once remarked, "Now, don't ask me what a breeder reactor is; ask Dr. Schlesinger [Atomic Energy Commission chairman]. But . . . unless you are one of those Ph.D.'s, you won't understand it either."*

Nuclear power had also remained a non-controversial issue because officials at the Atomic Energy Commission and the chieftains of the atomic industry withheld from the public adverse information for over twenty years. The unspoken rule was "doubts in private, assurances in public." Misrepresentations and overly simplistic statements by nuclear promoters kept the issue out of the public arena until relatively recently.

But, as the previous chapters have shown, the flaws of nuclear power are many, and there is even a serious controversy—from technical as well as non-technical standpoints—within the scientific community over nuclear power. Ordinary citizens, moreover, have demonstrated their ability to comprehend nuclear power's problems. For example, Daniel Ford, a Harvard-trained economist, mounted an accurate challenge to the adequacy of the Emergency Core Cooling System (ECCS), and June Allen, a Virginia music teacher, found that Virginia Electric and Power Company had built its North Anna plant over a geologic fault. Individual citizens have also brought damaging documents to light through agency proceedings in which they intervened and through Freedom of Information Act requests and suits. This once-secret information, along with information from leaks and whistle-blowers within the atomic establishment, has made nuclear power a controversial issue whose flaws are being subjected to closer and closer scrutiny.

Since nuclear power can affect all members of society, as well as future generations, the question of whether it should continue must

* Richard M. Nixon, remarks at the Atomic Energy Commission's Hanford works, Hanford, Washington, September 26, 1971.

be decided by the citizenry. The decision cannot be left to the so-called experts, who for too long have subserviently followed their employers or muted their voices. Policy-makers in government are now experiencing increased civic pressure for a nuclear moratorium and for national energy policies which emphasize safe, economical, and practical energy alternatives—such as energy efficiency and solar power.

Atomic power opponents presume that this technology cannot withstand thorough public scrutiny. There are a number of routes by which citizens can challenge the government and industry to publicly justify each step in the development of nuclear power and, in the process, raise the level of public awareness of its deficiencies.

The following chapters suggest ways to tackle each of the major decision-makers. Chapters XVIII and XIX review action at the federal level—in Congress, the Nuclear Regulatory Commission, and the courts. Chapter XX considers the possibility for action at state and local levels—through initiatives and referenda, state legislatures, and administrative agencies. Chapter XXI examines challenges to the nuclear industry, particularly to the utility companies building nuclear plants.

Grappling with these institutions is not simple, but the experience of the past several years indicates that citizen action can indeed affect public policy. In some ways, individual citizens can make the difference. Often, organized activities by citizens who join together have a greater impact; and there are always activities which many citizens can pursue on a very small scale which combined have a massive influence. In reading the following chapters, readers should keep in mind their own resources and goals. The suggestions given can cost thousands of dollars or can cost little in time and money. It is for each citizen or group to decide which activities are most suited to their abilities and desires. But it should be remembered that citizen action is essential if nuclear power's replacement is to be accomplished.

XVIII

Action at the Federal Level: The Congress

THE MOST EXPEDITIOUS ROUTE to limiting nuclear power is action by the legislature, which writes the laws to regulate nuclear power and to disburse taxpayer dollars in subsidies to the industry. The federal legislature, the United States Congress, could end nuclear power immediately by simply decreeing that nuclear plants will no longer be licensed. Or, it could take a multitude of lesser steps, such as removing some federal subsidies, placing minimum limitations and standards on the development of nuclear power, restricting the transportation of radioactive materials, or returning to the states some of the authority preempted by the Atomic Energy Act of 1954. None of these actions are likely to occur until members of Congress, who have been conditioned for years to endorse atomic power and who are now pressured by the industry and some labor unions to continue their support, experience countervailing pressure from citizens.

Votes are the ultimate currency in the political process, and the legislator's motivation to remain in office or succeed to higher elected office is the keystone to his behavior. The insecurity of his office demands that he be cognizant of the prevailing views of his constituents. As an elected public figure, the legislator is most responsive to public pressure, which is why he is solicitous of citizens or groups who know how to raise a legitimate public fuss.

With endless opportunities and public forums available, citizen groups can organize themselves to catapult their issue and their legislator's votes on it into public focus. Imagination and a modicum of funds are the only limitations on what citizens can do, individually or collectively, to generate a public debate on a serious national issue such as nuclear power. Only through public pressure will a change of policy occur and will decisions on continuation of nuclear technology be removed from the control of the corporate "specialists" and be placed in the hands of the voters.

There are two basic steps in influencing a legislator's view and vote: first, politicizing the issue in his or her constituency, and, second, publicizing his or her votes on the issue, so that large numbers of voters in the district become aware of his position. Because voters, once educated on the issue of nuclear power, often become strongly opposed to it, public exposure of the issue will be an integral element in the citizen strategy. If representatives or senators are generally sympathetic to the views of organized business representatives, it will be more difficult to persuade them to reassess their support for nuclear manufacturers and utilities.

Politicizing the Issue

There are numerous ways to politicize an issue. Essentially, they all involve encouraging groups of people to publicly support a position. Methods include the following examples:

PETITIONS

At age twenty, weighing 120 pounds, Franklin L. Gage is a leading candidate for the title "America's toughest kid." Working out of a small, drafty bedroom-office in a row house a few blocks from Congress, Gage is organizing a national Clean Energy Petition drive against nuclear power and for solar energy.

As coordinator for the Task Force Against Nuclear Pollution (P. O. Box 1817, Washington, D.C. 20003), which was founded by citizen activist Egan O'Connor, Gage works each day sorting petitions, answering requests for information from all over the country, and telling members of Congress where people in their districts stand on nuclear power hazards.

The Clean Energy Petition calls for active support by members of Congress for legislation to "1) develop safe, cost-competitive solar electricity and solar fuels within ten years or less, and 2) phase out the operation of nuclear power plants as quickly as possible."

The soft-spoken Gage is a veteran of twelve years of struggle against nuclear power. "Until 1965 [when he was nine] I was for nuclear power," he explains. "Then I learned that Consolidated Edison was not telling the truth about its nuclear plant destroying fish in the Hudson River." At that point, Con Ed, the giant New York City utility, produced an adversary.

By the time he was fifteen, Gage had organized a consumer boycott of Con Ed electricity for Earth Day 1970. Throughout his home town of Croton, New York, people turned off all their electricity, which came from Con Ed's nuclear plants, for fifteen minutes.

By the time he was eighteen, Franklin Gage was representing two New York environmental organizations with 1,000 members and testifying before the Congressional Joint Committee on Atomic Energy. Early in 1974 he became the Task Force's Washington, D.C., coordinator.

The Task Force circulates petitions, and when they are returned with signatures Gage tabulates them by congressional district. He can tell a representative or senator exactly how many people in the member's district or state have signed petitions for a moratorium on nuclear power. As of January 1977, about 450,000 persons had signed petitions, and members of Congress are taking notice. Largely as a result of the petition drive, about thirty-five members of Congress in 1976 supported the Nuclear Reappraisal Act, introduced by New York Representatives Hamilton Fish, III (Republican) and Edward Pattison (Democrat). Their proposed bill would order a five-year halt on the licensing of new nuclear plants until a study of unresolved safety and reliability problems is completed. These petitions are also used at the state and local level. In New York State, for example, five county governments have endorsed a nuclear moratorium.

As for Gage himself, his immediate needs are indicative of his single-minded mission. "All I want," he said recently, "is more volunteers and petition gatherers." Meanwhile, the petition drive continues to roll on.[1]

Gathering signatures from individuals who are concerned about nuclear power takes much time but is relatively inexpensive, and it can have a big payoff. In swing districts where members of Congress are elected by small margins, petitions carry particular weight. When members of Congress see a petition signed by 3,000 or 5,000 constituents (some districts have up to 11,000 signatures), they know there is a movement in their districts which they cannot ignore. Petitions particularly lend themselves to large numbers of people doing a little bit each day, in their own time and pace. The petitions can also provide a vehicle for educating other members of the public.

ENDORSEMENT OF PROMINENT GROUPS OR INDIVIDUALS

Since politics is the art of the possible, members of Congress have particular respect for the views of prominent groups and individuals in their districts. Of special importance for many members are university professors or officers, scientists, ministers, political figures, and the press. Many representatives of these groups, individually or on behalf of their organizations, have publicly recommended caution in the further development of nuclear power. If a significant number of well-respected figures in the community suggest that nuclear power should not be pushed, a member of Congress will at least take a second look. Citizens should invest time and energy in educating and persuading prominent individuals to take a public stand on the issue.

For example, the August 6, 1975, declaration by more than 2,300 scientists called for the president and Congress to diminish the growth rate of the nuclear program and to develop alternative energy sources. This kind of statement indelibly affects a member of Congress who might be worried about nuclear power but is not sure whether it is entirely respectable to express significant concern.

If local politicians are strongly opposed to the development of nuclear power, they should be encouraged to obtain resolutions by the city council, county commission, state legislature, and other local government bodies to indicate the broad base of support for their concern.

INITIATIVES

Twenty-one states permit citizens to enact legislation through the initiative process. If an issue such as limitation of nuclear power is placed on the ballot, it automatically becomes a political issue which politicians are asked to endorse or oppose. The nuclear industry will inundate the media with slick, misleading advertising, and the public will at least become aware that nuclear power is a controversial issue. If the citizen effort is effective, many people will become advocates for deemphasizing nuclear power. Thus, even if the final vote is not successful, the opposition to atomic power will expand. (See Chapter XX.)

PUBLIC DEBATES

In connection with an initiative, a key vote in Congress, or a decision on the location of a local nuclear power plant, a public debate between an advocate and an adversary can be scheduled. The debate should be well publicized and will have more impact if public radio or a local station covers it live. A group organizing the debate should make sure its spokesperson is well prepared and should try to get a company official, not a hired public relations person for the utility trade association, for the other side.

MONITORING LOCAL POWER PLANTS

Contrary to the industry's public relations claims, nuclear power plants have many problems. Citizen groups should arrange to monitor nuclear plants closely and report the breakdowns that develop. The local utility periodically reports "incidents"—accidents and malfunctions. These may be reviewed at a plant's local Public Document Room (PDR), which is established by the Nuclear Regulatory Commission (NRC). Generally, the local PDR is a library near the plant. If there is no PDR, a group or individual can request the NRC to establish one.

Nuclear plants frequently have equipment malfunctions which may not be of safety significance but may require shutdown for repairs or reductions in power. Such plant breakdowns cost utility ratepayers money. Citizen groups concerned about nuclear power can obtain information on the efficiency of plant operation at the local PDR and pass it on to consumers.

CONFLICTS OF INTEREST

Members of Congress frequently receive campaign contributions or honoraria from utilities that own nuclear plants or companies that manufacture or service nuclear plants. It is a blatant conflict of interest for a member to receive money in any form from such an interested person and then to vote for the interests of that person in Congress. If a member of Congress or a senator engages in such behavior, local groups should be alert to the conflict and inform the press in either a press conference or press release.

MEDIA REPORTING

Until an issue becomes controversial, the press rarely will pay attention. Citizen groups should be cognizant of legitimate factors that make an issue controversial, such as: (1) conflicts of interest; (2) picketing or other lawful protests about a nuclear plant; (3) refusal by the utility to respond to citizen requests, or to complaints about safety violations or price increases; (4) release of previously secret materials about the operation of nuclear plants (sometimes materials released by national organizations are not picked up by the local press and can be released subsequently by the local citizen group); (5) reports or studies about nuclear power; (6) activities in connection with initiatives or intervention in licensing proceedings.

Local newspapers, weeklies, and community newspapers are valuable sources of communication to members of Congress because the district office staff is usually assigned to "feel the pulse at home" by keeping track of local events and commentary. A well-argued advocacy letter to the editor—whether by a prominent individual, a citizen group, or an ordinary citizen—can bring quick results. Editorial writers are often anxious for new material and should always be contacted, as should reporters—by mail, by telephone, or in person. Articles for the "Op Ed" (opinion-editorial) page, where available, should be written periodically about new trends or events in reference to the issue.

OPINION POLLS

Most members of Congress want to stay within a few degrees of the center of political opinion. Thus, when most political forces in the country talked about nuclear power as safe, cheap, and clean, the safe political action was to support nuclear power. With public opinion now shifting to a more skeptical view, politicians are also shifting their positions, particularly on the most volatile issues, such as storage of wastes and proliferation of nuclear materials. A public opinion poll by a reputable local pollster, by a trained local group (such as one associated with a university), or even by a citizen group or the member of Congress can have a significant impact on political decisions.

If the citizen group wants to conduct its own poll, a nearby college political science or sociology department might help to design a statistically accurate one which can be conducted either by telephone or in person. To save time and to get the most informed results, the poll should follow substantial public discussion of the issue.

PUBLIC EDUCATION

Educating the public is an important goal in advancing any policy issue. There are numerous activities which can help to achieve this goal. The public library and school curricula are basic sources for communicating with young people. Displays, lectures, speakers at assemblies, school debates, films, special projects, or field trips can all be geared to educating students on nuclear and other forms of power generation. At the college level these can be expanded to include specialized courses, clinical courses outside the campus, teach-ins, term papers, and so on.

Exhibits at fairs or holiday events on the issues of nuclear power (economics, safety, proliferation, security, capital usage, lack of reliability, limitation of liability, waste storage, etc.) bring thousands of people in contact with the nuclear power problems facing this industry. Information about alternative sources of energy must also be conveyed to the public. Three Colorado environmental groups—Environmental Action of Colorado, Colorado Open Space Council, and Rocky Mountain Center on the Environment—cooperated with the University of Colorado to organize a traveling solar exhibit which was financed by a wide range of supporters, including the National Science Foundation, two local oil companies and a Denver bank. A handbook explaining how to obtain technical assistance, products, and other information was made available, and attracted a wide variety of volunteers and audiences. More information on this project can be obtained from: Solar Energy Exhibition Program, University of Colorado, Denver, Colorado 80202.

Influencing a Member's Votes

Most members of Congress maintain a consistent view on an issue, but to protect themselves politically they may vote to weaken a bill with amendments and then vote for final passage. When an issue such as nuclear power starts to become controversial, a member's votes become erratic, with the vote on each bill depending heavily on the politics of the issue in his district or state. Thus, a member of Congress concerned about the budget is more likely to vote against the breeder reactor because of its high cost overruns than against continued licensing of nuclear plants. A member concerned about international relations is more likely to vote against export of nuclear plants than to vote to discontinue nuclear plants in the United States.

Thus, citizens concerned about nuclear power must develop two strategies for influencing nuclear votes in Congress: (1) framing the issue so that it will attract the largest number of votes, and (2) exerting continuous pressure on members of Congress to heighten the exposure of their position. There are a number of techniques for exposing a member's views to the light of public scrutiny.

POLITICAL CAMPAIGN

There is nothing to equal a political campaign for forcing a member of Congress to become concerned about an issue. If the member is opposed by a candidate who continually raises the problems of nuclear power and criticizes the member throughout the district for failing to be concerned, the member will be forced to expose his positions and in many cases to change them. This result can also occur if an initiative or referendum is on the ballot and all major political figures are questioned on their positions.

An excellent example of this was the nuclear proliferation issue in the 1976 presidential race. President Ford and his Republican predecessor, Richard Nixon, had fostered the sale of nuclear plants and technology abroad. Just prior to and following his nomination, Democratic presidential candidate Jimmy Carter made two major speeches on the problems of nuclear proliferation. One month before the election, Ford announced that he would begin a new program to curb proliferation of atomic weapons. For two years prior to this announcement Ford had been a proponent of nuclear power technology export. Subsequent details showed that Ford's program was not a serious effort to curb proliferation. But candidate Carter's pressure had forced Ford to address the proliferation problem.

CONFRONTING THE MEMBER

When members of Congress or other politicians make speeches in their districts, they do not like to be asked embarrassing questions. One sure way to make a member pay attention to an issue is to ask him questions about it in public each time he appears. If he is unsure of the answer, he will brief himself before he goes on a public stage again. Unless his position is a hard and firm one, if his position is publicly criticized he will think carefully about it.

Because it is difficult to get a politician's attention, questions from citizens at public forums such as speeches are a basic tactic for focusing interest and thought on a matter. If a citizen group wants to use the opportunity to educate the people attending the speech, it can print a one-page flyer on nuclear power and the member's positions for distribution at the meeting. This technique—whether done at a meeting the member is attending or in a shopping center—really influences members of Congress because they do not like to be on the defensive. Since most members never tell their constituents how they vote, widespread distribution of his voting record often worries the member and makes him very cautious about his position.

When the member makes a speech, a group should make sure local reporters are informed on the issue and prepared to ask searching questions at the speech or press conference. Radio and television interviews and talk shows are also important opportunities for reporters and citizens to raise the issue.

CONTACT WHEN THE ISSUE IS HOT

Letters, telegrams, or telephone calls to members of Congress just before key votes in committee, on the floor, or in conference committee are very influential. Citizen groups should set up telephone trees to get the word out fast about imminent action on important nuclear legislation. Critical Mass, the Public Citizen group concerned about nuclear power, publishes a monthly newspaper and informs its subscribers about action on legislation in Congress. It needs and wants active citizens who are willing to develop a citizen network and contact members of Congress on important votes. Other groups publishing newsletters which include information on congressional action include the Center for Science in the Public Interest, Environmental Action, Friends of the Earth, the Natural Resources Defense Council, and the Sierra Club. Their addresses are all given in the Appendix.

Ordinarily, members of Congress work through their staffs. Thus, one should contact the member's legislative assistant or staffperson for a particular issue. Other effective actions include visits to the member's district office or Washington office (if possible), and encouraging other respected politicians (such as the mayor or governor), to lobby the member of Congress. An expensive but worthwhile effort is testimony by constituents at hearings on nuclear power in Washington or at regional or local hearings. This takes much work and travel expense but is frequently productive in making the member focus on the issue.

ASKING THE MEMBER FOR HELP

Over the years members of Congress have become ombudsmen for their constituents in their relationships with the federal government because this activity is very helpful in getting reelected. Most constituents ask their member for help in matters such as Social Security or veterans' benefits. But some ask a member to use the resources available to his office to develop policies on issues of public importance. Citizen groups should not forget to ask a member for a report from the Library of Congress Congressional Research Service or the General Accounting Office (GAO) in reference to a difficult nuclear power matter, if this is pertinent.

For example, the Energy Research and Development Administration (ERDA) distributed a misleading pamphlet which was used by opponents of the June 1976 California Nuclear Initiative. Mark W. Hannaford (D., Calif.) and other congressmen asked the GAO to investigate this matter. The subsequent critical GAO report and adverse publicity caused ERDA to remove its pamphlets from circulation and made Congress more skeptical of the agency's atomic advocacy.[2]

Constituents can also ask a member of Congress to have them invited to testify before a congressional committee, or to send them bills, reports, or committee hearings on nuclear power. These kinds of communications are important because they begin to influence a member and his staff to devote attention to the subject. Of course it is always appropriate for constituents to ask their member of Congress to co-sponsor legislation of interest to them, although most members will not do so unless they are sincere advocates of the matter.

RAISING A PUBLIC FUSS

If a member of Congress is recalcitrant and absolutely refuses to meet with or consider views in opposition to nuclear power, some citizen groups decide to make a public issue out of it and picket his district office or his public appearances. For most people this step is one of last resort because it tends to totally alienate the member from the constituents raising the fuss. However, if the member is isolated from his voters and does not reflect their views, it is important to raise the issue in a direct public manner so that large numbers of voters will become aware of his obstinacy. Since most members of Congress do not like to be isolated at one end of the political spectrum on any issue, this tactic, while difficult to accomplish with grace, can rapidly induce a member of Congress to reconsider or temper his positions.

XIX

Action at the Federal Level: The Nuclear Regulatory Commission

BESIDES THE CONGRESS, the major federal agency which presents opportunities for citizen action is the Nuclear Regulatory Commission (NRC), which licenses nuclear plants. Citizens may "intervene" in the licensing process to attempt to stop individual plants. But groups should be aware that there are severe obstacles to the process. The chances of winning are not great, although small victories, such as disclosure of critical information, and heightened community understanding, usually occur. Other ways of dealing with the NRC include taking the agency to court, or filing rule-making petitions, which request the commission to change its regulations on a generic (meaning applicable to several plants) basis. These options are described in more detail below.

The Licensing Process

In the United States, nuclear reactor construction and operation are licensed and regulated by the Nuclear Regulatory Commission. The licensing process has two stages. First, a utility applies for a construction permit, which would allow a nuclear power plant to be built. Later, the utility must apply for an operating license, which allows placement of nuclear fuel in the reactor and operation of the reactor to produce power. Figure 1 presents a simplified outline of the process that would be followed by a utility obtaining a construction permit. The same process is repeated, with minor differences, for an operating license.

In its application for a construction permit, a utility must submit several documents to the NRC. One is a Preliminary Safety Analysis Report (PSAR). The PSAR, as the name implies, is an initial analysis of the safety of the planned plant, and must include the following information:[1]

1. a summary description of the facility and its design;
2. an analysis of the plant's safety systems and their ability to prevent or mitigate accidents;
3. a description of the quality assurance program to ensure that the plant is constructed under adequate standards and that equipment is tested for defects;
4. the applicant's plan for training and conducting operations.
5. the applicant's preliminary plan for coping with radioactive emergencies.

The utility must also submit an environmental impact report outlining the effects that plant construction and operation will have on the environment. The utility must present information to allow the U.S. Department of Justice to determine whether construction or operation of the proposed plant would violate any antitrust laws. In addition, the utility must demonstrate that it has the financial capability to complete construction of the power plant safely.

Each application is reviewed by the NRC staff to determine whether the application contains sufficient information. If information is considered lacking or if the quality assurance program is unacceptable, the NRC staff will return the application. If the application is satisfactory to the staff, it will be formally accepted for a more detailed review.

The NRC is required by law to hold a public hearing on the construction permit before a three-member Atomic Safety and Licensing Board (ASLB). After the application is accepted, the NRC issues a public notice indicating that a hearing is to be held. Members of the public may participate in the hearing process in two ways. They can make a "limited appearance," which involves little more than making a brief statement opposing or supporting the plant. If an individual or a group of citizens enters the licensing process with the intent of preventing a nuclear plant's construction, the individual or group must become an "intervenor."

The intervening group or individual must petition the licensing

Figure 1. *Licensing of Power Reactors*

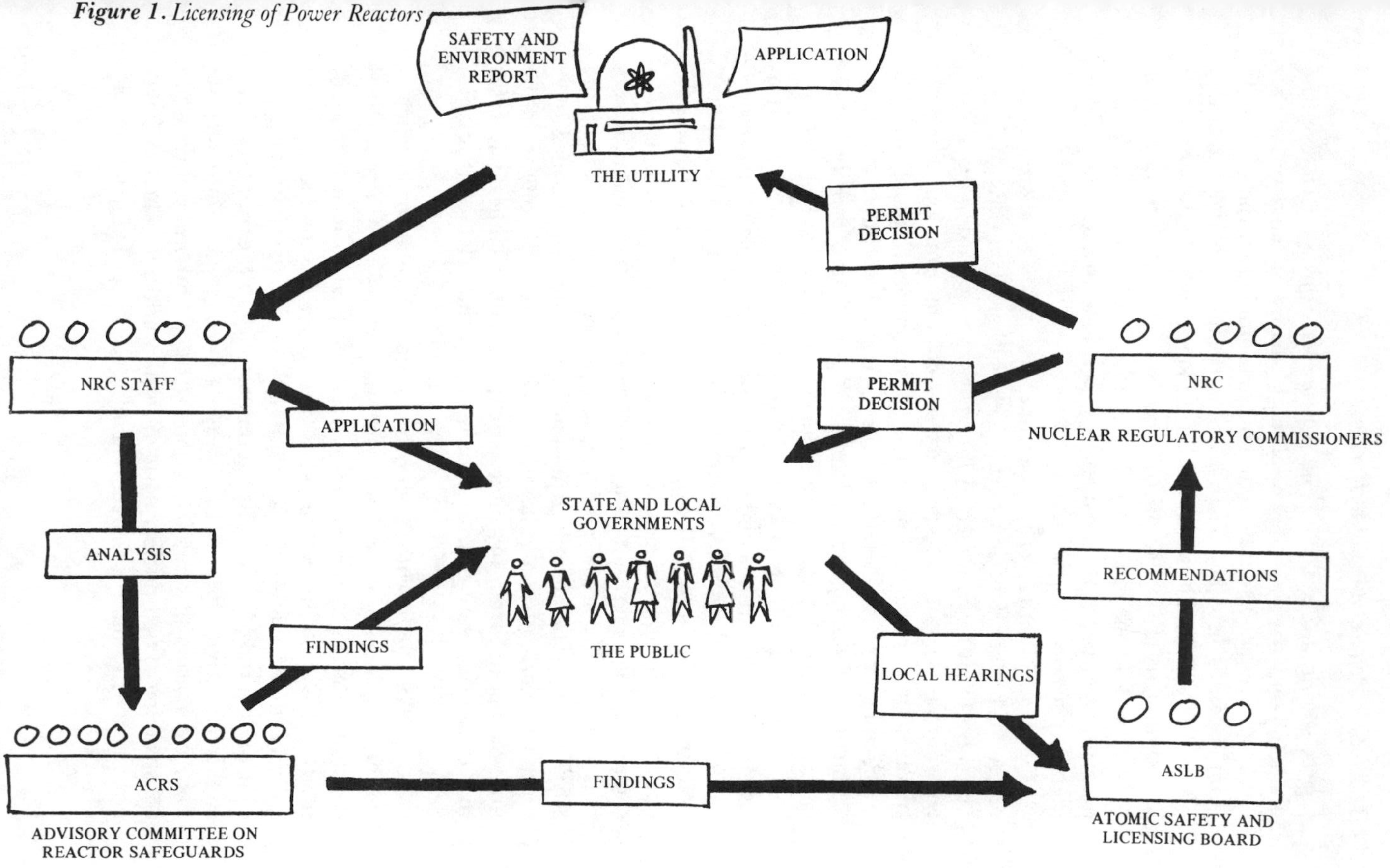

board, stating in detail how the petitioners will be affected by the plant and on what basis they oppose plant construction. An individual or group may also intervene on the grounds that they favor the plant, but this usually does not occur because it is not necessary. Since the ASLB will usually approve plant construction, almost all inventions are in opposition to the plant. About sixty days after public notice of the intent to hold a hearing, the ASLB holds a "prehearing conference" to consider any petitions to intervene and to identify the controversial issues.

The NRC staff, in the meantime, has begun its more detailed review of the construction permit application. The NRC will also prepare a Draft Environmental Impact Statement, using the utility's environmental report as its primary source of information. The draft statement is circulated for review and comment among federal, state, and local agencies and members of the public. Public notice, as with notice of hearings, is made in the *Federal Register* (a daily compilation of official statements by federal government agencies) and in local newspapers. The NRC, after receiving comments, will then issue a Final Environmental Impact Statement which will become part of the record in the environmental portion of the public hearing.

Ordinarily the ASLB would next schedule a public hearing on both the environmental and safety aspects of the proposed plant. In some cases, however, the utility may request a provisional construction permit and an accelerated licensing procedure, in which case the ASLB would schedule a hearing on the environmental issues only. Once the hearing begins, the ASLB will accept testimony from the utility applicant, from any intervenors, and from technical witnesses that either side may call. Statements from the NRC staff are also heard.

The ASLB will rule on all matters introduced at the hearing. If the hearing reviews only environmental matters, and if the ASLB concludes that such matters have been resolved to its satisfaction, the board will authorize the NRC director of reactor regulation to issue a Limited Work Authorization (LWA). This permits the applicant to undertake initial construction activity at the proposed plant site. Such activity may include preparation of the site for construction, installation of temporary construction support facilities, and construction of non-nuclear parts of thr power plant. The LWA will not ordinarily allow construction of nuclear-related parts of the plant.

Prior to the safety hearings, the NRC staff evaluates the safety aspects of the application, setting forth its conclusions in a Safety Evaluation which is released to the public and sent to the NRC's Advisory Committee on Reactor Safeguards (ACRS). The ACRS, which consists of technical persons from universities, industry, and the federal laboratories, forwards its comments to the licensing board and the NRC. Following these reviews, the ASLB will convene a hearing on plant safety issues. (As previously noted, both environmental and safety matters are addressed at this hearing, unless the applicant has requested an LWA).

During the safety hearings, the ASLB again receives the contentions of any intervening groups. If the ASLB decides that the plant can be built, it will authorize the director of reactor regulation to issue a construction permit. Objections to the decision may be filed with a three-member Atomic Safety and Licensing Appeal Board (ASLAP).* After the ASLAP releases its decision, only the NRC commissioners, on their own motion, can review particular licensing issues. The administrative review process is then exhausted, and the permit is granted. The only remaining appeal would be to the courts.

As construction of a nuclear plant nears completion, the utility will file an application with the NRC for an operating license. This must be accompanied by a Final Safety Analysis Report (FSAR) that covers the same subjects as the PSAR, but in greater detail. Any design changes, new information on safety equipment, or updated safety analysis are discussed. The FSAR must also include:

1. information on the facility's environmental and meteorological monitoring programs;

2. detailed description and safety analysis of the plant's equipment;

3. proposed methods for limiting discharge of radioactive materials;

4. description of the quality control system intended to assure safe plant operation;

5. pre-operational testing program planned for the plant;

*Members of the ASLB and the ASLAP are chosen from special panels established by the Nuclear Regulatory Commission. Their permanent members are NRC employees. Other members include university professors and employees of the federal laboratories. These panels have separate membership: the same individual cannot be eligible to serve both on an ASLB and an ASLAP.

6. more detailed plans for coping with radiation emergencies;
7. the physical security plans that are intended to protect the reactor from sabotage.

The utility must also demonstrate its continued financial capability to operate reactor safely.

After receipt of the applicant's request for an operating license, the NRC initiates a procedure similar to the licensing process for the construction permit. The Final Safety Analysis Report is reviewed by the NRC staff and the ACRS. A public hearing is not mandatory at this stage, but an ASLB will convene a hearing, usually in the area near the reactor, if requested by affected members of the public or by the commission itself. If the ASLB rules in the utility's favor, it will authorize the Director of Reactor Regulation to issue an Operating License. As with the construction permit, the operating license can be appealed to an ASLAP, and eventually the courts.

Problems with the Process

The above capsule summary of the licensing process does not mention the severe problems faced by citizen groups who decide to intervene. Many of these problems have been examined in a 1974 study by Steven Ebbin and Raphael Kasper, of George Washington University, who concluded that the licensing process could be aptly characterized as a "charade."[2] To begin with, citizens enter the licensing process at the eleventh hour. By the time public hearings begin, the NRC staff, the utility, and (often) the Advisory Committee on Reactor Safeguards have already made the decision that the plant should be built. The intervenors thus find themselves opposed by both the utility and the NRC staff.

Because they come so late to the process, and because technical expert witnesses cost money, the intervenors are often reduced to challenging the utility and the NRC staff via cross-examination. That is, because it is usually too difficult to mount direct testimony, in a short time period, on why a particular nuclear plant is not needed or inappropriate, the intervenors are forced to make their case in cross-examination. This, of course, results in large legal fees.

Intervenor challenges must also be limited to attacking the particular plant in the licensing hearing. Contentions on limited liability, proliferation, emergency core cooling, or other larger issues are not allowed. The intervenors cannot challenge the regulations of the NRC in a licensing hearing—they can only attempt to show that the plant will not meet the regulations. Changes in the commission's regulations can be considered only in generic rule-making proceedings.

One further problem is that it is often difficult to obtain documents to challenge a utility's statements and conclusions. The safety data which are supposedly the basis for some utility and vendor statements are often classified as "proprietary data" and thus not disclosable to the intervenors.

In addition to the procedural problems, the largest obstacle is cost. With fees for lawyers, expert witnesses, and travel expenses, an intervenor can spend $100,000 or more—without a prayer of stopping the plant. Thus, citizen groups who initially decide to work "within the system" by stopping a nuclear plant in the most obvious forum—the NRC licensing process—find that the odds are hopelessly stacked against them. Interventions have resulted in the addition of safety systems or use of cooling towers, against the utility's original desires, but rarely will intervention stop a plant.

Still, there are many groups who claim that intervention has value. For one, the cross-examination can result in the release of information on nuclear safety problems which might otherwise have remained hidden. Many intervenors have also succeeded in raising the awareness of local residents by publicizing the issues at public hearings. Chicago attorney Myron Cherry, who has represented several intervenors, believes that the intervention process can delay licensing and allow public opposition to the plant to develop further. In only a few cases, public opposition and the prospect of continued opposition have been sufficient to force a utility to abandon a plant or move to another site. But in most cases the positive value of intervention has been only to publicize nuclear power and educate the community.

With this in mind, groups have begun to examine less costly ways of making use of the intervention process. "Limited appearances" in licensing cases are short statements, oral or written. Usually five to ten minutes, they simply become part of the record, with any supporting documentation. While persons using this privilege are not entitled to present testimony, have access to documents, or cross-examine witnesses, the demonstration of local concern evidenced by large numbers of limited appearances, especially if specific hazards are detailed, can be reported by newspeople covering the hearings.

Another option is "mini-intervention": filing as a full intervenor, making a presentation—which can be longer than the few minutes allotted to a limited appearance—then withdrawing. This is a technique for making concerns known without going deeply into debt. Holding counter-hearings at key times and places also serves both to dramatize the citizens' plight and to present their case to the public. These strategies aim for media impact rather than persuading the agency through the intervention process itself.

In summary, the beneficial effects of an intervention are to make nuclear power a local issue and educate the public, but citizen groups must be aware of the severe problems with this process. A citizen group, if unsuccessful in its intervention before the NRC, has the option of taking the agency to court.

The Courts

Three recent cases representing follow-up activities to intervention provide some insight into the possibility of tackling nuclear power in court. These cases involve reactors in Bailly, Indiana; Vernon, Vermont; and Midland, Michigan. Another court case, which challenged the NRC's activities regarding plutonium fuel, is also instructive.

Intervention against the Bailly plant in Porter County, Indiana, allowed time for other opposition to mature. After two years of licensing hearings, the Atomic Energy Commission granted a construction permit to the Northern Indiana Public Service Company for Bailly. The intervenor groups, contending that the commission was violating its own guidelines by allowing a nuclear site to be located too near a major population and resort area, the Indiana Dunes National Lakeshore, filed a lawsuit in 1972. The suit was brought by the Porter County chapter of the Izaak Walton League and Business and Professional People for the Public Interest (BPI), based in Chicago.

The groups' challenge, initially successful in the court of appeals, eventually reached the Supreme Court, which reversed the lower court's decision and remanded the case for reconsideration.[3] The court of appeals refused to revoke the license. After four years of dispute and the expenditure of many thousands of dollars, the groups had exhausted all of their legal remedies.

But, an unexpected twist to the challenge developed. The appeals court noted parenthetically in its second ruling that, because the plant potentially could interfere with the viability of the 8,300-acre Indiana Dunes park, the Department of the Interior had authority to stop construction through legal action. The Interior Department had opposed the nuclear plant from the start because of its proximity to the Dunes but has yet to exercise its authority.

In the spring of 1976, Interior Secretary Kleppe wrote to NRC Chairman Marcus A. Rowden, pointing out that Bailly is an inappropriate site for the plant. Rowden rejected Secretary Kleppe's views, but the citizen groups urged the Interior Department to bring suit against the NRC. Interior decided, however, that it would first conduct an investigation to determine if additional data were available which would warrant further opposition to the plant.[4] As of March 1977, Interior had made no further decision, but determined citizen opposition had made possible the opportunity for action.

Another decision, reviewed in the chapter on the Plutonium Fuel Cycle, was the Second Circuit Court of Appeals order prohibiting "interim licensing" of plutonium in commercial power reactors. It is generally recognized that this ruling will effectively delay commercial use of plutonium for several years.[5]

But the victory may be a narrow one. The groups bringing the suit—led by Natural Resources Defense Council (NRDC), and the Sierra Club, *delayed* the NRC decision to license plutonium as a fuel. In order to *prevent* its commercial use, the groups must show in a series of lengthy hearings (the equivalent of an agency trial) why the highly toxic material represents a major hazard and should not, therefore, be used as a fuel. Toward that end NRDC will have to present expert testimony on plutonium toxicity, on problems of waste management, on reprocessing economics, on the threat to civil liberties created by plutonium, and on the relation of plutonium reprocessing to international nuclear proliferation. The cost to citizen groups of such a proceeding will be hefty. So while the initial court expenses were relatively low in this case, the required follow-up in agency proceedings will be quite expensive.

In other court cases, the Federal Court of Appeals for the District of Columbia in July 1976 handed down four interrelated decisions which produced a de facto nuclear moratorium for five months.[6] In these decisions, the court reversed two licenses granted by the NRC on the grounds that the commission had not adequately considered disposal of radioactive waste, nor energy conservation as an alternative to nuclear plants. These court suits had been brought by groups intervening against the Consumers Power, Inc., plant in Midland, Michigan, and the Vermont Yankee plant in Vernon, Vermont.

In two opinions dealing with the Vermont Yankee plant, the court disagreed with the commission's contention (1) that the effects of the nuclear fuel cycle were too remote and speculative to be given consideration in an individual licensing case and (2) that the National Environmental Policy Act permits these kinds of issues to be considered by the agency's expertise. The court ruled that the NRC must reassess the environmental effects of nuclear reprocessing and waste disposal, and that these matters—as part of the nuclear fuel cycle supporting each reactor—are relevant in individual nuclear plant licensing cases.

In a third decision on the Midland plant, the NRC was criticized for failing to consider energy conservation as an alternative to the building of a nuclear power plant. In its fourth decision, the court raised the same reprocessing matters at Midland as in the Vermont Yankee case, and ordered that the Advisory Committee on Reactor Safeguards letter of approval explain in each case and in clear language what is meant by "other generic problems." It had been the practice of the ACRS, in acceding to the issuance of a *construction permit*, to cite "other generic problems" which had to be corrected before the *operating license* was granted. But the ACRS rarely specified what these other problems were, a practice which the court found vague and unacceptable.

Even these court cases provided a hollow victory, however. By October 1976 the Nuclear Regulatory Commission had taken the minimally acceptable steps which it believed necessary to comply with the court order. The agency was back to issuing licenses in November.[7]

Court actions, then, at best can only win temporary delays in construction activities or implementation of NRC policy. Like interventions, they can be costly. Also, in the Bailly case and in the July 1976 court of appeals cases the groups bringing the suits were forced to exhaust their administrative remedies within the NRC before they could even bring their cases to court. Their resources thus were doubly taxed—by the intervention process and by the court suits.

Citizen groups should be aware of these difficulties before deciding to file a lawsuit. Furthermore, at least two or three attorneys and other experienced citizen organizations should be consulted beforehand. The most successful cases are those handled by experienced groups able to hire a full-time professional staff. For other groups, it may be that time and money can be better spent on different projects. Activists seriously considering court fights should write NRDC, 917 15th Street, N.W., Washington, D.C. 20005. Business and Professional People for the Public Interest (BPI) can be contacted at 109 North Dearborn, Chicago, Illinois 60602.

NRC Rule-Making: Evacuation as a Case Study

Another option for action is rule-making, which requires petitioning the NRC to amend its regulations on a generic basis. The best-known rule-making activity was the Emergency Core-Cooling System proceeding, but in 1972 and 1973 the AEC, responding to citizen concerns, held three other major rule-makings—on routine radioactive emissions, environmental effects of the nuclear fuel cycle, and transportation of radioactive material. These rule-making proceedings presented many of the same pitfalls for citizens as did the licensing process: they were expensive and time-consuming, and offered only limited likelihood of any major changes in atomic energy policies. The ECCS hearings, for example, resulted in 22,000 pages of testimony and cost citizen parties more than $100,000. But the AEC's final action was only to make minor changes in its ECCS Criteria.

But other rule-makings have been filed on less technical questions. In August 1975 the Public Interest Research Group (PIRG), our Washington, D.C., consumer organization, asked the NRC to require utilities to distribute emergency evacuation plans to persons living within a forty-mile radius of any nuclear facility, and to disseminate general information on evacuation procedures through the media. The petition also requested that the utilities be required to conduct actual drills in specified areas near nuclear plants so that citizens could become familiar with the evacuation plans. One little-known fact is that each nuclear plant is required to have plans to evacuate citizens in the event of a catastrophic accident. But if citizens are unaware of the plan, it is essentially useless. PIRG was joined on the petition by thirty-one consumer and environmental groups from twenty states. The NRC invited public comment on the petition but by March 1977 had made no decision.

One basis for the petition was the government's Reactor Safety Study. In its final version, the study predicted that 3,300 people would be killed quickly in a major atomic plant accident. But the study also estimated that evacuation within twenty-four hours of 70 percent of the people within a twenty-five-mile quadrant around the power plant could cut early fatalities in half.[8] Whether evacuation can actually be effective, however, can be seriously questioned, and depends on the efficiency and speed with which it is carried out.

Under the Nuclear Regulatory Commission requirements utilities are responsible for the safety of plant employees; and they are also under a legal obligation to give the NRC reasonable assurance that the public health and safety are protected beyond the plant gates. Federal regulations require utilities to show that plans prepared by local authorities for nuclear accident emergencies are workable on a continuing basis. However, the local jurisdiction has the final responsibility for designing emergency plans and making sure local officials are able to carry them out effectively. In actual fact, little attention has been paid to emergency planning; the plans are vaguely written and poorly publicized; most citizens have no idea what precautions or actions are necessary to protect against radiation contamination; and all the plans evidence severe weaknesses when actually tested.

The ineffectiveness of the emergency plans has been demonstrated in several states by drills which have been little more than tests of the telephone network by which the power plant would call local officials, local officials would alert others, and so on. The chief purpose of these drills has been to determine that officials recognize their position in the communications network and are aware of their responsibilities under the emergency plans. Even at that, the drills have uncovered major flaws.

In Minnesota, when the plans were tested, the communication systems failed and some persons were ignorant of their responsibilities. Roy Aune, deputy director of the state's Division of Emergency Services, said after a 1975 drill that "under the present system it is doubtful that the state EOC [Emergency Operating Center] could handle a major disaster adequately with the facilities presently available."[9]

An Oregon drill, which has been called the "most realistic" nuclear emergency drill, is perhaps the only one in which members of the general population were actually evacuated. Twenty-one volunteers were removed from the area. But this drill, too, had problems. The evacuation team leader for the NRC observed that the test showed a lack of communication and coordination among the local agencies, too few trained technicians, and confusion at evacuation centers. Persons assigned to simulate radiation monitoring, lacking essential automobile telephones, were forced to report their findings from roadside pay telephones.[10]

As a spin-off from rule-making petitions, several local groups filed "show-cause" petitions which demanded that the NRC and the reactor owners justify the continued operation of power plants that are in apparent violation of NRC standards. For example, the New York and Maine Public Interest Research Groups (NYPIRG and Maine PIRG) filed show cause petitions in February 1976 seeking to shut down the Indian Point (New York) and Maine Yankee reactors on grounds that the emergency and evacuation plans for nuclear accidents were extremely inadequate.

According to NYPIRG Director Donald Ross, the major failure in existing plans was "the absolute lack of planning for anything but very small accidents." The NYPIRG petition listed such other major flaws in the emergency plans as (1) confusion and a lack of coordination among state agencies, (2) unsatisfactory plans for notifying public agencies and officials of an accident, and (3) inadequate procedures for emergency drills. The petition called the plans "a satire on bureacratic thinking."[11] Maine PIRG Director Michael Huston, commenting on the Maine Yankee evacuation plan, pointed to a utility claim that in five towns near the reactor, specific sirens or alarms would be sounded to alert the public. "In fact, none of the towns has such a special signal. Plans for notifying the towns' 6,000 residents are based on a falsehood." Huston also noted the lack of evacuation plans for the seven schools and many summer camps in the area, the lack of hospital facilities for treating patients contaminated with radioactivity, and the failure to keep an updated list of local officials assigned to initiate the evacuation sequences. The Maine PIRG, in its report, *Helplessly Hoping*, by Robert Burgess, detailed the inadequacies of the state of Maine's emergency plan, which the utility explicitly incorporated into the plans it filed with the NRC.[12]

The evacuation petitions have been a qualified success. The NRC rejected both the show-cause petitions. But, with the Maine petition the NRC, the state, and the utility took significant action to improve the plan—between the time the petition was filed and the date the NRC rejected it. Although the NRC would not admit it, the petition was probably responsible for improving the evacuation plan.

It was probably too much to expect that the NRC would approve the show-cause petitions—such an action could eventually force the shutdown of several nuclear plants due to poor evacuation plans. But the groups were successful in publicizing the problems with nuclear evacuation plans and in obtaining media coverage of their petitions. Moreover, the evacuation plans provided a vehicle for national and local groups to work in conjunction. Filing the petitions was also a relatively inexpensive action—although if the NRC decides to hold hearings on the rule-making petition the resulting proceedings could be long and costly. For further information on the petitions, contact Public Interest Research Group, P.O. Box 19312 Washington, D.C. 20036.

In summary, processes within the Nuclear Regulatory Commission are available for citizens to use. But the procedures, particularly intervention, are costly, time-consuming, and very unlikely to end in victory. But until the nuclear Regulatory Commission has a change in leadership and liberalizes its rules for Citizen intervention, petitions before the NRC face formidable obstacles. The intervention process is worthwhile only for its legitimate news value in educating the general public on nuclear power hazards. Court actions can cause headaches for the industry, but bringing a successful suit is costly and requires the services of a skilled attorney.

Citizens would be well advised to look for methods which accomplish the same result as intervention—public education—but which bypass the intervention process. One alternative route within the NRC could be by simpler and less costly rule-making or show-cause petitions. Other chapters in this Citizen Action section give other ideas for informing the media and public.

Another possibility may be challenging state or local bodies

—utility or environmental commissions—which grant permits for nuclear plants but which may be less set in their ways than the NRC. Or perhaps the more likely answer lies in forming state and regional coalitions to lobby legislatures and the U.S. Congress—which, because they are elected bodies, may be more responsive to constituent concerns. Some of these possibilities are examined in the next two chapters. Certainly, the entrenched obstacles to citizen participation in the regulatory process argue powerfully for legislation to overcome these barriers to public scrutiny of the nuclear technology.

Action at the State and Community Level

CITIZEN GROUPS, recognizing the difficulties of working through federal agencies to stop the spread of nuclear power, have been examining other ways to discipline the federal role. Groups have begun to demand that state and local governments assume a greater role in the nuclear decision. As of August 1976, legislators in more than thirty states had introduced bills that would in one way or another restrict nuclear power plant expansion.[1] Vermont and California have already enacted measures which give their legislatures partial responsibility for the decision to site nuclear facilities in those states.

However, the atomic industry and its allies in the federal government maintain that the states do not have a legitimate role in regulating nuclear power unless particular responsibilities are delegated to the states by the Nuclear Regulatory Commission (NRC). According to the nuclear establishment, the legal doctrine of federal pre-emption assigns the power to regulate atomic energy to the U.S. government. Furthermore, the industry claims that the courts would invalidate any state action that supersedes or encroaches on the federal government's authority.[2]

The federal pre-emption doctrine was designed to ensure that *minimum* standards are enforced for ultra-hazardous technologies, such as atomic energy, that cannot be effectively regulated by the states. Industry apologists insist that federal standards should be the *maximum* level of safety required—implying that states should not regulate the design or operation of reactors more stringently than the NRC.

By its continual assertion of the pre-emption doctrine, the

atomic establishment has intimidated some state governments. A number of legislators have allowed moratorium bills to languish in committee because the nuclear industry has convinced them that the federal courts would overturn such legislation.

However, the federal pre-emption doctrine is neither all inclusive nor definitive. States can regulate nuclear power in a myriad of ways if citizens demand such actions.

The Northern States Power Case

The most significant court case used by atomic power promoters to support their argument that the states cannot, by law, regulate nuclear power is *Northern States Power Company* v. *The State of Minnesota* (1971).[3] The issue was joined when the state of Minnesota set standards, about 100 times more restrictive than those of the Atomic Energy Commission (AEC), for radioactive emissions from nuclear plants operating in Minnesota.[4] Minnesota attempted to enforce its stricter standards in 1968 when the AEC granted a construction permit to Northern States Power Company to build a nuclear plant in Monticello, forty miles upstream of Minneapolis. The company admitted that it could meet the Minnesota standards but claimed the cost would be too high.[5]

Northern States Power Company attempted to invalidate Minnesota's action on the grounds that federal law gave the AEC the exclusive right to regulate nuclear power plants, thus "pre-empting" any state authority. Since the outcome of the case would affect other states, Michigan, Wisconsin, and Maryland filed briefs in support of Minnesota's argument that it did have the right to set radiation standards more stringent than those set by the AEC.[6] In March 1971, the federal district court in St. Paul ruled in favor of the utility company. Both the appeals court and the U.S. Supreme Court upheld the lower court's decision.

The decision in the case centered on a clause in the Atomic Energy Act of 1954 which stated that no agreement could "provide for the discontinuance of any authority," and that the AEC "shall retain authority and responsibility with respect to regulation of" the construction and operation of "production and utilization facilities," including nuclear reactors.[7] This clause prompted the court's decision: "Accordingly, for the reasons stated, we hold that the federal government has exclusive authority under the doctrine of pre-emption to regulate the construction and operation of nuclear power plants, which necessarily includes regulation of the levels of radioactive effluents discharged from the plant."[8]

However, citizen activists were not halted by the adverse court decision. The Minnesota Environmental Control Citizens Association (MECCA) and other environmental groups encouraged the state legislature to enact a moratorium on nuclear power plants until the AEC promulgated more stringent radiation standards.

In April 1971, one month after the U.S. district court in St. Paul had released its decision in the Northern States Power case, Governor Wendell R. Anderson recommended a moratorium on construction of new nuclear plants in Minnesota until "such time as the Minnesota Pollution Control Agency certifies that new development can safely begin."[9]

In order to undercut the governor's recommendation and bills before the state legislature, Northern States Power announced that it would modify its Monticello plant to reduce radioactive discharges by 80 percent or more. The modifications were similar to the original demands of the Minnesota Pollution Control Agency. So, while the Northern States Power Company won its court case, it was a dubious victory, since the utility eventually agreed to operate under the stricter radiation standards similar to the ones it had opposed in Court.[10]

The Northern States Power case also spurred a move in Congress to overhaul the pre-emption doctrine. Legislation was introduced in both houses to amend the Atomic Energy Act so that states have authority to set more stringent standards than those of the AEC for operating nuclear facilities. However, the legislation was defeated in the Senate in 1972 and in the House in 1973.[11] Congressman Morris K. Udall (D., Ariz.) introduced similar "nuclear states rights" legislation in 1977.

State Action

It is important to understand that the Northern States decision only addressed federal and state roles in regulating the construction and operation of nuclear power plants, particularly in setting standards for emission of radioactive effluents. The Atomic Energy Act of 1954, on which the court relied in the Northern States Power case, also contains the following clause: "Nothing in this section shall be construed to affect the authority of any state or local agency to regulate activities for purposes other than protection against radiation hazards."[12]

Indeed, the states retain the authority to regulate nuclear power plants in a variety of ways under the principle of the Tenth Amendment to the Constitution—that rights not specifically delegated to the federal government are reserved to the states. Among state powers relevant to nuclear regulation are the protection of the general welfare and safety of the public, land use and zoning authority, utility regulation, and the development of alternatives. Some examples follow:

PROTECTION OF THE GENERAL WELFARE

The state of Vermont has found two ways to exercise its authority over nuclear power. In 1974, intervenor groups and Vermont Yankee, a consortium of utilities operating a nuclear power plant in Vermont, signed a contract with the state requiring the facility to conform with Vermont's radiation standards, which were tougher than those of the federal government.[13] The Vermont Board of Health later used this agreement to establish radiation standards for "turbine shine," which is radiation from the turbine building at the plant. The Board of Health action was especially significant because the NRC has never set any standards for turbine shine.[14]

In 1975 the Vermont legislature, again exercising the authority to protect the general welfare and safety of the public, enacted a law which was called at that time "the toughest state nuclear power control measure in the nation."[15] The major citizen organization which worked for the bill was the Vermont Public Interest Research Group, directed by Scott Skinner. The Vermont bill, signed into law on April 3, 1975, stipulated that both houses of the state legislature must voice approval before the state Public Service Commission can issue a permit for nuclear power plant construction. The legislation does not amount to an out-and-out ban on nuclear power in the state, but it does place part of the decision-making responsibility for nuclear power with elected representatives. Since the legislation does not affect the existing plant in the state and because the utilities have yet to build a second reactor, no court challenge has been initiated to strike down the law.

The Vermont bill was preceded by a series of town meetings

throughout the state in March of 1975 in which voters rejected nuclear power by large margins. The most dramatic rejection came in the town of Barnet, across the Connecticut River from a proposed nuclear power plant in New Hampshire. Barnet voiced opposition to that plant by a margin of 190–2.[16] In March 1977, nuclear power was again the topic at Vermont town meetings. Thirty-one towns voted to prohibit the siting of nuclear plants, or the disposal or storage of nuclear wastes, within their borders.

LAND USE AUTHORITY

In June 1976, the California legislature, citing the state's authority to regulate land use, enacted three bills which regulate the construction of nuclear power plants within the state. The bills, signed into law by Governor Brown, give the legislature the power to prohibit the construction of any new atomic power plants in the state, unless it decides by majority vote that:

> The United States through its authorized agency has approved and there exists a demonstrated technology or means of disposal of high-level nuclear waste.
>
> The United States, through its authorized agency has identified and approved, and there exists a technology for the construction and operation of nuclear fuel rod reprocessing plants.
>
> The [California Energy] Commission has undertaken and completed a study of the necessity for, and effectiveness and economic feasibility of, undergrounding and berm containment of nuclear reactors. . . .[17]

These bills, which apply only to future plants, were passed as an alternative to a citizen initiative that would have mandated more stringent requirements for both present and future nuclear facilities. However, they also represent a significant challenge by the most populous state in the country to the federal pre-emption doctrine.

ALTERNATIVES

State actions can also be addressed to alternatives to nuclear power. Any act which promotes energy conservation—in particular, conservation of electricity—will reduce the need for nuclear plants. Individual states have already taken action to require more energy-efficient buildings, to improve energy efficiency in state-owned buildings, and to ban the use of non-returnable containers.[18] Another alternative is to restructure the pricing systems for electricity and other energy sources, which traditionally have charged large users less per unit of energy consumed, thus reducing the incentive for energy conservation. In California, "Lifeline" concepts were instituted in 1975 by the state legislature and the Public Utilities Commission.[19] The California Lifeline concept would require higher residential rates for higher consumption levels.

Furthermore, states can reduce their need for nuclear energy by encouraging alternative means of supplying energy. The state of Minnesota now has a policy of encouraging the use of solid waste as fuel.[20] Several states have also taken action to encourage solar energy. As of August 1976, twenty states had passed laws granting tax advantages for solar energy. Five of these states had established income tax incentives, and the others had allowed sales and property tax exemptions for solar hardware.[21] A complete description of state legislation on solar energy, energy conservation, and other alternatives to nuclear power is available from the National Conference of State Legislatures, 1150 17th Street, N.W., Washington, D.C. 20036.

Influencing the State Legislative Process

Citizens can have a role in the enactment of legislation at the state level because their representatives may very well be receptive to assistance. State legislators typically have too few staff members or other resources to properly write, introduce, and work for passage of significant legislation. In a given year they may have to consider as many as 1,000 separate pieces of legislation. Until recently legislators were dependent upon an overworked legislative counsel, limited committee staff or industry lobbyists to prepare bills for them. Now, however, increasing numbers of citizens are contributing their time and effort to ensure that legislation in the public interest has a chance to pass.

Any citizen group can become involved with the legislative process by offering assistance to a sympathetic lawmaker. Specific recommendations for action follow:

1. After talking with persons who are knowledgeable about legislative affairs, activists can determine the best candidate who might consider introducing bills on nuclear power. At the time of the meeting, it would be helpful to have: a) a well-defined proposal, including a discussion of the problem and a variety of options from

which the legislator can choose; and b) some indication of hometown concern, whether it be a petition from constituents, a number of letters of endorsement, or recent publicity about the issue in the home district. Offers to provide research assistance and other help may be appreciated and accepted.

The more "homework" a group can do ahead of this initial meeting the better impression it can make. For example, a group that has already obtained copies of legislation in other states, can specifically show its representative the kinds of steps taken by other legislators. One or two concise articles available for the legislator's reading, with suggestions for additional references, are also helpful.

2. After a bill has been drafted and introduced, it will probably be assigned to a committee for study. Citizens should visit with committee members and ask for public hearings. A bill's lead sponsor, if he/she has been carefully selected, can best advise what specific tactics should be used to promote the bill's passage.

3. A group should obtain as many endorsements for the legislation as possible from groups and individuals throughout the state. Endorsements provide credibility for the legislation; a large enough number of significant endorsements helps to create a "bandwagon effect" that can significantly improve the prospects for passage.

4. Pressure must be put on the committee. It is not enough just to request public hearings. Innumerable pieces of legislation have never come out of committee, or have been assigned to "special commissions" for further study. The committee must report the bill to the floor, if the entire membership is to vote on it.

Frequently, the committee chairperson will have complete power over legislation assigned to the committee. Unless the chairperson is favorable, or at least neutralized, a specific bill may never see the light of day.

Although committee chairpersons are less susceptible to pressure than ordinary members, they are not immune. A steady barrage of letters, mailgrams, phone calls, and visits from constituents can influence even the most recalcitrant Solon.

Other "pressure points" are party workers in the committee chairperson's home district. The chairperson, no matter how powerful, must rely on these people when running for re-election or seeking higher electoral office.

A group should obtain a list of precinct captains or ward leaders from the local party headquarters. If one can be convinced to support the legislation, that person can then write to other precinct officials in the district to recommend enactment of the bill.

In all cases it is preferable that people who actually live in the ward, precinct, legislative district, etc. contact the relevant officials. Requests from constituents almost *always* carry more weight with party officials.

Other community leaders, such as a member of Congress, might be able to sway the chairperson. Prominent members of the business community, religious leaders, or well-known educators should also be approached.

5. If the chairperson is determined to block the legislation, then a majority of committee members will have to be persuaded that the bill deserves a vote. This will not be easy. Committee members are in general unwilling to try to reverse a decision of the chairperson, especially if they are of the same party. Committee chairpeople can make life difficult for members who don't "go along." The chairperson frequently controls allocation of staff time and office space. More important, the Chair is in a position either to thwart or expedite the passage of a legislator's pet bill. Many committee chairpersons are quite willing to use their powers to retaliate against a member who defies their authority.

The techniques described earlier—constituent pressure, a push from the party faithful and the like—are at least as effective with individual members as they are with chairpersons. All should be used to bring the issue to a vote.

Legislation reported favorably out of committee has a significantly greater chance of passing than does a bill which must be forced out. However, a favorable report does not guarantee passage. Once again constituents have to be mobilized, party workers contacted and the like to help ensure that the bill will be enacted. Efforts should be concentrated on "swing" members—those who are not yet committed on the bill. By now, it should be clear that effective use of the legislative process is hard work. It requires time, patience, persistence and dedication. Nevertheless, it is frequently the most effective tool available for citizen action on the nuclear issue.

The Initiative

"If we can give to the people the means by which they may accomplish such other reforms as they desire, the means as well by which they may

prevent the misuse of the power temporarily centralized in the Legislature and admonitory and precautionary measures which will ever be present before weak officials, and the existence of which will prevent the necessity for their use, then all that lies in our power will have been done in the direction of safeguarding the future and for the perpetuation of the theory upon which we ourselves shall conduct this government.

"This means for accomplishing other reforms has been designated the 'Initiative and Referendum,' and the precautionary measure by which a recalcitrant official can be removed is designated the 'Recall.' And while I do not by any means believe the Initiative, the Referendum and the Recall are the panacea for all our political ills, yet they do give to the electorate the power of action when desired, and they do place in the hands of the people the means by which they may protect themselves. I recommend to you, therefore, and I most strongly urge, that the first step in our design to preserve and perpetuate popular government shall be the adoption of the Initiative, the Referendum and the Recall.

". . . The opponents of direct legislation and the Recall, however they may phrase their opposition, in reality believe the people cannot be trusted. On the other hand, those of us who espouse these measures do so because of our deep-rooted belief in popular government, and not only in the right of the people to govern, but in their ability to govern; and this leads us logically to the belief that if the people have the right, the ability and the intelligence to elect, they have, as well, the right, ability and intelligence to reject or recall; and this applies with equal force to an administrative or judicial officer."

—California Governor Hiram Johnson,
Inaugural Address, January 4, 1911

The man whose energy and enthusiasm helped to ignite the nuclear safeguards initiative campaigns around the country had not even registered to vote before he was thirty-eight. Former California Governor Ronald Reagan's cuts in spending for education and mental hospitals spurred Ed Koupal's involvement into politics. Before his immersion in public interest issues, Koupal was a used car salesman, a bartender, a chicken rancher, and an Air Force drum major. Ed Koupal and his wife, Joyce, founded the People's Lobby, a grass-roots organization that sought to strengthen the political power of citizens. In the early seventies, their efforts spawned the reintroduction of the initiative process as a major tool for citizen action. The recent focus of the initiatives was nuclear power since the Koupals believed that no other issue exemplified the systematic exclusion of citizens from the decision making process in an issue that so intimately affected their well-being.

Tragically, Ed Koupal never lived to see his dream become reality. He died of cancer on March 30, 1976—barely two months before Californians voted on California's Nuclear Safeguards Initiative, Proposition 15. Even though the initiative was defeated by a 2–1 margin, the result was a vindication of Ed Koupal's vision. It was the first time that citizens were given the chance to break the government/industry stranglehold on the nuclear issue.

The initiative process enables the people, through petition, to propose new legislation directly. If a specified number of registered voters signs a petition to put a proposed law to a vote, the issue qualifies to appear on a statewide ballot to be approved or disapproved by the electorate. In most states, laws created by initiative cannot be changed by the legislature. Energy policies, political reform, and tax laws are among the issues that people can shape directly.

If California's Proposition 15 had been approved in 1976, it would have established the most comprehensive and stringent nuclear safeguard standards in the nation. First, the owners of nuclear power plants would have become fully liable for the consequences of a nuclear accident. Utilities would have been required to waive the liability limitations established by the Price-Anderson Act. If any operating plant did not waive the liability limitation, it would have been required to reduce operating power levels to 60 percent capacity within one year, and 10 percent per year for each additional year that the provision had not been met. Thus, if an operating plant had not waived the liability limitation after six years, it would not have been permitted to operate. Additionally, no new plant could be constructed unless the operators agreed to waive any liability limitations.

Second, the legislature would have to find, by a two-thirds vote, that all nuclear reactor safety systems (including ECCS) are effective. Such a finding could be made only after comprehensive testing of substantially similar systems had been undertaken on operating reactors.

Third, the legislature would have had to certify, by a two-thirds vote, that radioactive wastes could be stored and disposed of with no reasonable risk to the land or people of California.

After five years, the legislature would have been required to make a final determination that safety systems and waste disposal provisions had been met. If two-thirds of each house was unable to make such a finding, then nuclear power would be phased out of operation in California over a period of five more years.

Proposition 15 would also have established the machinery for a full-fledged public debate on the nuclear power issue. A fifteen-member commission, consisting of representatives with different points of view on nuclear safety, was proposed to advise the legislature before a vote was taken. The commission was empowered to make its recommendations only after conducting extensive public hearings into the safety and waste disposal questions.

In discussing the defeat of Proposition 15, Larry Levine, a spokesperson for the YES on 15 Campaign Committee, stated, "We suspected during the last weeks of the campaign that the massive amounts of money poured into the campaign (by utility companies and other concerned business interests across the country) was going to lose it for us. But the nuclear industry has a constituency to deal with that it's never had before.[22] Despite the fact that more than $4 million was spent to defeat Proposition 15, almost two million persons voted for the California initiative.

In November 1976, nuclear initiatives in six other states—Oregon, Washington, Colorado, Montana, Arizona, and Ohio—were defeated by the same industry tactics and resources that won in California. The November initiatives had been patterned after the California proposition, with the exception that operating reactors were excluded: the initiatives would have affected only future plants.

Most of the measures lost by 2–1 margins, with the Montana and Oregon initiatives (42 percent each) coming closest to majority approval.[23] Most of the initiatives had enjoyed early leads in the polls, with the Colorado and Washington measures reportedly favored by 2–1 in September. But once the atomic industry began its media blitz against the measures, they lost support. The industry outspent initiative proponents by 80–1 in Ohio, by 90–1 in Arizona, and by large margins in other states. In Oregon, where initiative proponents were outspent by "only" 5–1, the industry still spent forty cents per voter—more than three times as much per voter as the industry spent in California.[24]

In addition to the spending issue, the initiatives forced the industry to tip its hand in another way. CBS-TV had scheduled for October 21 an episode of "Hawaii Five-O," a detective series, dealing with stolen plutonium fabricated into nuclear warheads by a clandestine operation. The Atomic Industrial Forum, the industry's trade association, pressured CBS to postpone the program until after the election. Although one Seattle station did not run the show, CBS refused to yield to the industry, and aired the program in most regions on schedule.[25] The AIF's heavy-handed attempt to manipulate media coverage of a fictional account showed the depths of the atomic industry's desperation.

In two states, Missouri and Michigan, November 1976 initiatives relevant to nuclear power were successful. Missouri citizens approved, with 62 percent of the vote, a measure that would prevent construction work in progress (CWIP) from becoming part of a utility's rate base. In most states, only plants producing electricity are allowed to be considered as part of the rate base. CWIP would encourage needless utility construction and force today's consumers to pay for tomorrow's power. Although the Missouri proposition will apply equally to all power plants, two nuclear plants are directly affected. Prior to the election, Union Electric Company, which serves the St. Louis area, had indicated that the measure could delay the construction of its large Callaway nuclear power plants, the first two in the state. Both the *St. Louis Post-Dispatch* and Joseph Teasdale, the successful Democratic challenger for the governor's seat, had endorsed the Missouri proposition because of its obvious economic fairness.

The Michigan counties of Charlevoix, Alpena, and Oscoda voted overwhelmingly—by 8–1 margins—against the idea that they should be used for a nuclear waste repository. The non-binding referenda in these counties, which are in the northern tip of lower Michigan, had been prompted by the Energy Research and Development Administration's (ERDA) activities. ERDA planned test drilling into Michigan salt deposits, and had informed the state government that Michigan was under consideration as a site for a federal nuclear waste facility.[26]

A unique factor in Missouri was a gubernatorial candidate who supported the initiative. Even though the initiative proponents were heavily outspent, candidate Teasdale's endorsement assured favorable coverage and credibility for the measure. In Michigan, since the initatives were on a countywide basis, canvassing could practically be done on a door-to-door basis. Major media coverage was not necessary—and even if it had been, the industry would have been hard pressed to endorse the concept of Michigan as the nation's nuclear dumping site.

In short, experience shows that some initiatives can win. But the fact that other state measures were beaten indicates that more groundwork will have to be done for future campaigns. The initiative stands a better chance of being approved if legislation has been passed curbing corporate campaign contributions. Voters will also have to be better informed in order to recognize the flaws in the industry's arguments. As in Missouri and Michigan, it may be prudent to focus the initiative on specific issues, such as economics or waste. Activists should also consider measures to encourage development of positive alternatives to nuclear—such as solar energy or conservation. Lastly, qualifying the initiative for the ballot is an important step, but it is only a first step. There must be a strategy for getting the measure approved, once the initiative has qualified to be on the ballot.

Other Avenues

There are still other ways in which the states can regulate nuclear power, including areas which the NRC already has delegated. The "Agreement State" program was established to relieve the NRC from the burden of regulating small quantities of radioisotopes (such as are used for medical or research purposes). The NRC gives a state the authority to regulate certain nuclear materials, if it is satisfied that a given state has an adequate radiation protection program.

Agreement states have regulatory powers in at least two significant areas apart from supervision of small quantities of materials (medical and research). For example, uranium mills in New Mexico, at which radioactive drinking water was found, were regulated by the State. Mills in Texas, Colorado, and Washington State also fall under the agreement state program.[27]

Agreement states also regulate "low-level" waste dumps, disposal sites for waste which contains radioactive material of low enough concentration to be defined as "suitable for burial" by the NRC. Private companies presently operate burial facilities at Maxey Flats, Kentucky; Beatty, Nevada; Barnwell, South Carolina; West Valley, New York; and Richland, Washington. Each is regulated by the applicable state under federal contract. Another facility in Sheffield, Illinois, is regulated directly by the NRC. These facilities were initially established to receive waste from hospitals, research facilities, and industries using small quantities of radioisotopes. But the sites have increasingly been used for low-level waste from the nuclear power industry.[28]

The U.S. Environmental Protection Agency (EPA) is conducting studies at the Maxey Flats, Kentucky, and West Valley, New York, sites and has reached some alarming conclusions. Roger Strelow, assistant administrator for air and waste management of EPA, has testified: "These locations were licensed by the respective States in the early 1960's based on analyses of site hydrology, geology, and meteorology. It was believed from these analyses that the buried radioactive wastes *would not migrate from these sites.* That is, they would be retained on the site for hundreds of years. In ten years or less, however, radioactivity has been detected offsite."[29] (Emphasis in original.)

Waste disposal at these sites is merely a euphemism for burial in trenches. During heavy rainfall water pours into the trenches and radioactive material flows outside the burial ground. One EPA study at Maxey Flats found plutonium contamination as far as one kilometer (three-fifths of a mile) outside the burial site.[30]

The Maxey Flats site is sparsely populated, so relatively few persons have been exposed to radioactive runoff. At Beatty, Nevada, however, radioactive material from that town's dump site turned up in individual dwellings. Over several years, workers had pilfered shovels, hammers, other tools, and lumber from the dump site for their own use or sale. Lax security at the waste dump, managed by the Nuclear Engineering Company, had permitted the wholesale theft of materials that were radioactively contaminated and should have been buried. In March 1976, state and federal investigators found more than twenty truckloads of contaminated items, including a cement mixer used to lay a patio in a local saloon.[31]

Although federal officials believe there were no serious exposures from the material at Beatty, that is a matter more of luck than of strong federal regulation. In fact, it was the Nevada state agency which first ordered the dump site to shut down for investigation.[32] The occurrences at Maxey Flats and Beatty make it clear that if the states are going to assume the responsibility for low-level waste dumps, they need to ensure that the dumps are closely monitored. States could also pressure the NRC and the nuclear industry to develop better techniques for containing wastes, or they could demand veto power over where the dumps will be sited. The State of Kentucky has implemented a use tax at Maxey Flats. A charge of 10 cents per pound of waste handled is assessed, the revenues being set aside for environmental monitoring or corrective action.

State and local goverments are increasingly concerned about transportation of nuclear materials by air, sea and land. For example, New York City refused to accept the government's assurances and drafted its own proposal—which is tied up in court proceedings with the NRC—to regulate nuclear shipments through the city. A compilation of regulations has shown that shipments of radioactive materials are prohibited from such roadways as the Delaware, Maine, and West Virginia turnpikes, as well as from specified bridges and tunnels in other states.[33]

State and local governments can intervene in NRC licensing hearings. Citizens can pressure local governments to pass resolutions expressing their opposition to nuclear plants in the area. Even if such resolutions have no authority, they are a clear signal to the local utility that an aroused citizenry and their elected officials are determined to oppose siting a power plant in their area. Since utilities value their public relations images, an anti-nuclear resolution may dissuade a company from building a nuclear plant. Nonbinding initiatives and referenda could have similar results.

In summary, authority over nuclear power is not precluded for the states. The Northern States Power case in Minnesota established federal pre-emption in some areas, but that pre-emption is certainly not unlimited. In a host of ways states can challenge the limits of pre-emption. There are some areas where states already have jurisdiction that could be applied vigorously to nuclear plants, and in others the NRC has placed regulatory authority with the states. States thus have the authority for some effective regulation, and federal law can always be amended to provide for more state authority. Vigorous citizens can provide local officials with the will to act and the support they need to sustain the struggle against the nuclear establishment.

Challenging Electric Utilities

Citizen activists have many opportunities to challenge the electric utilities that build and operate nuclear power plants. At the state level one specialized form of action is through the state agencies which regulate utilities. For example, challenges before the Public Utility Commissions (PUCs) which set rates can focus attention on the economic problems of nuclear power which damage consumers in their wallets and pocketbooks. Utility ratepayers and nuclear power activists can unite not only in contesting rate increases but also in the formation of Residential Utility Consumer Action Groups (RUCAGs). Citizens can also investigate promotional and other aspects of utility advertising. Stockholder and municipal power challenges can be mounted for consumer justice as well.

Utility Administrative Agencies

Utility executives often complain that they must obtain permits from a multitude of different agencies before they can begin construction of a power plant. A 1974 federal study on energy regulation concluded: "Virtually all energy projects require more state and local authorizations than federal permits."[1] A utility can often obtain most state and local permits fairly simply, but in most states, one or two permits are crucial. A utility generally needs certification that a particular power plant is necessary, and a determination that the plant can be built without unacceptable environmental effects. At least forty states require some type of authorization for power plant construction, and at least twenty-five have specific laws for siting power plants.[2] In most states, the utility commission is responsible for determining the need for a plant, and state siting boards or environmental agencies have jurisdiction to deal with environmental and land use issues.

Today, the most opportune time to halt a nuclear power plant may be before it is built, by requiring a utility to demonstrate that the plant is needed. At the end of 1977, the electric utility industry nationwide was drastically overbuilt, with a total generating capacity 32 percent greater than the industry's peak load for the year.[3] This excess capacity, along with a slowdown in electrical demand growth, has made it increasingly difficult for power companies to justify their ambitious construction programs. Public utility commissions need not accept a utility's contention that another plant must be built to satisfy the public's desire for electricity. Frequently, a careful analysis will show that the utility has inflated projected demand or ignored the potential for energy efficiency in order to justify its own desire to expand.

Such was the case in Wisconsin. On November 8, 1976, the state Public Service Commission (PSC) ordered Madison Gas and Electric to sell its 7 percent share of two reactors to be constructed by Wisconsin Electric Power Company on Lake Koshkonong, in the southeastern part of the state. The PSC rejected the Madison utility's projections and decided that with a limited service area and a fairly stable number of customers, Madison G&E would not need the generating capacity from its share of the Koshkonong plants. The PSC also ruled that the Madison utility could ill afford the large investment required. Madison G&E's share was $100 million, an amount not easily replaced by other Wisconsin Electric investors.

Two days later, the Wisconsin Department of Natural Resources (DNR), which must issue permits for power plant water use, released its evaluation of the Koshkonong reactors. In a letter to the Nuclear Regulatory Commission, DNR Secretary Anthony Earl called the Koshkonong project "environmentally unacceptable," and stated that rather than issue the required permits, the DNR would appear "at future hearings in opposition" to the proposed plant. Moreover, the DNR revealed that one week before Wisconsin Electric announced in June 1974 that it had chosen the Koshkonong site, the DNR had informed the utility that of five alternate sites, Koshkonong environmentally was the worst. Lake Koshkonong, a large but very shallow body, could not provide the large amounts of cooling water the reactors require. The DNR attacked the NRC's environmental impact statement, which depended heavily on utility data, as containing "frequent examples of inaccurate information."[4]

The PSC and DNR decisions had been preceded by over two years of opposition from a coalition of utility ratepayers, farmers, and environmentalists. On August 4, 1977, Wisconsin Electric informed the NRC that it would abandon Koshkonong as a nuclear plant site.[5] Although the utility will attempt to relocate its reactors to Sheboygan County, Wisconsin Electric will be forced to justify construction in more proceedings before state agencies. Whatever the outcome of those proceedings, Koshkonong residents for the time being have prevented their area from becoming a nuclear plant site. Groups interested in emulating the Wisconsin experience should first assess their own state agencies: if the agencies have a long history of decisions favorable to the utilities, actions before the legislature or other forums might bear more fruit. Additional information about the Koshkonong case may be obtained from Concerned Citizens of Wisconsin (CCOW), P.O. Box 1105, Madison 53703; or Wisconsin Coalition for Energy Alternatives, 114 North Carroll, Madison 53703.

In Iowa, a coalition of consumer organizations pursued a strategy to require that electric utilities demonstrate need for generating plants before the Iowa Commerce Commission, the state agency regulating utilities. The groups found that the Iowa Code specifically stated that commission approval was not required for the construction or operation of electrical generating facilities. According to this statute, a utility had unlimited discretion to build electric plants. In fact, the law even stated that the Commission's only obligation relative to plant construction was to the utility stockholders—for whom the commission must set rates to assure a certain return on investment—and not to the public. Thus, the commission's interpretation of this law effectively prevented citizens from confronting utilities on any basis other than ratemaking.

In response to this situation, the Iowa Student Public Interest Research Group (ISPIRG), a student-financed consumer group, assisted state representatives in drafting legislation to repeal this section of state law and to require a utility to obtain a "certificate of necessity" from the Commerce Commission. The legislation would have required utilities to demonstrate clearly to the commission that a new plant was needed, whether it be fossil-fueled or nuclear. The utility would have been required 1) to provide evidence that future electrical demand could not be lowered by conservation; 2) to show that the company has the financial capability to build the facility; and 3) to offer evidence that the new plant will not become an economic burden to the consumer. Citizens would have an opportunity to participate in a hearing on the certificate. The commission, on the basis of the record developed in the case, could reject, accept or modify the utility's application.[6]

The Commerce Commission, responding to the legislative impetus, reversed its previous rulings and decided in August 1975 that Iowa utilities must obtain approval from the commission before constructing any power plants expected to cost more than $250 million. Large nuclear plants, which typically cost at least $1 billion, would of course require the commission's approval. In explaining the commission's ruling, Chairman Maurice Van Nostrand noted that he was concerned about nuclear plant economics as they related to uranium fuel prices and the cost of nuclear waste storage.[7]

The commission's ruling was followed in 1976 by the legislature's enactment of the ISPIRG-sponsored bill into law, although in somewhat weakened form. The legislature also instructed the Commerce Commission to develop regulations outlining the technical and economic information required before construction of a nuclear plant would be allowed. The commission subsequently demanded information on the costs of reprocessing and waste storage, which no utility could supply. As a result, nuclear plant construction in Iowa has been halted.[8]

Rate Fights and RUCAG

The last five years have witnessed unprecedented increases in the cost of electricity. Utility rates have soared for a number of reasons: the 1973 oil embargo and the subsequent cartel price arrangement, the impact of inflation on wages and prices, the automatic fuel adjustment clause, and the insatiable desire of electric utilities to build more and more power plants. Nor is the end in sight. No sooner does one increase become effective than a utility will file with the regulatory commission a request for more of the consumers' hard-earned dollars.

In response to the ever-escalating costs of the basic necessities of life, consumer groups throughout the country have organized to challenge rate increases. They have had some success, such as partial rate reform in California and Michigan, but mostly consumers have been frustrated.

Their desire is simple: electricity at prices they can afford, and a governmental regulatory agency responsive to their needs, rather than the local electric utility's penchant for transferring the costs of waste, mismanagement, and energy monopolies to the final consumer instead of trying to contain them. The issue is *strictly* economic, especially for senior citizens on fixed incomes and low-income people. The strategy of consumers is oriented to rate reform, such as lifeline rates, where the residential utility consumer pays a substantially lower monthly charge for the first several hundred kilowatt-hours, than for a larger amount of consumption. However, rate reform is only part of the issue. Rates are increasing because utilities want money to build more unneeded nuclear plants.

Utility rate setting is essentially a two-step process. The first step is to determine the amount of money, or revenues, the utility will be allowed to earn. The second step is to determine which customers pay which part of the total revenues. Total revenues are based on a simple rule: if the utility is properly managed, it is permitted to earn sufficient revenues to cover its operating expenses plus a return on investment adequate to compensate existing investors and to attract new capital. For the anti-nuclear activist, the last step is most significant. For the low-income consumer, it is equally significant, suggesting a similarity of interests.

In almost all rate cases over the last five years, each electric utility has asked for a higher rate of return on investment so that it can attract new capital to pay for the new power plants under construction or in planning. Coincidentally, an increase in the rate of return means an increase in total revenues and therefore an increase in everyone's rates. It is, therefore, in the interests of consumer and anti-nuclear groups to combine their efforts in a rate case for the purpose of demonstrating the lack of need for new plants and the concomitant reduction in requested additional revenue.

An appropriate focus for the anti-nuclear movement is an increased emphasis on the economic realities facing so many people. Anti-nuclear groups can build coalitions with those consumers who dispute the projected growth of electrical consumption and the burdens such growth has placed on their incomes. It is time to join hands with blue-collar workers and explain to public utility commissions that continued construction of nuclear and other power plants will diminish the standard of living and, most important, employment opportunities.

A broad-based coalition is waiting to be formed. Utility economics can provide the connecting issue around which different organizations can rally. The arena for action is the public utility commission, where the utilities must ask for those higher revenues to attract the capital to construct more generating stations.

Interventions before utility commissions have already met partial success—for example, by limiting a utility's rate increase to less than the full hike requested. Even such partial victories are important. They limit increases in consumer bills, and they limit the cash available to utilities to continue unneccessary construction programs. Intervention before the NRC, in contrast, is an all-or-nothing affair, with the government and utility arrayed together consistently to defeat the intervenors.

The Connecticut Citizen Action Group (CCAG, Box G, Hartford 06106) is a citizen-supported consumer group that has had particular success in challenging utilities. In December 1976 the group's efforts contributed to a decision by the Connecticut utility commission that not only rejected a $56-million rate increase request by Northeast Utilities, but also ordered the company to *lower* rates by $21 million.[9] In 1977, CCAG testimony helped the state commission to order Northeast to dispose of its interests in a nuclear plant in Seabrook, New Hampshire, and scuttle its plans for two reactors in Montague, Massachusetts, because the units were unneeded. The commission also ordered Northeast to revise its rate structures to encourage energy conservation, which would reduce further the need for future power plant construction.[10]

Although CCAG scored these victories, consumers should recognize that utility rate fights can be long, costly, and frustrating affairs. If citizens are to be successful, they need a funded organization, accountable to the public, with a full-time staff of organizers, lawyers, economists, accountants, and engineers, and the stamina to challenge utilities again and again. The concept of a Residential Utility Consumer Action Group (RUCAG) would meet these criteria. Such an organization could be established by state law or utility commission regulation and would be financed by a "check-off." Included with each month's utility bills sent to residential users would be a payment envelope or room on the bill for listing a voluntary contribution, however small, to support a utility-users action group. This contribution would be over and above the charge for the utility bill. The utility would be obligated, under strict audit, to pass such monthly contributions on to the Residential Utility Consumer Action Group.

A RUCAG would be organized and run by democratic election procedures. Its membership—all those residential consumers contributing a minimal amount to the fund—would each have one vote in the periodic election of a council of directors. The council would be responsible for hiring the staff which would intervene in rate and other proceedings, handle consumer complaints, and insure that the public was kept constantly informed of its activities before the utility commission, the legislature and in the courts.

Membership would be limited to residential consumers, those least able to fund organized opposition to rate cases on an ongoing basis. There should be one exception. Residential customers either employed by or holding stock in a local utility should not become members of the council of directors. These people would have an inherent conflict of interest and might act against the interests of the vast number of residential consumers.

The first step in creating an organization such as RUCAG is to organize a coalition to urge passage of legislation. Groups working on nuclear power should contact local utility rate groups and labor unions, and explain to them the similarity in each organization's goals. In turn, these other groups can contact their friends, civic and religious leaders, and still more citizens to arrange an organizational meeting and plan action strategy. With a solid consumer base, the campaign to establish a utility action group can begin. Activists working to establish the RUCAG concept are already lobbying their state legislators. Bills have been introduced in several states, and the concept has passed one house in Maryland, New York, Wisconsin, and California, but utility industry opposition thus far has blocked any legislation. Citizens who want more information on electric utility issues should contact RUCAG or the Environmental Action Foundation. Both groups are listed in the Appendix.

Challenging Advertising

The ability of utilities to mount aggressive advertising campaigns, often at the consumer's expense, gives the utilities a significant advantage over citizen groups in their ability to influence public policy. To attack this imbalance, many organizations are challenging both the content and the ratepayer funding sources of excessive and misleading utility advertisements on atomic energy. Such challenges can be mounted before state regulatory commissions or federal agencies such as the Federal Communications Commission (FCC) or the Federal Energy Regulatory Commission (FERC).

The legal basis for FCC regulation of media advertising is that the airwaves belong to the public. The Communications Act of 1934 states that broadcasters must not restrict their programming to one point of view on "controversial issues of public importance." This concept, known as the "fairness doctrine," requires any broadcaster presenting one side of a controversial issue to afford reasonable opportunity for the presentation of contrasting viewpoints. Citizen groups have filed fairness doctrine complaints on many issues, including utility advertising.

Before filing a complaint with the FCC, a citizen organization must first make a reasonable request to the station for free time. This should be in the form of a letter to the broadcaster containing the following, (1) citation of specific advertisements; what they said and when they were run; (2) identification of the specific controversy addressed by the ad(s); (3) explanation of why the issue is controversial and of importance within the broadcaster's signal range (evidence such as newspaper clippings and references to coming hearings may be included); (4) indication of evidence that the station has failed to present the other side of the issue.[11]

This effort alone may prompt a broadcaster to offer free air time to citizen groups. For example, the Vermont Yankee Nuclear Power Corporation ran radio advertisements featuring "Atom-Man," a fictional character promoting atomic power. When the Vermont Public Interest Research Group (VPIRG) requested equal time in 1977, radio stations in the state responded by running VPIRG ads, free of charge, rebutting the claims of "Atom-Man."[12] Other stations, rather than offer free time, may choose to cancel the controversial advertising. If a fairness doctrine request to individual stations fails, then a formal complaint to the FCC is warranted.

Seven public interest groups in California filed just such a complaint with the FCC in September 1974. The groups claimed that pro-atomic energy radio advertisements sponsored by the Pacific Gas & Electric Company (PG&E), a northern California utility, were controversial and warranted "equal time."

In a decision released May 18, 1976, the Federal Communications Commission found that each of the PG&E announcements "directly addressed the issue of the desirability of nuclear power and addressed the interrelated issues of the safety and environmental cleanliness of nuclear power." The FCC also ordered nine California radio stations to submit information showing how they planned to meet their fairness doctrine obligations.[13]

Roger Hickey, the Washington, D.C. director of the Public Media Center, one of the seven petitioners, commented that the ruling "guarantees that those opposed to nuclear power because of safety and economic concerns will have the right to be heard." Hickey added that the FCC ruling requires radio stations to show that they provide opponents of atomic energy with a "reasonable opportunity to be heard by the same huge audience as is reached by utility advertising campaigns. A boring talk show on the Sunday night midnight shift won't do."[14]

Groups considering a fairness doctrine complaint should first consult attorneys familiar with FCC regulations. The Media Access Project, 1609 Connecticut Avenue, N.W., Washington, D.C. 20009, or the Public Media Center, 2751 Hyde Street, San Francisco, California 94109, can give some helpful advice although groups will have to retain their own attorney to handle the petition.

The practice of charging customers for utility advertising has also been challenged before a federal agency. In April 1975, the Chief Accountant of the Federal Power Commission determined that nuclear power advertising was "political" in nature. Such a ruling would prohibit utilities from charging ratepayers for the cost of nuclear ads. Rather, the cost would have to be borne by utility stockholders. The following month, over forty utilities requested the FPC Commissioners to overturn the Chief Accountant's ruling.

In June 1975, the Media Access Project, on behalf of eleven organizations and four members of Congress, petitioned to intervene before the FPC in opposition to the utilities' request. The petition also requested public hearings before the commission decided the matter.

The petition argued that, because of the numerous bills relating to nuclear power that have been filed in Congress and state legislatures, all advertising which discusses nuclear power is, in effect, an effort to influence public sentiment and thus, to influence legislative votes. The petitioners, who have publicly opposed construction of additional power plants, stated further that the power companies' massive pro-nuclear advertising forces great additional expense, which the petitioners may not be able to afford, to counter the advertising. This imbalance in spending capability, as it affects public communication on the nuclear controversy, would result in grave First Amendment injury to the petitioners. The FPC has since become the Federal Energy Regulatory Commission (FERC), and as of June 1978 had not acted on either the utilities' appeal or the petition to intervene.

FERC's authority is generally limited to interstate distribution of electricity, but its rulings often establish precedents for the commissions regulating intrastate power. In addition to the forthcoming FERC ruling on the Media Access Project petition, other groups have challenged atomic energy advertising through state bodies.

For example, the Connecticut Citizen Action Group has relieved consumers in its state from subsidizing utility advertisements. Through the group's efforts, the Connecticut legislature in 1976 enacted a law prohibiting utilities from including most advertising in the rate base. (Ads giving energy conservation tips, or for normal business activities such as employment solicitation, were exempted from the act.) The utilities may still run ads puffing up their images or promoting atomic power, but those must be paid for by stockholders, not ratepayers. Utility commissions in California, Massachusetts, and New Jersey have also prohibited utilities from charging consumers for pro-nuclear advertising.[15] As a result of citizen complaints, the New York Public Service Commission completely outlawed utility bill inserts promoting the company's position on atomic energy or other controversial issues. The New York commission considered such advertising "tantamount to taking advantage of a captive audience."[16]

The utility funds involved in advertising are ordinarily a small part of the company's expenditures—a few hundred thousand dollars, compared to the billion dollars or more for a single new power plant. But the practice of charging customers for a utility's hype adds insult to injury, and certainly should be challenged on principle. This is also a relatively inexpensive effort for a citizen group to mount. The Media Access Project (MAP) petition before the FPC provides a useful example for organizations wishing to challenge advertising before their state agencies. A copy of the petition can be obtained from MAP by sending $2.00 and asking for *Petition in re Alabama Power Company, before the Federal Power Commission, Docket No. E–9478.*

Stockholder and Consumer Challenges

Utility policies have recently come under attack from none other than the companies' own stockholders. While a few large banks and other financial institutions usually control utilities, individuals, colleges, and churches hold a substantial amount of company stock. Some stockholders are concerned about electric rates or about electric company performance on social issues such as nuclear power.

By purchasing a few shares of a utility's stock, an individual or organization can attend the annual stockholders' meeting and may be able to gain access to information not normally available to the public. For example, a stockholder investigating the Public Service Company of Colorado turned up expense vouchers for golf balls, country club memberships, and other questionable items. He embarrassed company officials by asking them to explain the vouchers at a stockholders' meeting.[17]

Virtually all issues at annual company meetings are decided in favor of management. Therefore, persons who organize stockholder challenges should not expect to win a majority vote. The purpose of a stockholder action is to bring certain social issues to public view. At times, though not frequently, the public exposure may induce a company to change its practices. In 1978, for example, the University of Minnesota agreed to withdraw stockholder resolutions against ten companies, including General Electric, when those corporations agreed to follow certain guidelines for improved labor practices by U.S. businesses with operations in South Africa.[18]

Regulations of the Securities and Exchange Commission (SEC) make it possible for stockholders to have certain issues placed on a proxy ballot. Corporations such as General Motors and Gulf Oil have been publicly embarrassed by stockholder challenges of their policies regarding minorities, pollution, and other social issues. Utilities and other nuclear corporations are also vulnerable to such challenges.

In 1978, at least three companies faced stockholder resolutions on atomic energy. Thirteen individuals asked the Iowa Power and Light Company to develop alternate energy sources to atomic power, particularly energy conservation. Ecology Alert of Bloomsburg, Pennsylvania asked Pennsylvania Power and Light Company to cancel construction of its Susquehanna nuclear reactors. And three church groups introduced a resolution asking Westinghouse Electric to report on its nuclear operations and policies.[19] These church groups were affiliated with the Interfaith Center on Corporate Responsibility, which has sponsored many stockholder challenges. Other organizations can benefit from this previous experience by contacting the Center at 475 Riverside Drive, New York, New York 10027.

One of the most publicized utility stockholder challenges was a 1973 effort in Minnesota to elect a public representative to the board of directors of the Northern States Power Company (NSP). The Coalition to Advocate Public Utility Responsibility (CAPUR), composed of eleven environmental, consumer, and religious organizations, began by asking NSP to place a public representative on its board. When the company refused, CAPUR forced NSP to put the name of the coalition's candidate, former state legislator Alpha Smaby, on the proxy ballot.

Smaby's chances looked good because NSP's stockholders could cast "cumulative" votes. Cumulative voting, which some corporations use, allows a stockholder to cast all votes, multiplied by the number of board members, for one candidate, rather than voting separately for each position on the board. Smaby, campaigning on consumer and environmental issues, seemed assured of receiving the 7 percent vote she needed to win one of the fourteen board seats.

But Northern States Power reorganized its voting procedures, reducing the impact of cumulative voting. CAPUR challenged the company before the SEC and federal court, and eventually reached an out-of-court settlement restoring the earlier voting ground rules. Following the settlement, CAPUR mailed its own proxy cards to all stockholders, soliciting support for Smaby, and the returns gave her an apparent victory. However, a few days before the election deadline NSP sent a confusing proxy card to each stockholder. Most of

Smaby's supporters signed and returned the card (which did not mention her name), and unwittingly revoked their original votes.

So NSP retained its original fourteen directors. CAPUR, which had spend $15,000 on its campaign, was no match for the utility, which spent an estimated $500,000 on extra proxy mailings and telephone calls to every stockholder. However, CAPUR's effort was hardly a failure; nationwide publicity about the campaign damaged NSP's image, and the company added two of its own "public" representatives to the board a few months later. CAPUR spokespersons say it was worth every penny they spent.[20]

For public power utilities, a variation of the stockholder challenge with a much higher chance of success is the voter challenge. Public utilities—owned by municipal or regional governments—are usually controlled by a popularly elected board of directors or by the elected city or regional council. Activists can work to place consumer-oriented candidates in the decision-making agency.

The Jacksonville Electric Authority provides a case in point. Through the efforts of Joe Cury and others, the City Council deferred approval of JEA's purchase of two floating nuclear plants. The Eugene, Oregon, Water and Electric Board, a public authority, abided by a May 1970 referendum in which consumers voted against spending money to construct a nuclear power plant.

In Nebraska, the boards of directors of the Omaha Public Power District (OPPD) and the Nebraska Public Power District (NPPD) are elected by consumers in their respective service areas. Some Nebraskans have for several years been working to elect consumer-oriented persons to the boards. Although activists have not been successful with the NPPD, the OPPD gained a 4–3 majority of progressive directors. In the 1976 elections, a coalition of the Nebraska Public Power Project (a citizens' group not to be confused with NPPD), the Gray Panthers, the Quality Environmental Council, and community organizing groups were successful in replacing the OPPD chairman with a consumer-minded candidate, to establish the majority.

In February 1977, the OPPD Board voted 4–1 to abandon its 50-percent interest in the Fort Calhoun 2 nuclear plant. (The older members, in an unaccustomed minority position, did not bother to attend board meetings, and the one dissenting member resigned in a huff.) One board member cast her vote against the plant for environmental reasons, but the other three identified as their main concern the plant's questionable economics. The NPPD retains its half-interest in Fort Calhoun 2 but will be hard-pressed to find partners to pick up the share OPPD abandoned.[21] For further information, contact Nebraska Public Power Project, P.O. Box 4524, University Place Station, Lincoln, Nebraska 68504.

Public power is not a panacea. Public utilities, such as the Tennessee Valley Authority, which President Carter has criticized as "just another utility,"[22] can be unresponsive to many public concerns. But a utility directed by elected officials can, with vigilant citizen oversight, both provide more sensible electric rates and respond to environmental concerns. Persons who want to learn more about public power, or about converting private utilities to public entities, should contact the Environmental Action Foundation for *Taking Charge*, their report on public power (see the Appendix).

As the tide of public opposition to atomic power mounts, the industry increasingly will attempt to neutralize opponents with misleading arguments that atomic energy is necessary for jobs and economic stability. The proper response to such arguments is to meet them head-on, pointing out why they are mythical and why consumers must prevent government and industry from pouring more good money after bad. Economic pressure on the industry may in the end halt atomic power, if a major accident does not. All voters should fight future state and federal taxpayer subsidies to keep atomic power going, by reminding their representatives of atomic power's economic drawbacks. All ratepayers should fight the continuous utility rate increases necessary to fuel nuclear power expansion, by reminding public power authorities and public service commissions that nuclear power will contribute to higher electrical bills. Citizens need not allow the atom industry to play them off against each other as "environmentalists," "consumers," and "workers." They can all unite to fight nuclear power when they understand that it is a pocketbook issue.

Afterword

THE NUCLEAR INDUSTRY is crumbling technically and economically. Pressures from a deteriorating position on the inside are feeding the pressures of citizen action from the outside. But, it is ultimately in the political arena where the verdict will be rendered to discontinue the nuclear fuel cycle. This is so whether the decision is a matter of political foresight or a reaction to a catastrophic atomic power disaster somewhere in the country.

The market will affect but not determine the final judgment for two reasons. First, utilities have superseded market disciplines with their monopoly status and their government subsidies and bailouts—a process which may yet jar the ideology of economic conservatives. Second, the concentrated energy conglomerates, by continuing to push up the price of alternative fuels and to avoid the development of cheaper energies, protect atomic power from a truly competitive price comparison.

Nuclear power is in many ways this country's "technological Vietnam." The reactor vendors, utilities, engineering construction firms, energy companies, and others have invested many billions of dollars in nuclear technology. So have government officials invested billions of tax dollars subsidizing the development of this energy form over the years. Many scientists, engineers and administrators have invested their careers in highly specialized nuclear operations. Except for a growing number of courageous dissenters, most of these individuals will not renounce a technology they have spent years promoting, *even* if that technology requires massive taxpayer subsidies and quasi-military controls to keep it going. While the news of the industry's problems and hazards gets worse, the façade is optimistic and cover-ups breed faster than they are exposed. At what point will the industry and its government patrons relinquish their intransigence and cut their losses? Will it take a continuation of leaks and spills and losses of radioactive material around the country? Or will it take the horrific devastation of an atomic meltdown or plutonium terrorism? They will not say, for they dare not contemplate what they refuse to envision.

But citizens must envision. For they will be the victims should this technological fury be unleashed beyond its uncertain containment. Since consistent public scrutiny began in the seventies, the number of new and revealed problems of atomic power has grown while the solutions have lagged. The serious problems are rooted in fact, while the important solutions are steeped in faith. This state of affairs surely warrants pause. Yet the pro-nukes rush on as if only a costly radioactive energy can be the salvation of a society hooked on an ever-wasteful orgy of Btus. However, the jargon has been breached and the truth is beginning to tell. The energy crisis is one of gluttony, greed, and gigantism. It is not one of a dearth of safer and renewable alternatives to power our economy.

Under the psychological and economic jolt of the 1973 embargo, energy conservation and many forms of solar energy were rediscovered and found either promptly applicable or practical for development in the relatively short term. The diffusion of this knowledge to millions of Americans is generating a countervailing civic movement against the hegemony of the energy monopoloids. The emerging realization among people that more voice means more choice is stirring the grass roots. Atomic power has become an issue in local, state, and national elections, and, at last, a recurrent story for the mass media. Both sides have had political gains and defeats, but the momentum on a diversity of nuclear issues (such as waste disposal) and superior energy alternatives is with the opponents of fission. Public awareness of nuclear power's unknown risks and known faults is at a higher level, although it is far from peaking.

The stopping of a major technology through public opposition is both difficult and new. In the past, new technologies displaced existing ones, as the railroad displaced the stagecoach. But, then, what technology has had the potential for both inadvertent and willful mass destruction to present and future generations, to present and future natural resources, that the nuclear technology has? What technology has the potential for wiping out cities and contaminating states after an accident, a natural calamity, or a successful sabotage?

What technology has been so unnecessary, so avoidable by simple thrift or by the deployment of renewable energy supplies? What technology has such an attraction and a vulnerability to unstable forces both here and abroad? What technology has the Moloch-like capacity to sacrifice the capital, the tranquility, and the liberties of a society?

The purpose of this book is to invite citizens of all ages and stations to contemplate and move toward a democratic shaping of energy policy—one that embraces renewable energy forms that tolerate human error and do not jeopardize the earth's descendants. This is an eminently realizable journey. Its destination is fraught with attributes of peace and well-being for the world's peoples. The journey is one of study and effort made pleasurable by the importance of its success. The pages of this book, covering the unacceptability of atomic power, the alternatives that are available, and some strategies for a consumer-sensitive energy policy, are meant to encourage individuals and groups to develop their own policies, tailored to their own objectives in their community or region. As the descriptions in this volume of valiant citizen efforts reveal, the exercise of our liberties makes for imaginative diversities in approaches toward common goals.

Many persons working in groups or at their own individual pace can together constitute a movement that government and industry cannot ignore. The atomic energy industry may wither away, or it may suddenly collapse due to a catastrophe or to a reversal of governmental policy. How much more preferable would it be to resolve the awesome risk by the foresight of democracy than by the hindsight of disaster. Though some of the past insults from nuclear power are irreversible, the successful practice of citizenship on this industry should provide the experience and wisdom to foresee and forestall other mindless technological juggernauts.

And what of those in the nuclear establishment who have to be saved from themselves? Solar energy offers so many opportunities for redemption.

Update 1978

SINCE MARCH 1977, when the main text was completed, many new developments in the nuclear arena have confirmed and validated this book's major points. Nuclear power's serious unresolved problems have remained unresolved, or have been recognized as more serious. The timetables may have changed, but the issues have not.

Price-Anderson

As the book was going to press last year, a federal court struck a blow against the atomic industry's limited liability protection. The Carolina Environmental Study Group, a citizens' organization represented by the Public Citizen Litigation Group, challenged the Price-Anderson Act as unconstitutional, and the court agreed. Judge James B. McMillan, who heard the case, found that the act violated the Fifth Amendment by allowing the destruction of life and property "without reasonable certainty that the victims will be justly compensated."[1] Duke Power Company, the utility directly affected by the decision, appealed to the U.S. Supreme Court, noting that the district court's ruling "may place a cloud over relations" between atomic corporations and the financial community.[2] Although the Burger Supreme Court saved the industry by overturning the district court decision in June 1978, Duke's warning provides convincing support for those who believe that without the limited liability of Price-Anderson, the atomic industry might never have been established.

Plutonium

In April 1977, President Carter expressed his desire to "defer indefinitely" plutonium reprocessing and to terminate the Clinch River Breeder Reactor project, for reasons of health, safety, and economics, and to reduce the spread of nuclear weapons.[3] With President Carter's announcement, the idea that "the plutonium economy" was not necessary, economical, or sensible—which had been a minority position five years before—became presidential policy.

But the final resolution of plutonium issues remains uncertain. In December 1977, the Nuclear Regulatory Commission (NRC) terminated its proceedings on plutonium fuel, but the atomic industry has taken the agency to court to keep the proceedings alive. Breeder forces were successful in inducing Congress to authorize and appropriate funds for the Clinch River project, against the President's wishes. Although President Carter vetoed the authorization bill, Congress attached to the appropriations bill riders—such as funding for environmental protection and authority to terminate the B–1 bomber—which prevented a veto.

Carl Walske, president of the Atomic Industrial Forum, the industry's trade association, has advanced his belief that without plutonium, atomic power at best is viable "only for the near term."[4] Because its future is intertwined with plutonium fuel, the industry will continue its insistent lobbying of federal agencies and members of Congress to push a plutonium future. Although the plutonium economy may no longer present an immediate threat, the discussion of its dangers is very much relevant, both now and for the future.

Nor is the thorium-uranium breeder, which has gained acceptance in some academic circles, a solution. That breeder, as any other, requires reprocessing of fissionable material—a procedure which can be expected to be costly and dangerous, and which provides the proliferation connection. There are no proliferation-proof reprocessing methods: The country that controls the reprocessing plant can manipulate the process to produce weapons material.

Reactor Safety

While President Carter has been right on plutonium, he and his administration remain infatuated with light water reactors and the uranium fuel cycle, which are not nearly safe enough. Late in April

1977, the Union of Concerned Scientists (UCS) released its analysis of a large number of government documents obtained under the Freedom of Information Act, showing the serious institutional weaknesses of the NRC's Reactor Safety Study, which underpins the safety propaganda of government and industry. Although the safety study has been widely and severely critiqued for technical flaws, the released documents revealed that even these critiques were too kind to the study, because they assumed that it was independent, scientific, and unbiased. In fact it was none of these.

The documents UCS released showed that:

• The Reactor Safety Study was conceived in letters dated March 1972 from Manson Benedict and Norman C. Rasmussen, nuclear engineering professors at the Massachusetts Institute of Technology (MIT), to AEC Chairman James Schlesinger and his special technical adviser, Stephen Hanauer. Benedict had served on the Board of Directors of the Atomic Industrial Forum, the atomic industry's trade association. In their letter to Hanauer, the MIT professors noted that, "The report to be useful must have reasonable acceptance by people in the industry."[5]

• Though he represented himself as a nuclear safety expert, Professor Rasmussen, who directed the safety study, had previously specialized in a field unrelated to reactor safety. Nonetheless, he has consulted for six nuclear industry organizations.

• As the safety study was conducted, atomic industry companies such as Westinghouse, General Electric, and the Bechtel Corporation were not only asked to provide raw information, but also were given responsibility for performing reactor safety analyses. Industry representatives—from Stone & Webster Engineering Corporation, General Electric, Virginia Electric and Power Company, Bechtel, Westinghouse, and Philadelphia Electric Company—participated with Atomic Energy Commission (AEC) officials in an internal review of the Reactor Safety Study's Preliminary Draft. No members of the public, including the growing number of technical critics of atomic power, were invited to this internal review.

• An early draft of the study contained an appendix discussing problems with industry and AEC inspection programs, which were supposed to detect equipment and construction defects before they became a problem. AEC officials were concerned that discussing these numerous real-world defects would undercut government assurances about reactor safety. One AEC internal memo noted that a "disadvantage" of publishing the appendix was that "the facts may not support our pre-determined conclusions."[6] The offending appendix never made it into print.

In summary, the internal documents revealed that the Reactor Safety Study was conceived with the idea that it would be acceptable to the industry, and was conducted by the industry, for the industry, and with the atomic industry's approval. The study provides no foundation for claims that reactors are safe.

Department of Energy

To carry out his energy policies, the President appointed James R. Schlesinger as Secretary of the Department of Energy, which was formed in October 1977 through a consolidation of the Energy Research and Development Administration (ERDA), the Federal Energy Administration, the Federal Power Commission, and parts of the Department of the Interior. Schlesinger, former Atomic Energy Commission Chairman, has managed to surround himself with pro-nuclear advisers. His Deputy Secretary is John O'Leary, who served under Schlesinger as the AEC's Director of Licensing. The Department's Undersecretary, third in command, is Dale Myers, who as vice-president of Rockwell International presided over projects to build the B–1 bomber and develop breeder reactor technology.

Schlesinger has also done little to reduce the pro-nuclear bias of the officials he inherited from ERDA, which in turn came from the AEC. Quite the contrary, Schlesinger instigated the appointment of Robert D. Thorne as his Assistant Secretary of Energy Technology, an office which retains most of ERDA's programs. Thorne is a former ERDA Assistant Administrator for Nuclear Energy, a breeder supporter, and as Manager of ERDA's San Francisco office, he carried out the agency's campaign against the California Initiative in 1976.

When Richard Nixon dedicated a breeder research facility at Hanford, Washington, he admitted that "all this business about breeder reactors and nuclear energy is over my head," and told reporters, "don't ask me what a breeder reactor is, ask Dr. Schlesinger," who accompanied him.[7] Secretary Schlesinger's actions and past association with the breeder lead many to suspect that not only is he wedded to atomic energy, but he is no supporter of the President's plutonium policy. Schlesinger was the apparent architect of a "compromise" with the House Committee on Science and Technology, the most pro-nuclear congressional body since the Joint Committee on Atomic Energy, which was formally abolished in 1977. Schlesinger offered to begin a design study for a larger (600–900 MWe) breeder, if the Science and Technology Committee would terminate funding for the 380 MWe Clinch River reactor. But the committee responded in April 1978 by authorizing funds for *both* the design study and Clinch River, making Schlesinger look foolish in the process.[8]

Schlesinger's influence extends outside the Department of Energy. He succeeded in convincing Carter to appoint as Chairman of the Nuclear Regulatory Commission Joseph M. Hendrie, who served on the AEC Staff under Schlesinger and O'Leary. A second Schlesinger-endorsed NRC appointee, Kent Hansen, told the Senate committee considering his nomination that he had consulted extensively for the atomic industry, and saw no reason not to return to industry consulting the day he left the NRC. By rejecting Hansen's nomination in October 1977, the Senate served notice that it will not look kindly on the idea that the NRC should provide on-the-job training for the atomic industry.

In March 1978, Schlesinger submitted to Congress nuclear licensing legislation which is supposed to speed the construction of atomic power plants, and which will govern the procedures by which NRC operates. At about the same time, Schlesinger also drafted for President Carter's signature an executive order which would give the Secretary of Energy authority to set schedules for the NRC and other federal agencies to follow in licensing nuclear plants.[9] Although the draft executive order has not been enacted, these intrusions by Secretary Schlesinger subvert the will of Congress, which split the AEC to keep regulation and promotion of atomic energy in separate agencies.

Occupational Hazards

New evidence has also indicated that the effects of low-level radiation, particularly on workers, have been underestimated. Thomas F. Mancuso, scientist at the University of Pittsburgh School of Public Health, in September 1977 released his final report on the mortality of workers in federal atomic programs at Hanford,

Washington. Mancuso found a correlation between radiation exposure and cancer as a cause of death. Moreover, he found this correlation at an average exposure far less than the 5-rem-per-year standard considered acceptable for workers by the federal government. Mancuso concluded that radiation caused about a 6 percent increase in cancer in the worker population, over what would have been expected without the exposure.

The reaction of the atomic establishment to Mancuso's findings recalled memories of the Gofman-Tamplin controversy. In 1974 when Samuel Milham, an epidemiologist working for the state of Washington, found a higher incidence of certain cancers among Hanford workers, the AEC, which was funding Mancuso's work, asked the Pittsburgh scientist to publish his findings. At the time, Mancuso saw no correlation between radiation exposure and cancer, but he knew that his analysis was incomplete. The AEC, and then ERDA, increased their pressure, and in 1975 decided to terminate Mancuso's grant, on the grounds that he was 62 years old and his retirement was "imminent." Unfortunately for ERDA, its strategy became transparent because the agency did not realize that Pittsburgh's mandatory retirement age was 70.[10]

When Mancuso published his final report after twelve years of research, he noted the cancer-radiation correlation, and added that ERDA's decision to terminate his grant was "not in the best interest of science."[11] ERDA's contracting officer for Mancuso's project, Sidney Marks, had since left the agency to work for Battelle Northwest Laboratories, near Hanford. Shortly thereafter, the work Mancuso was doing was transferred to Battelle Northwest.[12] Federal occupational radiation standards, meanwhile, remain the same.

Radioactive Waste

The intervening months have also seen no resolution of the nuclear waste problem. In October 1977, the Department of Energy (DOE) offered to take title to spent reactor fuel for storage, which represented the logical extension of the President's no-reprocessing fuel cycle. Under the DOE proposal, spent fuel rods would be stored deep underground in geological formations, and DOE would charge utilities a fee to cover the costs of waste management.[13] The fee, however, would be assessed on a fixed, one-time basis, which could end up harming taxpayers. There would be no provision for reassessing utilities if DOE, in an attempt to subsidize the atomic industry, underestimated the costs of storing the waste.

Specific details of the DOE policy have not been formulated, and will probably require a lengthy rulemaking proceeding before they are put into place. As of this writing, DOE has no regulations, no facility, and no disposition for the spent fuel beyond temporary storage. Nor has the Department given much confidence that it can avoid the history of mismanagement, false steps, and mistakes of the AEC and ERDA in dealing with nuclear waste. Attempts by atomic promoters to brush away concerns over geological uncertainties were also shaken in early 1978 by a candid U.S. Geological Survey circular, which concluded that:

> This Circular has dealt largely with the difficulties and uncertainties connected with the geological disposal of high-level radioactive waste because, from our viewpoint, these are significant potential stumbling blocks that need critical attention.[14]

> A concluding statement concerns the uncertainties involved in geologic prediction for long timespans, discussed earlier. These uncertainties need to be faced candidly in public discussions of radioactive-waste disposal. Earth scientists can indicate which sites have been relatively stable in the geologic past, but they cannot guarantee future stability. Construction of a repository and emplacement of waste will initiate complex processes that cannot, at present, be predicted with certainty.[15]

Congress is also becoming less tolerant of the atomic establishment's historical inability to deal with its waste. In April 1978, the U.S. House Committee on Government Operations released a report noting that:

> Radioactive waste is a significant and growing problem . . . yet there is still no demonstrated technology for permanently and safely disposing of this waste.

> Neither the Federal Government nor the nuclear industry has prepared reliable cost estimates for the ultimate disposal and perpetual care of radioactive wastes and spent nuclear fuel.[16]

The committee report ended with the recommendation that "Congress and the executive branch should consider requiring that further licenses for nuclear powerplant construction be conditioned upon the timely and satisfactory resolution of radioactive waste and spent nuclear fuel permanent disposal and storage problems,"[17] which is the closest a U.S. congressional committee has come to endorsing a moratorium on civilian nuclear plant construction.

Economics

The time since March 1977 has also seen heightened recognition of atomic energy's economic woes. The electric utility industry remained seriously overbuilt—by the end of 1977, generating capacity was 32 percent greater than the year's highest electrical demand.[18] As a result, the atomic industry's doldrums continued. There were only four new nuclear plant orders in 1977, and none in the first half of 1978, and reactor vendors were warning that their manufacturing capabilities might "disintegrate" without rapidly expanding demand.[19] The notion that uranium fuel would be cheap was dealt another blow in August 1977 when Jerry McAfee, chairman of Gulf Oil, admitted to a U.S. House of Representatives subcommittee that his company willingly participated in the international uranium cartel's efforts to drive up prices.[20] In June 1978, Gulf pleaded "no contest" to Justice Department charges that the company's cartel involvement violated antitrust laws, and paid a fine of $40,000[21]—a trivial sum, considering the millions of dollars gouged out of consumers from spiraling uranium prices.

The Ford Foundation assembled a Nuclear Energy Policy Study Group, composed chiefly of scientists and academicians, which in March 1977 concluded that atomic energy was neither crucial nor even very important to the nation's economic well-being, contrary to the industry's propaganda. Using scenarios that consciously tried to err on the side of overestimating energy demand, the Study Group concluded that the economic effects of a 15-year moratorium on nuclear plant construction would be "essentially insignificant in this century," reducing the Gross National Product only by fractions of a percent.[22] Of course, if the money not invested in atomic plants were instead invested in energy efficiency, more energy could be saved than the nuclear plants would supply, more jobs would be created, and the economic results would benefit more of the population.

Early in 1978, important representatives of minority groups voiced their understanding that atomic energy is a poor buy. On January 20, Vernon E. Jordan, president of the National Urban League, noted in a speech that:

Reactors are the most expensive way to meet energy needs. Their construction and development impose price restraints that must be considered. Expanded nuclear energy carries with it capital costs that encourage inflation.[23]

Three days later, the U.S. Congressional Black Caucus affirmed that:

Building new electric generating facilities is one of the most ineffective ways to create jobs. . . . Similarly, nuclear energy is low labor intensive and produces less energy per dollar invested than other sources. The development of alternative energy sources such as energy conservation and a solar energy industry will provide more permanent jobs and will be more environmentally sound than nuclear energy.[24]

Solar Energy

Energy efficiency is already a much more cost-effective option than atomic power, and solar power's appeal is gaining greater recognition. The California state government has set a goal of 500,000 solar homes in the state by 1985 and has legislated a large tax credit to help meet that goal. The program for California, with one-tenth the nation's population, is more energetic than President Carter's energy plan, which set a goal of 2.5 million homes for the entire nation by 1985.

Enthusiasm for photovoltaic cells has also increased, and even old-line nuclear supporters at the Department of Energy are optimistic about solar-electric options. In July 1977, a study conducted for the Federal Energy Administration and Department of Defense (DOD) found that photovoltaics were already cost-effective for some remote applications at DOD bases. Moreover, the report found that a DOD photovoltaic purchase of less than $500 million over five years would save the Department $1.5 billion in energy costs over twenty years. At the same time, the market provided by DOD could induce mass production, which would bring the costs of photovoltaics down to a point where they were competitive with conventional electric generating stations.[25] The development of cheap photovoltaic cells with storage would be truly revolutionary, for it would give consumers direct access to solar-powered electricity, allowing them to divorce from central generating stations and transmission lines.

In September 1977, a study from the University of California found that by the year 2020, with a doubling of its population and a quadrupling of its economy, California could supply all its energy needs from indigenous, renewable sources, chiefly solar power in all its forms.[26] This report was released to the press not by the U.S. Department of Energy, which funded the study but left it on the shelf to gather dust, but by the Critical Mass Energy Project. On a national level, the President's Council on Environmental Quality in April 1978 estimated that a strong commitment to solar development could allow the U.S. to be generating one-quarter of its energy from solar power by the year 2000.[27] But in spite of the unquestioned potential of solar energy, the inherently democratic ability of the sun to bypass energy corporations causes those whose investments solar energy would devalue to construct barriers to its development. Thus there is a need for active citizens to inject themselves into solar policy.

Denis Hayes, who organized Earth Day in 1970, now sits on the board of Solar Lobby (1028 Connecticut Avenue, N.W., Washington, D.C. 20036), a freshly established group which organized Sun Day for May 3, 1978. As Earth Day increased environmental awareness among citizens and politicians, so Sun Day's organizers hope to establish a national network of citizen activists dedicated to removing the obstacles to solar development.

International

Lastly, there is greater recognition that the nuclear issue is an international issue. This is so because of the proliferation problem: if President Carter's plutonium policies are to succeed, then other nations must also turn back from the plutonium economy. By the same token, if other nations follow the U.S. example, the resolve of the Carter Administration to stick by its "indefinite deferral" of plutonium fuel will be increased. Atomic power is an international issue because the technology's dangers know no geographical barriers. Although reactor types may change from nation to nation, the dangers of the nuclear fuel cycle—catastrophic accidents, occupational hazards, the need to manage atomic waste, the hazards of sabotage and terrorism, and the dependence of atomic power's future on more dangerous breeders—do not.

Atomic power is an international issue because the industry cannot exist anywhere without subsidies from taxpayers. Although tax loopholes already pay for 20 percent of every U.S. nuclear power plant, congressional conferees considering the National Energy Plan are also considering an *additional* 15 percent credit for nuclear plants. England's Central Electricity Generating Board (CEGB) has 20 percent more capacity than it needs, and did not plan to order new plants until 1980 or later. In order to prop up the nation's heavy power industry, including Babcock & Wilcox, a foreign division of the U.S. reactor vendor, the CEGB may order $1.5 billion in new generating plants, *before they are needed.*[28]

When Australia opened its uranium reserves for exploitation, the Australian Atomic Energy Commission agreed to provide 72 percent of the capital requirements for uranium mining. Industry will meet 28 percent of the capital requirements, but will share the profits 50–50.[29] France is also trying to save its faltering atomic industry. The French Atomic Energy Commission bought the nuclear technology subsidiary of Saint-Gobain,[30] and the government has established a cabinet-level committee to consider financing nuclear power plant exports.[31] Atomic power is an international issue because the reactor manufacturers, unable to sell their products to glutted or declining markets at home, will increasingly have to find their markets in developing nations.

Citizen Action

But most importantly, atomic power is an international issue because citizens and scientists in democratic nations world-wide are opposing this ultrahazardous technology in increasing numbers. In this country, the state of California has seen dramatic developments in the aftermath of the California Initiative. Although the defeat of that and other state measures in 1976 was widely viewed as a setback, the initiatives helped lay the groundwork for a heightened institutional skepticism of atomic energy. As an alternative to the initiative, the California legislature passed three bills in 1976, two of which would prohibit construction of new nuclear power plants unless the California Energy Resources Conservation and Development Commission found that technology for reprocessing and high level waste disposal existed.

The first nuclear facilities to be affected by these bills were San Diego Gas and Electric Company's two Sundesert reactors, in Riverside County. The energy commission in fact held in January 1978 that reprocessing and waste disposal technology did not exist, and that practical alternatives could eliminate the need for Sundesert. Attempts by San Diego G&E to exempt Sundesert legislatively were beaten in the California Assembly, and the energy commission's ruling stood.

But the blow that stopped Sundesert was economic. The California Public Utilities Commission in May 1978 adopted an order prohibiting San Diego G&E from financing the reactors, whose costs had soared to $2.3 billion, through rate increases. On May 4, 1978, the utility's board of directors announced they were abandoning plans for Sundesert, as a result of the utility commission ruling.[32]

The residents of Kern County, California had earlier dealt the atomic industry a blow on March 7, 1978 by objecting to a proposed nuclear plant by a 2–1 margin. Although the Kern County referendum carried no legal force, the Los Angeles Department of Water and Power had pledged beforehand that it would not build the plant without voter approval. A coalition of ranchers, farmers, local politicians, and public interest organizations were successful in stopping the reactors.

These developments also provided support for California governor Jerry Brown, who is campaigning for reelection on an avowed pro-solar/anti-nuclear platform. If he emerges from his campaign victorious, and if he decides to challenge Jimmy Carter for the Democratic Party's presidential nomination in 1980, Governor Brown will have established a clear-cut difference between his energy platform and President Carter's.

In Australia, citizens are marching against the government's decision to exploit uranium, and unions have refused to mine and handle uranium. Canada's citizens, stung by the revelation that India used a Canadian reactor to produce its nuclear weapon, have more recently been shocked by allegations of bribery in CANDU reactor sales abroad. The nuclear debate was joined in Britain by the Flowers report, directed by a member of the United Kingdom Atomic Energy Authority, which highlighted the dangers of plutonium, reprocessing, and the breeder reactor. Protests in Japan kept a nuclear ship at sea and have led the government to seriously reduce its projections for a nuclear-powered future.

In Sweden, Prime Minister Olof Palme attributed his party's defeat to voter dissatisfaction with atomic energy. In Switzerland, new nuclear plant licenses have been halted pending the outcome of a national referendum. West German court decisions on nuclear waste have stymied atomic development in that country. And France, Germany, Italy, and other nations are experiencing citizen protests not seen since the Vietnam War.

But merely protesting is not enough. Citizens must use the tools of democracy to challenge atomic energy promoters: before agencies; in the courts; before bodies that set utility rates, where applicable; in referenda, for those states and countries that sponsor them. But perhaps most importantly, citizens must challenge atomic energy through their elected representatives. The elected official is most sensitive to citizen concern, because widely perceived insensitivity can lose an election. Since turnover and change of opinion in elected representatives is more frequent than for appointed officials in an entrenched bureaucracy, voters must depend on their representatives to ride herd on bureaucrats.

The obstacles to an enlightened energy policy are great. Citizens must overcome giant corporations with multibillion dollar investments, as well as governments determined to prop up an expiring technology. But the stakes are also great. There are two choices for future energy systems. One is a future increasingly dependent on energy supply options that are expensive, in need of massive subsidies, complicated, dangerous, destructive to workers and the environment, and which, by requiring large amounts of capital investment, lend themselves to control by large corporate entities. The second future would turn to energy options that are simpler, safer, economical, job-creating, decentralized, local, socially and environmentally beneficial, and inherently more democratic and susceptible to citizen control. Because different interest groups will benefit from the two futures, they are probably mutually exclusive. The world's citizens and future generations, who will benefit from the second path, cannot afford to lose this struggle.

RALPH NADER
JOHN ABBOTTS

June 10, 1978
Washington, D.C.

Appendix
Notes
Index

Appendix

PERSONS INTERESTED *in obtaining information on nuclear energy issues or on strategy should contact the following sources (prices may change).*

CITIZEN GROUPS

Critical Mass Energy Project, P.O. Box 1538, Washington, D.C. 20013. Publishes a monthly newspaper on the citizen movement to stop nuclear power. Annual subscription costs $7.50 for individuals, $37.50 for business-professional-institution. Write for a sample copy.

Center for Renewable Resources/Solar Lobby, 1028 Connecticut Avenue, N.W., Washington, D.C. 20036. This group, which organized Sun Day in 1978, plans to maintain a continuing presence, educating the public and encouraging government officials to develop solar energy. Write for information.

Citizens' Energy Project, 1413 K Street, N.W., Washington, D.C. 20005. Publishes *People & Energy*, a monthly newsletter on citizen activity on different energy issues. Also distributes several publications. Write for a list and prices.

Environmental Action Reprint Service (EARS), 2239 East Colfax Avenue, Denver, Colorado 80206. A large selection of information on nuclear, solar, and other energy sources. Write for a recent catalog.

Environmental Action Foundation, Dupont Circle Building, Washington, D.C., 20036. Distributes *Countdown to a Nuclear Moratorium* ($3.00), *How to Challenge Your Local Electric Utility* ($3.50), *Utility Scoreboard* ($3.50), *Taking Charge* ($3.50), and *The Power Line*, a monthly newsletter on electric utility issues and activism.

Friends of the Earth, 124 Spear Street, San Francisco, California 94105. Distributes *Not Man Apart*, a newsletter including a section on nuclear energy. $25 for membership, $10 for a subscription. Also write for a list of books.

Media Access Project, 1609 Connecticut Avenue, N.W., Washington, D.C. 20009; or Public Media Center, 2751 Hyde Street, San Francisco, California 94109. Distribute information on how to challenge advertising, including utility advertising.

Natural Resources Defense Council, 917 15th Street, N.W., Washington, D.C. 20036. Distributes information on the breeder reactor, plutonium, and nuclear proliferation. Write for a list.

Natural Resources Defense Council, 2345 Yale Street, Palo Alto, California 94306. Publishes a citizen's handbook on nuclear waste. Write for a list of other publications.

Public Interest Research Group, P.O. Box 19312, Washington, D.C. 20036. Distributes free general information on energy policy, nuclear and solar energy, specific reports and petitions before the Nuclear Regulatory Commission at a cost. Write for a list. Send a self-addressed stamped enevelope.

Residential Utility Consumer Action Group (RUCAG), P.O. Box 19312, Washington, D.C. 20036. Distrubutes a practical proposal on how state governments can facilitate the organization of residential utility consumers into an effective and well-staffed consumer protection group. Send a self-addressed stamped envelope.

Sierra Club, 530 Bush Street, San Francisco, California 94108. General environmental organization. Publishes a monthly magazine and weekly news report which includes information on congressional activity. Write for price and membership information.

Task Force Against Nuclear Pollution, Inc., P.O. Box 1817, Washington, D.C. 20013. General information. Conducting a nationwide petition drive for a nuclear moratorium.

Union of Concerned Scientists, 1208 Massachusetts Avenue, Cambridge, Massachusetts 02138. Distributes general and technical information. Write for a list.

U.S. GOVERNMENT SOURCES

Office of Senator Gravel (D., Alaska), U.S. Senate, Washington, D.C. 20510. Has an extensive list of reprints on nuclear power problems and alternative energy sources.

Office of Senator Ribicoff (D., Conn.), U.S. Senate, Washington, D.C. 20510. Reprints on security, safeguards, and international nuclear proliferation issues.

Office of Senator Abourezk (D., S.D.), U.S. Senate, Washington, D.C. 20510. Reprints on solar energy, solar legislation, energy issues in general.

U.S. Nuclear Regulatory Commission, Office of Public Affairs, Washington, D.C. 20555. Distributes a weekly compilation of press releases on its activities.

U.S. Department of Energy, Office of Public Affairs, Washington, D.C. 20545. Distributes a weekly compilation of press releases on its activities.

Notes

1. How It Started and How It Works

I. THE NUCLEAR COMMITMENT

1. Jacob Bronowski, *The Ascent of Man* (Boston: Little, Brown, 1973), pp. 368–370.
2. See, for example, Charles F. Zimmerman, "Competition in the Nuclear Industry" (Ph.D. dissertation, Department of Agricultural Economics, Cornell University, Ithaca, N.Y., 1975), p. 86.
3. Quoted by Wilson Clark, *Energy for Survival* (Garden City, N.Y.: Anchor Press, 1975), p. 277.
4. Joint Committee on Atomic Energy, U.S. Congress, *Government Indemnity*, hearings, May and June 1956, pp. 248–250.
5. Peter Milius, "Major Battle Is Brewing on A-Plants," *Washington Post*, June 23, 1975, A4.
6. Materials Policy Commission, *The Promise of Technology*, Resources for Freedom, Vol. 4 (Washington, D.C., 1952), p. 220.
7. Joint Committee on Atomic Energy, U.S. Congress, *Possible Modification or Extension of the Price-Anderson Insurance and Indemnity Act*, hearings, January and March 1974, Phase I, p. 220.

II. NUCLEAR POWER OVERVIEW

1. Environmental Action Foundation, *How to Challenge Your Local Electric Utility* (Washington, D.C., 1974), p. 12.
2. Environmental Action Foundation, *The Power Line*, May 1976, p. 2.
3. "Why Atomic Power Dims Today," *Business Week*, November 17, 1975, p. 98.
4. Adapted from Ralph Nader, "America's Unsung Heroines," *Ladies Home Journal*, July 1976, pp. 22 ff.
5. U.S. Atomic Energy Commission, WASH-1250, *The Safety of Nuclear Power Reactors and Related Facilities* (Washington, D.C., July 1973), p. 1–6.
6. George Masche and Don L. Testa, *Systems Summary of a Westinghouse Pressurized Water Reactor Nuclear Power Plant* (Westinghouse Electric Corporation, PWR Systems Division, 1971), p. 35.
7. Union of Concerned Scientists, *The Nuclear Fuel Cycle* (Cambridge, Mass.: MIT Press, 1975), p. 71, 84.
8. See, for example, U.S. Nuclear Regulatory Commission, WASH-1400, *Reactor Safety Study* (Washington, D.C., October 1975), Appendix V, pp. V-74, V-98.
9. See, for example, "The Dwindling Orders at General Atomic," *Business Week*, October 6, 1975; and "General Atomic, Delmarva P & L Cancel Contracts," *Wall Street Journal*, October 29, 1975, p. 15.

10. Peter T. Faulkner, letter to Public Interest Research Group (Washington, D.C.), June 10, 1975.
11. David Burnham, "Power Reactors Face Safety Test," *New York Times,* September 22, 1974.
12. David Burnham, "Safety an Issue at Indian Point," *New York Times,* January 21, 1976, p. 58.
13. David Burnham, "Three Engineers Quit G.E. Reactor Division and Volunteer in Antinuclear Movement," *New York Times,* February 3, 1976, p. 12.

2. The Case Against Nuclear Power

III. OVERVIEW OF THE ISSUES

1. Union of Concerned Scientists, press release (Cambridge, Mass., August 6, 1975).
2. Committee of Inquiry: The Plutonium Economy, "The Plutonium Economy: A Statement of Concern," background report, submitted to the National Council of Churches of Christ in the U.S.A., New York, September 1975.
3. Keith Miller (University of California at Berkeley), letter and attached memorandum to Stan Fabic (Nuclear Regulatory Commission), May 7, 1976; and statement on "CBS Evening News with Walter Cronkite," May 12, 1976.
4. Carl H. Builder (director, Division of Safeguards), memorandum to Ronald A. Brightsen (assistant director for licensing, Division of Safeguards, U.S. Nuclear Regulatory Commission), January 19, 1976.
5. "Summary of GAO Report Evaluating ERDA's Safeguards Systems," *Congressional Record,* 122: H7838–H7840, July 27, 1976.
6. David E. Lilienthal statement before Senate Committee on Government Operations, U.S. Congress, January 19, 1976.
7. U.S. Atomic Energy Commission, WASH-1327, Draft, *Generic Environmental Statement Mixed Oxide Fuel* (Washington, D.C., August 1974), Vol. 4, p. V-7 and passim.
8. Russell W. Ayres, "Policing Plutonium: The Civil Liberties Fallout," *Harvard Civil Rights–Civil Liberties Law Review,* Vol. 10, No. 2 (Spring 1975), pp. 369–442.
9. David G. Snow, *The Uranium Stocks: Nuclear Industry Kaleidoscope Coming Together* (New York: Mitchell, Hutchins, January 30, 1976).
10. "Why Atomic Power Dims Today," *Business Week,* November 17, 1975, p. 98.
11. Denis Hayes, *Energy: The Case for Conservation,* Worldwatch Paper 4 (Washington, D.C.: Worldwatch Institute, January 1976).
12. American Institute of Architects, *Energy and the Built Environment: A Gap in Current Strategies,* and *A Nation of Energy Efficient Buildings by 1990* (Washington, D.C., 1975).
13. See, for example, Ron Lanoue, *Nuclear Plants: The More They Build, the More You Pay* (Washington, D.C.: Center for Study of Responsive Law, 1976), Chap. I.
14. John Holdren testimony, Ph.D. (Energy and Resources Group, University of California at Berkeley), Before Subcommittee to Review the National Breeder Program, Joint Committee on Atomic Energy, U.S. Congress, June 10, 1975.
15. Farno L. Green, *Energy Potential from Agricultural Field Residues* (Detroit: General Motors Corporation, June 1975).

IV. RADIATION EFFECTS

1. The story of Gofman and Tamplin is adapted from *Whistle Blowing,* ed. Ralph Nader, Peter J. Petkas, and Kate Blackwell (New York: Bantam Books, 1972), Chap. 6.
2. John W. Gofman and Arthur R. Tamplin, *Poisoned Power* (Emmaus, Pa.: Rodale Press, 1971), p. 26.
3. Ibid., p. 296.
4. Ibid., p. 97.

5. Ibid., p. 259.
6. Ibid., pp. 307–308.
7. National Academy of Sciences, National Research Council, *The Effects on Populations of Exposure to Low Levels of Ionizing Radiation* (Washington, D.C., November 1972), p. 2.
8. U.S. Nuclear Regulatory Commission, *Federal Register*, 40: 19439, May 5, 1975.
9. Office of Radiation Programs, U.S. Environmental Protection Agency, *Environmental Radiation Protection Requirements for Normal Operation of Activities in the Uranium Fuel Cycle* (Washington, D.C., May 1975), p. 2.
10. Ibid., p. 7.
11. Ibid., p. 143.
12. Ibid., p. 82.
13. K. Z. Morgan, "Suggested Reduction of Permissible Exposure to Plutonium and Other Transuranium Elements," *American Industrial Hygiene Association Journal*, August 1975.
14. Ibid.
15. W. J. Bair and R. C. Thompson, "Plutonium: Biomedical Research," *Science*, 183: 715–722 (February 22, 1974).
16. John W. Gofman, "The Cancer Hazard from Inhaled Plutonium," reprinted in *Congressional Record*, 121: S 14610, July 31, 1975.
17. Roger Rapoport, *The Great American Bomb Machine* (New York: Dutton, 1971), pp. 41–49.
18. Edward A. Martell, "Radioactivity Of Tobacco Trichomes and Insoluble Cigarette Smoke Particles," *Nature*, 249: 215–217, May 17, 1974; and "Radioactive Smoke Particles: New Link to Lung Cancer?" *NCAR Quarterly* (Boulder, Colo.: National Center for Atmospheric Research, May/August 1974).
19. Edward A. Martell, letter to Public Interest Research Group (Washington, D.C.), May 22, 1975.
20. John W. Gofman, "Estimated Production of Human Lung Cancers by Plutonium from Worldwide Fallout," *Congressional Record*, 121: S 14616, July 31, 1975.
21. Ibid., S 14619.
22. Irwin J. Bross, "Cumulative Genetic Degradation," statement presented at Critical Mass '74, Washington, D.C., November 1974; also *Preventive Medicine*, 3: 361–367 (September 1974).
23. Ibid.
24. Lauriston S. Taylor, "History of the International Commission on Radiological Protection (ICRP)," *Health Physics*, 1958, Vol. 1, pp. 103–104.

V. THE FRONT-END OF THE FUEL CYCLE

1. NUREG-0002, U.S. Nuclear Regulatory Commission, *Final Generic Environmental Statement on the Use of Recycle Plutonium in Mixed Oxide Fuel in Light Water Cooled Reactors* (Washington, D.C., August 1976), Vol. 3, pp. IV.F-7–IV.F-12.
2. H. Peter Metzger, *The Atomic Establishment* (New York: Simon and Schuster, 1972), p. 162.
3. Union of Concerned Scientists, *The Nuclear Fuel Cycle* (Cambridge, Mass., October 1973), p. 97.
4. Ibid.
5. Ibid., p. 98.
6. *The Atomic Establishment*, p. 165.
7. Natural Resources Defense Council (Palo Alto, California), letter to Nuclear Regulatory Commission *re* Generic Environmental Impact Statement on Uranium Milling, March 28, 1975, p. 9.
8. Joint Committee on Atomic Energy, U.S. Congress, *Use of Uranium Mill Tailings For Construction Purposes*, hearings, October 1971, p. 106.
9. Ibid., p. 15.
10. Natural Resources Defense Council, letter, p. 7.

11. Union of Concerned Scientists, *Nuclear Fuel Cycle*, pp. 94, 102.
12. U.S. General Accounting Office, *Controlling the Radiation Hazard from Uranium Mill Tailings*, May 21, 1975, p. 10.
13. Environmental Protection Agency, EPA-520/1-76-001, *Potential Radiological Impact of Airborne Releases and Direct Gamma Radiation to Individuals Living Near Inactive Uranium Mill Tailings Piles*, January 1976.
14. Robert O. Pohl, "Nuclear Energy: Health Effects of Thorium-230" (Physics Department, Cornell University, Ithaca, New York, July 1975), pp. 2–4.
15. Natural Resources Defense Council, letter, p. 1.
16. Ibid., pp. 1–2.
17. Ibid., pp. 3–4.
18. U.S. Nuclear Regulatory Commission, "Uranium Milling" *Federal Register* 41: 22430, June 3, 1976.
19. U.S. Atomic Energy Commission, WASH-1248, *Environmental Survey of the Uranium Fuel Cycle* (Washington, D.C., April 1974), pp. C-2, C-4.
20. Ibid., p. D-7.
21. According to Joint Committee on Atomic Energy, U.S. Congress, hearings, *Future Structure of the Uranium Enrichment Industry*, July and August 1973, pp. 78, 79: U.S. enrichment plants in 1973 supplied 3.5 million separative work units for U.S. reactors and 2.1 million separative work units for foreign reactors.

    According to U.S. Energy Research and Development Administration, ERDA-1543, Draft Environmental Statement, *Expansion of U.S. Uranium Enrichment Capacity* (Washington, D.C., June 1975), p. 2.1-15, ERDA expects to supply 35 percent of the total foreign demand for enriched uranium until the year 2000.

    A separative work unit (SWU) is a measure of the work required to enrich a given amount of uranium to a given percentage of enrichment. Separative work units are generally expressed in units of mass to give the same dimensions as material quantities. For example, about 110 metric ton-SWU is required to produce enough enriched uranium—47 metric tons—for the annual requirements of a 1,000 Megawatt-electric plant, which is the size of a typical modern nuclear plant. A metric ton is 1,000 kilograms (kg), equal to 2,200 pounds. See WASH-1248, table S-1, p. S-5.
22. WASH-1248, table S-1, p. S-5, states that 250 metric tons of UF_6 is fed into the enrichment plant and 47 metric tons of enriched UF_6 is produced.
23. Ibid., p. D-7.
24. Environmental Action Foundation, *How to Challenge Your Local Electric Utility* (Washington, D.C., 1974), pp. 82–83.
25. This is based on the following calculations: WASH-1248, pp. D-1–D-4, states that the equivalent of a 45-MWe coal plant is necessary to supply the electricity for the 116,000 kg-SWU that annually are used for the enrichment of fuel for a 1000 MWe reactor. ERDA's three enrichment plants can annually supply 4.7 million, 5.2 million, and 7.3 million kg-SWU, respectively (ERDA-1543, p. 2.1-11).

    This means that to supply the 5.2 million SWU plant, for example,

    $$45 \text{ MWe} \times \frac{5.2 \text{ million}}{.116 \text{ million}} = 2000 \text{ MWe}$$

    are needed, which is the equivalent of two large, modern, coal-fired plants.
26. WASH-1248, p. D-10.
27. Ibid., p. D-4.

    Robert O. Pohl, in "Nuclear Energy," p. 4, states that a 1000 MWe coal plant annually burns 3 million tons of coal. The equivalent for 45 MWe is then:

    $$\frac{45}{1000} \times 3{,}000{,}000 = 135{,}000 \text{ tons/year}$$
28. WASH-1248, p. D-11.

29. See note 25, above. $\frac{17.2 \text{ million}}{.116 \text{ million}} \times 1000 \text{ MWe} = 148{,}000 \text{ MWe}$
30. "The President's Plan for a Competitive Nuclear Fuel Industry," White House Fact Sheet (Washington, D.C., June 26, 1975), p. 3.
31. U.S. General Accounting Office, *Evaluation of the Administration's Proposal for Government Assistance to Private Uranium Enrichment Groups* (Washington, D.C.: October 31, 1975), p. 32.
32. "The President's Plan," p. 8.
33. "Evaluation of the Administration's Proposal," p. iv.
34. Ibid., p. v.
35. *Congressional Record* 122: S 17065, September 29, 1976.

VI. REACTOR SAFETY: THE TECHNICAL CONTROVERSY

1. David Dinsmore Comey, *The Incident at Browns Ferry* (Washington, D.C.: Friends of the Earth, 1975), p. 5. See also Union of Concerned Scientists, *Browns Ferry: The Regulatory Failure* (Cambridge, Mass., June 10, 1976).
2. "Inoperative A-Power Plant Said to Cost TVA $18 Million," *Washington Post*, August 19, 1976.
3. Nuclear Energy Property and Liability Association, *TVA's Browns Ferry Nuclear Plant* (Farmington, Conn., May 23, 1975); *The Incident at Browns Ferry*.
4. "The Incident at Browns Ferry," p. 8.
5. U.S. Atomic Energy Commission, *Theoretical Possibilities and Consequences of Major Accidents in Large Nuclear Power Plants*, WASH-740 (Washington, D.C., March 1957). The conclusions of the report on possible accident consequences are summarized on p. viii.
6. Dr. John Gofman (professor of medical physics, University of California at Berkeley), "The Fission-Product Equivalence Between Nuclear Reactors and Nuclear Weapons." Reprinted in the *Congressional Record* by Senator Mike Gravel (D., Alaska), July 8, 1971.
7. Idaho Nuclear Corporation, Nuclear Safety Program Division, Monthly Report, November 1970, p. 18.
8. Ibid., December 1970, p. 8.
9. Ibid., January 1971, p. iii.
10. Ibid., February 1971, p. iii.
11. "ACRS Questioning PWR Emergency Core Cooling; Plant Delays Possible," *Nucleonics Week*, May 6, 1971, p. 1.
12. Union of Concerned Scientists, *Nuclear Reactor Safety: An Evaluation of New Evidence* (Cambridge, Mass., July 1971).
13. Union of Concerned Scientists, *A Critique of the New A.E.C. Design Criteria for Reactor Safety Systems* (Cambridge, Mass., October 1971).
14. Steven Ebbin and Raphael Kasper, *Citizen Groups and the Nuclear Power Controversy* (Cambridge, Mass.: MIT Press, 1974), p. 125.
15. Daniel F. Ford and Henry W. Kendall, "Nuclear Safety," *Environment*, September 1972, p. 2.
16. *Citizen Groups*, p. 126.
17. Ibid., p. 134.
18. Ibid., pp. 135–136.
19. Those interested in a more detailed evaluation of the ECCS hearings may refer to *Citizen Groups*, pp. 122–138; and Daniel F. Ford and Henry W. Kendall, *An Assessment of the Emergency Core Cooling Systems Rule-Making Hearing* (San Francisco: Friends of the Earth, 1974).
20. The statements cited in notes 21–24 are from Daniel F. Ford and Henry W. Kendall, *The Nuclear Power Issue: An Overview* (Cambridge, Mass.: Union of Concerned Scientists, 1974). Notes 21–24 give the sources as cited in that publication.

21. M. Rosen and R. Colmar, AEC internal memorandum to ECCS Task Force, June 1, 1971.
22. George Brockett, et al., *Loss of Coolant/Emergency Core Cooling Augmented Program Plan*, August 1971.
23. J. Curtis Haire, testimony before Atomic Energy Commission, April 6, 1972, Docket No. RM-50-1, ECCS Hearings, transcript, pp. 7951–7953.
24. Milton Shaw AEC internal memorandum to R. E. Hollingsworth, AEC general manager, February 19, 1971.
25. MITRE Corporation, M76-53, *Public Participation in Energy Related Decision Making: Six Case Studies* (McLean, Va., September 1976), p. 52.
26. The statements cited in notes 27–33 are from "The Nuclear Power Issue: An Overview." Notes 27–33 give the sources as cited in that publication.
27. Andrew J. Pressesky, assistant director for nuclear safety, Division of Reactor Development and Technology, AEC internal memorandum to Milton Shaw regarding the German moratorium, May 8, 1972.
28. *Federation of American Scientists Newsletter*, February 1973.
29. R. D. Docker et al., R-1116-NSF/CSA, *California's Electric Quandary* (Los Angeles: RAND Corporation, September 1972).
30. H. G. Mangelsdorf (chairman, Advisory Committee on Reactor Safeguards), letter to Dixy Lee Ray (chairman, AEC), September 10, 1972.
31. Assembly Science and Technology Advisory Council, *Nuclear Power Safety in California*, report to California Legislature, May 1973.
32. Dr. Bjorn Kjellstron (AB Atomenergi, Miljocentrumkonferens), statement, Uppsala, Sweden, June 14, 1973.
33. Twenty-third Pugwash Conference on Science and World Affairs, Report of Working Group 5, September 4, 1973, p. 11.
34. Opinion of the Commission, AEC Docket RM-50-1, December 28, 1973.
35. U.S. General Accounting Office, *This Country's Most Expensive Light-Water Reactor Safety Test Faciltiy* (Washington, D.C., May 26, 1976), p. 18.
36. Third Water Reactor Safety Research Information Meeting, Gaithersburg, Maryland, sponsored by Nuclear Regulatory Commission, September 30, 1975, LOFT.
37. "Most Expensive Light-Water Reactor," pp. 45–46.
38. Exhibit 1013, ECCS Rule-Making Hearings, AEC Docket RM-50-1.
39. Union of Concerned Scientists, press release, August 6, 1975.
40. Keith Miller (University of California at Berkeley), letter and attached memorandum to Stan Fabic (Nuclear Regulatory Commission), May 7, 1976; and statement on "CBS Evening News with Walter Cronkite," May 12, 1976.

VII. REACTOR SAFETY: HIDDEN DOCUMENTS AND ACCIDENT STUDIES

1. Robert D. Pollard, letter to William A. Anders (chairman, U.S. Nuclear Regulatory Commission), second enclosure, "Report On the Nuclear Regulatory Commission Reactor Safety Review Process," February 6, 1976.
2. Ibid., second enclosure, p. 9.
3. Ibid., p. 11.
4. Ibid., p. 16.
5. Ibid., p. 17.
6. Ibid., pp. 18–19.
7. Robert D. Pollard, letter to William A. Anders, January 23, 1976.
8. Robert D. Pollard, letter, February 6, 1976, second enclosure, pp. 3–4.
9. Ibid., p. 4.
10. Robert D. Pollard, letter, February 6, 1976.
11. U.S. Atomic Energy Commission, WASH-740, *Theoretical Possibilities and Consequences of Major Accidents in Large Nuclear Power Plants*, (Washington, D.C., March 1957, p. viii).

12. Chad J. Raseman (Brookhaven National Laboratory), letter to Dr. Clifford K. Beck (deputy director of regulation, U.S. Atomic Energy Commission), Document 103, WASH-740 update, January 22, 1965, p. 27. The WASH-740 update files are available at the Nuclear Regulatory Commission public document room, Washington, D.C. The 45,000-death figure was confirmed by AEC Press Release R-252, June 25, 1973, which announced the release of the update papers.
13. D. N. Sunderman (Battelle Memorial Institute), letter to S. Z. Szawlewicz (U.S. Atomic Energy Commission), Document 136, WASH-740 update, April 13, 1965, p. 3.
14. S. A. Szawlewicz, memo to U. M. Staebler on Steering Committee Meeting on the Revision of WASH-740, held on December 16, 1974, Document 87, WASH-740 update, December 21, 1974, pp. 4–5.
15. C. K. Beck, memo to Commission on Brookhaven Report, Document 129, WASH-740 update, March 17, 1965, p. 1.
16. Ibid., p. 3.
17. Glenn T. Seaborg (chairman, AEC), letter to Chet Holifield (chairman, Joint Committee on Atomic Energy), Document 154, WASH-740 update, June 18, 1965.
18. Adapted from Sheldon Novick, *The Careless Atom* (Boston: Houghton Mifflin, 1969), Chap. III.
19. The following information relevant to Mr. Pesonen's requests are filed with the WASH-740 update but have not been assigned document numbers: David Pesonen, letter to John Palfrey, (commissioner, AEC) August 14, 1965, and Commissioner Palfrey's response; David Pesonen, letter to Commissioner Palfrey, September 13, 1965, and the commissioner's response, October 8, 1965; David Pesonen, "Atomic Insurance, the Ticklish Statistics," *The Nation*, October 18, 1965, pp. 242–245.
20. Duncan Clark (AEC), memo to the commission, Document 159, WASH-740 update, October 26, 1965.
21. Forrest Western, note to C. L. Henderson (AEC), point paper on WASH-740, cover sheet, Document 188, WASH-740 update, August 26, 1969.
22. Ibid., background paper, p. 18.
23. Ralph Lapp, *The Nuclear Controversy* (Greenwich, Conn.: Reddy Kilowatt, April 1975), p. 48.
24. AEC Commissioner C. E. Larson, letter to Senator Mike Gravel, Document 199, WASH-740 update, December 4, 1970, p. 20.
25. U.S. Atomic Energy Commission, WASH-1400, *Reactor Safety Study Draft* (Washington, D.C., August 1974).
26. Sierra Club–Union of Concerned Scientists, *Preliminary Review of the AEC Reactor Safety Study* (San Francisco and Cambridge, Mass., November 1974).
27. U.S. Environmental Protection Agency, *Comments by the Environmental Protection Agency on Reactor Safety Study* (Washington, D.C., November 1974).
28. U.S. Atomic Energy Commission Regulatory Staff, *Review of the Reactor Safety Study (WASH-1400) Draft of August 1974* (Washington, D.C., December 2, 1974).
29. "Report to the APS by the Study Group on Light-Water Reactor Safety," *Reviews of Modern Physics*, Vol. 47, Supplement No. 1 (Summer 1975).
30. Ibid., p. S5.
31. Ibid., pp. S51, S102.
32. Frank von Hippel, "Reactor Safety," talk delivered at the Spring Meeting of the American Physical Society, Washington, D.C., April 30, 1975, p. 1.
33. U.S. Nuclear Regulatory Commission, WASH-1400 (NUREG-75/014), *Reactor Safety Study* (Washington, D.C., October 1975).
34. Ibid., p. 107.
35. Phillip M. Boffey, "Rasmussen Issues Revised Odds on a Nuclear Catastrophe," *Science* 190: 640 (November 14, 1975).
36. WASH-1400, p. 255.
37. Ibid., p. 263.

38. Ibid., p. 265.
39. Ibid., p. 266.
40. Ibid., p. 266.
41. U.S. Environmental Protection Agency, EPA-520/3-76-009, *Reactor Safety Study (WASH-1400): A Review of the Final Report* (Washington, D.C., June 1976).
42. Frank von Hippel, "Looking Back on the Rasmussen Report," *Bulletin of Atomic Scientists*, February 1977, p. 47.
43. *Congressional Record* 121: S 22339 (December 16, 1975).

VIII. THE ENVIRONMENTAL EFFECTS OF REACTOR OPERATION

1. Christian Williams, "A Nuclear Family, a Sea of Trouble," *Washington Post*, September 15, 1974, p. F10. Other sources for information on the Oyster Creek shipworm problem include: Richard Phalon, "Boatmen Lay Shipworm Growth to Nuclear Plant," *New York Times*, July 22, 1974, p. 33; "Utility to Buy Three Marinas Damaged by Shipworms," *New York Times*, February 12, 1975, p. 81; "Shipworms Now Threaten Barnegat Bay," *New York Times*, May 17, 1975, p. 61.
2. "Court Ready to Act on Atom Plant Fine," *New York Times*, October 26, 1975, p. 77; "Court Vindicates a Nuclear Plant," *New York Times*, January 23, 1976, p. 65.
3. Robert Boyle, "A Stink of Dead Stripers," *Sports Illustrated*, April 26, 1965, p. 81.
4. Staff of *Environment* magazine, "A New River," January –February 1970, p. 37.
5. The Governor's Task force on Nuclear Power Plants, *Nuclear Power Plants in Maryland* (Annapolis, December 1969), pp. 23, 102.
6. Federal Power Commission, *Problems in Disposal of Waste Heat from Steam-Electric Plants* (Washington, D.C., 1969), p. 4.
7. "A New River," pp. 38–40.
8. U.S. Nuclear Regulatory Commission, NUREG-0001, *Nuclear Energy Center Cite Survey—1975* (Washington, D.C., January 1976), Part I, pp. 4-4–4-11.
9. David Freeman, *Energy: A New Era* (Cambridge, Mass.: Ballinger, 1975), p. 64.
10. Quoted by Mark Swann in "The Energy Park Battle," *Environmental Action Bulletin*, April 5, 1975, p. 2.
11. General Electric Company, *Assessment of Energy Parks vs. Dispersed Electric Power Generation Facilities* (Washington, D.C., May 30, 1975), p. ES-5.
12. See, for example, Steven Ebbin and Raphael Kasper, *Citizen Groups and the Nuclear Power Controversy* (Cambridge, Mass.: MIT Press, 1974), p. 21.
13. U.S. Atomic Energy Commission, WASH-1288, *Evaluation of Nuclear Energy Centers* (Washington, D.C. January 1974), Vol. 2, p. 7.33.
14. *Assessment of Energy Parks*, p. 1/2–74.
15. MITRE Corporation, Technical Report MTR-6811, *Volume I: Minutes and Summary of the First Meeting of the Advisory Committee on Energy Facility Siting*, November 20, 1974 (McLean, Va., January 1975), p. 34.
16. "Energy Parks Offer No Simple Solutions," editorial, *Electrical World*, November 1, 1975, p. 55.
17. *Assessment of Energy Parks*, p. ES-5, concluded that the cooling pond could cover 39,000 acres. 1 acre = .00156 square miles, so 39,000 acres = 60.84 square miles.
18. NUREG–0001, Part I, Table 4.2, p. 4–12.
19. "Rush for Riches on the Great Pipeline," *Time*, June 2, 1975, p. 20; Michael Rogers, "The Dark Side of the Earth," *Rolling Stone*, May 22, 1975, pp. 52 ff.
20. See, for example, Mike Goodman and William Endicott (*Los Angeles Times* staff), "Thieves Greasing Their Palms Threaten Alaska Oil," *Trenton Times*, November 23, 1975, pp. B1, B16, B17. (This article also appeared in other newspapers.)
21. Helena Huntington Smith, "The Wringing of the West," *Washington Post*, February 16, 1975, pp. B1, and B4.

22. *Assessment of Energy Parks*, p. ES-23.
23. Ibid.
24. *Assessment of Energy Parks*, p. ES-5.
25. Ibid., p. ES-11.
26. Barry Commoner, *The Poverty of Power* (New York: Knopf, 1976), p. 214.
27. Lee Schipper, *Energy Conservation: Its Nature, Hidden Benefits and Hidden Barriers* (Berkeley, Calif.: Lawrence Berkeley Laboratory, University of California at Berkeley, June 1, 1975), Table 4b, p. 31.
28. "Energy Parks," p. 55.
29. Steven Harwood, Kenneth May, Marvin Resnikoff, Barbara Schlenger, and Pam Tames, *Decommissioning: Nuclear Reactors* (Buffalo, N.Y.: New York Public Interest Research Group, January 21, 1976), p. 1.
30. U.S. General Accounting Office, *Selected Aspects of Nuclear Powerplant Reliability and Economics* (Washington, D.C., August 15, 1975), Appendix I, p. 8.
31. Ibid., Appendix I, p. 9.
32. Ibid.
33. Pacific Gas & Electric Company, Diablo Canyon 1 and 2, Environmental Report, Supplement 2, Volume I, July 28, 1972, Appendix A, Question E-1. Filed in Nuclear Regulatory Commission Docket 50-275.
34. Virginia Electric and Power Company, North Anna 1 and 2, Final Environmental Statement, April 1973, p. 8-8. Filed in Nuclear Regulatory Commission Docket 50-338.
35. MITRE Corporation, MTR-7010, *Proceedings of Quantitative Environmental Comparison of Coal and Nuclear Electrical Generation and Their Associated Fuel Cycles Workshop* (McLean, Va., August 1975), Vol. 2, pp. 18, 19.
36. *Decommissioning: Nuclear Reactors.*
37. Ford Foundation, Energy Policy Project, *A Time to Choose* (Cambridge, Mass., 1974), p. 210.
38. Ben C. Rusche (U.S. Nuclear Regulatory Commission), letter to Senator James Buckley (R., N.Y.), July 16, 1975, Docket 50-16, estimated Fermi's decommissioning costs at $7.1 million.
39. Title 10, *Code of Federal Regulations* (Washington, D.C.: U.S. Government Printing Office, January 1976), Section 50.82.

IX. THE BACK-END OF THE FUEL CYCLE

1. U.S. Environmental Protection Agency, *Environmental Analysis of the Uranium Fuel Cycle, Part III—Nuclear Fuel Reprocessing* (Washington, D.C., October 1973), p. 3.
2. Robert Gillette, "Plutonium (I): Questions of Health in a New Industry," *Science* 185: 1032 (September 20, 1974).
3. "Getty Oil Subsidiary Says It Won't Reopen Nuclear-Fuels Facility," *Wall Street Journal*, September 23, 1976.
4. Ralph W. Deuster (Nuclear Fuel Services), letter to N. Richard Werthamer (New York State Energy Research and Development Authority), April 29, 1976.
5. *Environmental Analysis, Part III*, p. 4.
6. Robert Gillette, "Nuclear Fuel Reprocessing: GE's Balky Plant Poses Shortage," *Science* 185: 770–771, August 30, 1974.
7. U.S. Energy Research and Development Administration, ERDA-76-25, *LWR Spent Fuel Disposition Capabilities* (Washington, D.C., May 1976), p. 9.
8. U.S. Nuclear Regulatory Commission, *Federal Register* 40: 42801, September 16, 1975.
9. Union of Concerned Scientists, *The Nuclear Fuel Cycle* (Cambridge, Mass., October 1973), Chap. VI, "Nuclear Fuel Reprocessing: Radiological Impact of West Valley Plant," pp. 149–207.
10. John Gofman, testimony before the Nuclear Study Committee, Legislature of the State of South Carolina, January 7, 1972.
11. U.S. Environmental Protection Agency, Draft Environmental Statement, *Environmental Ra-*

diation Protection Requirements for Normal Operations of Activities in the Uranium Fuel Cycle (Washington, D.C., May 12, 1975).
12. Ibid., p. 82.
13. Union of Concerned Scientists, *The Nuclear Fuel Cycle* (Cambridge, Mass., October 1973), Chap. I, "Storage and Disposal of High Level Wastes," p. 15.
14. Ibid., pp. 16, 47.
15. Hannes Alfvén, quoted in Natural Resources Defense Council, *Citizens' Guide: The National Debate on the Handling of Radioactive Wastes from Nuclear Power Plants* (Palo Alto, Calif., August 1974, revised November 1975), cover page. The *Guide* is an excellent in-depth analysis of the waste problem.
16. Dennis Farney, "Ominous Problem: What to Do with Radioactive Waste," *Smithsonian*, April 1974, p. 21.
17. U;S. Atomic Energy Commission, WASH-1539, Draft Environmental Statement, *Management of Commercial High Level and Transuranium-Contaminated Radioactive Waste* (Washington, D.C., September 1974), pp. 1.2-10–1.2-14).
18. ERDA, press release 75-51, April 11, 1975.
19. ERDA, press release 76–136, May 7, 1976.
20. U.S. Energy Research and Development Administration, ERDA-48, *Creating Energy Choices for the Future* (Washington, D.C., June 1975), Vol. 2, p. 120.
21. WASH-1539, p. 1.2-9.
22. *The Nuclear Fuel Cycle*, p. 27.
23. Ibid., pp. 28–29.
24. *Citizens' Guide*, p. 26.
25. WASH-1539, p. 1.2-10.
26. H. Peter Metzger, *The Atomic Establishment* (New York: Simon and Schuster, 1972), p. 158 (pp. 154–161 give a good summary of the Lyons situation); *Citizens' Guide*, p. 34.
27. "Ominous Problem," p. 25.
28. *The Atomic Establishment*, p. 157.
29. Ibid., p. 158.
30. Philip M. Boffey, "Radioactive Waste Site Search Gets into Deep Water," *Science* 190: 361 (October 24, 1975).
31. Meeting of the Environmental Subcommittee, Advisory Committee on Reactor Safeguards, U.S. Atomic Energy Commission, Washington, D.C., July 26, 1974, Transcript, pp. 84, 85. The exchange is between Dr. Parker, consultant to the subcommittee; Dr. Moeller, member of the subcommittee; and Mr. L. Joseph Deal, Division of Operational Safety, Atomic Energy Commission.
32. *Citizens' Guide*, p. 30.
33. Daniel F. Ford and Henry W. Kendall, *The Nuclear Power Issue: An Overview* (Cambridge, Mass.: Union of Concerned Scientists, 1974).
34. U.S. Atomic Energy Commission, Richland Operations Office, *Characterization of the Z-9 Trench* (Richland, Wash., December 1973), p. 1.
35. U.S. Atomic Energy Commission, WASH-1250, *The Safety of Nuclear Power Reactors and Related Facilities* (Washington, D.C., July 1973, p. 4-89).
36. *Citizens' Guide*, p. 23.
37. WASH-1250, p. 4-89.
38. WASH-1250, p. 4-72, states that one metric ton of spent fuel becomes 330 gallons of high-level liquid waste at a reprocessing plant. WASH-1250, p. 4-7, states that the annual removal of spent fuel from an 1100 MWe plant is 29–37 metric tons. The annual generation from such a plant would therefore be 33 × 330, or about 10,000 gallons of high-level waste.
39. *The Nuclear Fuel Cycle*, p. 17.
40. *Natural Resources Defense Council Newsletter*, Vol. 4, Issue 1 (Winter 1974–75), p. 3.
41. Marion Anderson, "Fallout on the Freeway" (Lansing: Public Interest Research Group in Michigan, January 18, 1974), p. 13.

42. Ibid., pp. 17–18.
43. U.S. Atomic Energy Commission, WASH-1238, *Environmental Survey of Transportation of Radioactive Materials to and from Nuclear Power Plants* (Washington, D.C., December 1972), p. 80.
44. "Fallout on the Freeway," p. 14.
45. Ibid., p. 19.
46. Senate Commerce Committee, U.S. Congress, *Transportation of Hazardous Materials*, hearings, June 1974, p. 226.
47. Ibid., p. 227.
48. See U.S. Interstate Commerce Commission, Docket 36312, public files, Washington, D.C.; and "ERDA, TVA Rap Rails' Proposal on Nuclear Waste," *Journal of Commerce*, December 24, 1975.
49. Friends of the Earth, press release (New York, March 24, 1975).
50. Special Panel to Study Transportation of Nuclear Materials, *Transportation of Radioactive Material by Passenger Aircraft*, Report No. 1 to the Joint Committee on Atomic Energy, U.S. Congress, September 17, 1974, p. 14.
51. New York Public Interest Research Group, background information (New York, n.d.).
52. *Congressional Record* 121: H 5895, June 20, 1975.
53. "Uranium Transport Through the City Is Banned by Bellin," *New York Times*, August 1, 1975, p. 23.
54. Hearings Before New York City Department of Health, November 6, 1975, Transcript, p. 290.
55. "U.S. Sues on Shipping Radioactive Goods," *New York Times*, January 16, 1976, p. 27.

X. WORKER SAFETY

1. "Former Dow Employee Seeks Benefits," *Denver Post*, April 25, 1975; "Ex-Dow Employee Seeks Compensation," *Rocky Mountain News* (Denver), April 25, 1975. Also, telephone conversation between David Eisner (attorney representing Albert Moon, Boulder, Colo.) and Public Interest Research Group, Washington, D.C., June 7, 1976.
2. Lauriston S. Taylor, "History of the International Commission on Radiological Protection (ICRP)," *Health Physics*, 1958, Vol. 1, pp. 103–104.
3. Most of the information in this section comes from four investigative reporters who did articles on Karen Silkwood: Howard Kohn, "Malignant Giant: The Nuclear Industry's Terrible Power and How It Silenced Karen Silkwood," *Rolling Stone*, March 27, 1975, pp. 43 ff.; B. J. Phillips, "The Case of Karen Silkwood," *Ms.*, April 1975, pp. 59 ff.; Barbara Newman, "Plutonium: A Question of Life and Death," Washington, D.C., National Public Radio, March 23, 1975. David Burnham was the *New York Times* reporter whom Karen Silkwood was to have met in Oklahoma City. He has written periodic articles on the case.
4. U.S. Energy Research and Development Administration, WASH-1174-74, *The Nuclear Industry 1974*, pp. 54–55.
5. "Plutonium: A Question of Life and Death."
6. Arthur R. Tamplin and Thomas B. Cochran, "Radiation Standards for Hot Particles" (Washington, D.C., Natural Resources Defense Council, February 14, 1974), p. 5.
7. W. J. Shelley (Kerr-McGee Nuclear Corporation, Oklahoma City), letter to D.C. Rouse (U.S. Nuclear Regulatory Commission, Washington, D.C.), March 22, 1975, Appendix J, table I, NRC Docket 70-1193.
8. David Burnham, "Hearing on Plutonium Plant Is Told of a Conflict Over Health Reports," *New York Times*, April 27, 1976, p. 12.
9. Thomas O'Toole, "Plutonium Worker Contaminated Self, AEC Aides Suspect," *Washington Post*, January 7, 1975, p. A8.
10. Gregory Curtis, "This Man Loves Car Wrecks More Than Anyone in the World," *Texas Monthly* (Austin), May 1975, p. 66.
11. U.S. General Accounting Office, "Federal Investigations into Certain Health, Safety, Qual-

ity Control, and Criminal Allegations at Kerr-McGee Nuclear Corporation" (Washington, D.C., May 30, 1975), pp. 2, 3.

12. Anthony Mazzocchi and Steve Wodka (Oil, Chemical and Atomic Workers International Union), statement before the House Committee on Small Business, U.S. Congress, April 26, 1976.
13. *Federal Investigations*, pp. 2, 3.
14. Ibid., pp. 17–18.
15. Ibid., pp. 18–19.
16. Ibid., pp. 20–21.
17. "A-Pellets Planted, Memo Indicates," *Washington Post*, June 5, 1975, p. A20.
18. "FBI: No Foul Play in Death," *Washington Star*, May 3, 1975, p. A2.
19. "The Case of Karen Silkwood," p. 66.
20. "Malignant Giant," p. 62.
21. Howard Kohn, "Shutdown at Oklahoma's Kerr-McGee: The End of a Plutonium Relationship." *Rolling Stone*, December 4, 1975, p. 32.
22. "Women Press U.S. on Silkwood Case," *New York Times*, August 27, 1975.
23. National Public Radio, press release, November 3, 1975.
24. Robert Gillette, "Plutonium (I): Questions of Health in a New Industry," *Science* 185: 1030–1031 (September 20, 1974).
25. Ibid., p. 1030.
26. National Resources Defense Council, "Comments on WASH-1327" (Washington, D.C., October 30, 1974), General Comments, pp. 12–14.
27. "Plutonium (I)," p. 1032.
28. AEC Directorate of Regulatory Operations, letter to NUMEC, August 12, 1974, AEC Dockets 70-135, 70-364.
29. AEC Directorate of Regulatory Operations, letter to NUMEC, June 5, 1974, AEC Dockets 70-135, 70-364.
30. NUMEC, letter to AEC Directorate of Regulatory Operations, July 5, 1974, reporting exposure of June 11, 1974; AEC Docket 70-364.
31. NUMEC, letter to AEC Directorate of Regulatory Operations, August 28, 1974, reporting exposure of August 27, 1974; AEC Docket 70-135.
32. "Comments on WASH-1327," pp. 15–16.
33. "Plutonium (I)," p. 1032.
34. Ibid., p. 1031.
35. Robert Gillette, " 'Transient' Nuclear Workers: A Special Case for Standards," *Science* 186: 125 (October 11, 1974).
36. Letter from James L. Liverman (U.S. Energy Research and Development Administration) to Public Interest Research Group, October 8, 1976.
37. C. C. Johnson (U.S. Public Health Service), cited in Union of Concerned Scientists, *The Nuclear Fuel Cycle* (Cambridge, Mass., October 1973), Chap. V, "Lung Cancer Among Uranium Miners," pp. 116–148.
38. Atomic Energy Act of 1954, as amended, Section 62.
39. Joint Committee on Atomic Energy, U.S. Congress, *Radiation Exposure of Uranium Miners*, hearings, May–August 1967, p. 159.
40. Some of the many studies performed are reviewed in *The Nuclear Fuel Cycle*, and by F. E. Lundin, et al., *Radon Daughter Exposure and Respiratory Cancer, Quantitative and Temporal Aspects* (Rockville, Md.: National Institute for Occupational Safety and Health, National Institute of Environmental Health Science, Joint Monograph No. 1, 1971). The first reference to excessive lung cancer deaths among European miners appeared in 1879.
41. H. Peter Metzger, *The Atomic Establishment* (New York: Simon and Schuster, 1972), pp. 120–121.
42. Study cited in *Radon Daughter Exposure*, p. xxi.

43. International Labor Office, *The Protection of Workers Against Ionizing Radiations*, report submitted to the International Conference on the Peaceful Use of Atomic Energy, Geneva, August 1955, pp. 39–40.
44. D. A. Holaday et al., *Control of Radon and Daughters in Uranium Mines and Calculations on Biologic Effects* (Washington, D.C.: Public Health Service Publication No. 494, 1957), p. ix.
45. Joint Committee on Atomic Energy, U.S. Congress, "Employee Radiation Hazards and Workmen's Compensation," summary-analysis of hearings, March 1959, p. 11.
46. Ibid., hearings, p. 193.
47. Joint Committee on Atomic Energy, U.S. Congress, *Radiation Exposure of Uranium Miners*, hearings, May–October 1967.
48. U.S. Environmental Protection Agency, "Underground Mining of Uranium Ore," *Federal Register*, 36: 12921.
49. Victor E. Archer (National Institute for Occupational Safety and Health, Salt Lake City), "Risks Associated with Mining and Processing of Uranium," paper presented at meeting of Health Physics Society, Atlanta, June 23–27, 1975.
50. Ibid., p. 7.
51. Ibid., p. 9.
52. U.S. Environmental Protection Agency, Denver, EPA-330/9-75-001, Draft, *Water Quality Impacts of Uranium Mining and Milling Activities in the Grants Mineral Belt* (New Mexico), July 1975, p. 5.
53. John R. Wright (chief, Water Quality Division, State of New Mexico Environmental Improvement Agency, Santa Fe), letter to George J. Putnicki (U.S. Environmental Protection Agency), September 25, 1974.
54. J. V. Rouse and J. L. Hatheway, *Preliminary Report on New Mexico Uranium Mine and Mill Survey, Grants Mineral Belt, New Mexico*, report to National Enforcement Investigations Center, U.S. Environmental Protection Agency, Denver, June 2, 1975; *Water Quality Impacts.*
55. Ibid., pp. 11–13.
56. *Preliminary Report*, p. 2.
57. *Water Quality Impacts*, p. 5.
58. Ibid.
59. *Preliminary Report*, p. 8.
60. The violations of the different companies as listed in EPA reports are summarized in letters from the Public Interest Research Group, Washington, D.C., to the Environmental Protection Agency and Nuclear Regulatory Commission, August 15, 1975.

XI. THE PLUTONIUM BREEDER REACTOR

1. Roger Rapoport, *The Great American Bomb Machine* (New York: Dutton, 1971), pp. 115–117; "Radiation Standards for Hot Particles" (Washington, D.C.: Natural Resources Defense Council, February 14, 1974), pp. 28, 29, and Appendix B.
2. H. Peter Metzger, *The Atomic Establishment* (New York: Simon and Schuster, 1972), p. 145.
3. Ibid., p. 287.
4. *Natural Resources Defense Council Newsletter*, Vol. 3, Issue 2 (Summer/Fall 1974), p. 4.
5. Ibid.; C. Peter Gall, "Inside Look at Soviet Breeders," *Chemical Week*, May 28, 1975, pp. 55–56.
6. *Natural Resources Defense Council Newsletter*, p. 3.
7. U.S. Atomic Energy Commission, WASH-1535, Proposed Final Environmental Statement, *Liquid Metal Fast Breeder Reactor Program* (Washington, D.C., December 1974), p. 4.2-146.
8. S. J. Board, R. W. Hall, R. S. Hall (Berkeley Nuclear Laboratories, Gloucestershire, United Kingdom), letter to *Nature* 254: 321 (March 27, 1975).
9. U.S. Nuclear Regulatory Commission, *Supplement No. 2 to the Safety Evaluation of the Fast Flux Project No. 448* (Washington, D.C., March 7, 1975), p 1-6.

10. WASH-1535.
11. U.S. Atomic Energy Commission, Argonne National Laboratory, *Liquid Metal Fast Breeder Reactor Program Plan*, August 1968, pp. 10-90, 10-92.
12. George L. Weil, *Nuclear Energy: Promises, Promises* (Washington, D.C., 1971), inside front cover.
13. George L. Weil, letter to National Intervenors (Washington, D.C.), July 26, 1973, reprinted in Joint Committee on Atomic Energy, U.S. Congress, *Nuclear Reactor Safety*, hearings (Washington, D.C., January 1974), Part 2, Vol. I, p. 378.
14. WASH-1535, pp. VI.38-256, 257.
15. Natural Resources Defense Council, *Bypassing the Breeder* (Washington, D.C., March 1975), p. 6.
16. *Radiation Standards for Hot Particles*, p. 5.
17. Sheldon Novick, *The Careless Atom* (Boston: Houghton Mifflin, 1969), pp. 155–156.
18. The "maximum credible accident" is described in *Technical Information and Hazards Summary Report* (Power Reactor Development Company, June 14, 1961), p. 603.15. The Fermi fuel melt accident was described in *Preliminary Report on Fuel Damage in Fermi Reactor* (Power Reactor Development Company, October 10, 1966); and *Report on the Fuel Melting Incident in the Enrico Fermi Atomic Power Plant on October* 5, 1966 (Power Reactor Development Company, January 8, 1969). All documents are filed in NRC Docket Number 50-16, at the NRC's document room in Washington, D.C.
19. John G. Fuller, *We Almost Lost Detroit* (New York: Reader's Digest Press, 1975), p. 231.
20. *Bypassing the Breeder*, Appendix, p. 1.
21. Ibid., p. 8.
22. WASH-1535, p. 11.2-6, gives the benefits as $19.4 billion. P. 11.2-32 gives the program costs as $4.7 billion. Both figures are in 1974 dollars.
23. *Bypassing the Breeder*, pp. 12–14, and Appendix, pp. 21–30. The Cornell researchers are Duane Chapman and Timothy Mount.
24. Ibid., Appendix, p. 40.
25. Ibid., Appendix, pp. 40–41.
26. Mark Sharefkin (Resources for the Future), *The Fast Breeder Reactor Decision: An Analysis of Limits and the Limits of Analysis*, prepared for Joint Economic Committee, U.S. Congress, April 19, 1976.
27. Brian G. Chow (Saginaw Valley State College), *The Liquid Metal Fast Breeder Reactor: An Economic Analysis* (Washington, D.C.: American Enterprise Institute, December 1975).
28. WASH-1535, pp. 1.1-7, 1.2-1.
29. Ibid., p. 11.2-134.
30. U.S. Energy Research and Development Administration, ERDA-48, *Creating Energy Choices for the Future* (Washington, D.C., June 1975), Vol. I, Figure 5-5, p. V-5.
31. Ibid., pp. VIII-3, B-8.
32. *Bypassing the Breeder*, Appendix, p. 6.
33. WASH-1535, p. 11.2-33.
34. U.S. General Accounting Office, Staff Study, *Fast Flux Test Facility Program* (Washington, D.C., January 1975), p. 9.
35. Ibid., p. 7.
36. Ibid., p. 1.
37. Thomas B. Cochran, Ph.D. (Natural Resources Defense Council, Washington, D.C.), statement before Joint Economic Committee, U.S. Congress, May 8, 1975, p. 3.
38. *Bypassing the Breeder*, Appendix, p. 10.
39. Thomas Cochran, statement, pp. 3–4.
40. Ibid., pp. 3–5.
41. "Energy: Receiving the Lion's Share," *Science News* 109: 52 (January 24, 1976).
42. Dr. Robert C. Seamans, Jr. (administrator, U.S. Energy Research and Development Ad-

ministration), statement before the Joint Committee on Atomic Energy, U.S. Congress, February 4, 1975.

43. *Energy: Receiving the Lion's Share.*
44. *Bypassing the Breeder*, p. 16.
45. Ibid., p. 15, estimates that solar could provide 5.5 trillion kilowatt-hours of electricity (tkwh), geothermal 1.7 tkwh, and organic wastes 0.6 tkwh in 2020. Use of organic wastes is an indirect form of solar energy. Lower growth in electrical consumption could account for 13.8 tkwh. The total of these sources would then be 21.6 tkwh, which is 79 percent of the 27.6 tkwh which the AEC estimated would be required.
46. Ibid., p. 14.

XII. THE PLUTONIUM FUEL CYCLE: THE IMMEDIATE THREAT

1. U.S. Nuclear Regulatory Commission, NUREG-0002, *Final Generic Environmental Statement on the Use of Recycle Plutonium in Mixed Oxide Fuel* (Washington, D.C., August 1976), Vol. I, p. S-43.
2. U.S. Atomic Energy Commission, WASH-1327, Draft *Generic Environmental Statement Mixed Oxide Fuel* (Washington, D.C., August 1974), Vol. 1, p. S-27.
3. Mason Willrich and Theodore B. Taylor, *Nuclear Theft: Risks and Safeguards* (Cambridge, Mass.: Ballinger, 1974). Biographical sketches of the authors are printed on the back cover.
4. Theodore B. Taylor, statement before Subcommittee on International Finance, U.S. Senate, July 15, 1974, p. 10.
5. "M.I.T. Student Designs Atomic Bomb," *Technology Review*, December 1975, p. 7.
6. John Darcy and Joe Shapiro, statement, Washington, D.C., June 4, 1975.
7. William A. Anders (chairman, U.S. Nuclear Regulatory Commission), letter to Ralph Nader (Washington, D.C.), October 31, 1975.
8. John M. Baer, "Intruder Eluded Guards in N-Plant," *Harrisburg Patriot*, February 20, 1976, pp. 1, 2.
9. U.S. Nuclear Regulatory Commission, press release 76-64 (Washington, D.C., March 17, 1976).
10. U.S. General Accounting Office, "Improvements Needed in the Program for the Protection of Special Nuclear Material" (Washington, D.C., November 7, 1973).
11. U.S. General Accounting Office, "Protecting Special Nuclear Material in Transit: Improvements Made and Existing Problems" (Washington, D.C., April 12, 1974).
12. Atomic Energy Commission, "A Special Safeguards Study," reprinted in *Congressional Record* 120: S 6622 (April 20, 1974).
13. WASH-1327, Vol. 4, pp. V-49, V-50.
14. Carl H. Builder (director, Division of Safeguards), memorandum to Ronald A. Brightsen (assistant director for licensing, Division of Safeguards, U.S. Nuclear Regulatory Commission), January 19, 1976.
15. U.S. Nuclear Regulatory Commission, press release 76-22 (Washington, D.C., February 4, 1974); Kenneth R. Chapman (director, Office of Nuclear Material Safety and Safeguards, U.S. Nuclear Regulatory Commission), letter to Natural Resources Defense Council, March 22, 1976.
16. Title 10: Energy, *Code of Federal Regulations* (Washington, D.C.: U.S. Government Printing Office, January 1, 1975), Section 70.51.
17. Thomas O'Toole, "Fear of Nuclear Theft Stirs Experts, AEC," *Washington Post*, May 26, 1974, p. A16.
18. U.S. Atomic Energy Commission, press release T-403 (Washington, D.C., August 14, 1974).
19. U.S. Nuclear Regulatory Commission, press release 76-86 (Washington, D.C., April 19, 1976).

20. "Summary of GAO Report Evaluating ERDA's Safeguards Systems." *Congressional Record,* 122: H7838–H7840, July 27, 1976.
21. David Burnham, "Report Says U.S. Cannot Account for 2 Tons of Atom-Bomb Material," *New York Times,* August 6, 1976, p. A14.
22. "A Special Safeguards Study," S 6624.
23. NUREG-0002, p. S-16.
24. WASH-1327, Vol. 4, p. V-4.
25. J. Gustave Speth (Natural Resources Defense Council), statement before House Subcommittee on Energy and Environment, U.S. Congress, May 2, 1975, p. 16.
26. Russell W. Ayres, "Policing Plutonium: The Civil Liberties Fallout," *Harvard Civil Rights–Civil Liberties Law Review,* Vol. 10, No. 2 (Spring 1975), p. 436.
27. U.S. Nuclear Regulatory Commission, NUREG-0095, *Joint ERDA-NRC Task Force on Safeguards* (unclassified version) (Washington, D.C., July 1976), pp. xiii, xiv.
28. "Policing Plutonium, p. 397.
29. Ibid., p. 441–442.
30. Ibid., p. 411.
31. Ibid., p. 374.
32. Ibid., p. 443.
33. "Atom Power Foe Object of Inquiry," *New York Times,* August 5, 1974, p. 26.
34. "Texas Agency Destroys Disputed Files," *New York Times,* August 25, 1974, p. 38. Other accounts of the Texas surveillance are printed in the *Dallas Times-Herald,* August 1, August 2, and September 19, 1974, and the *Dallas Morning News,* August 3 and October 15, 1974.
35. Paul G. Edwards, "Bill to Allow Va. Utilities to Name Own Police Sought," *Washington Post,* January 18, 1975, p. E3.
36. John G. Davis (U.S. Nuclear Regulatory Commission), letter to James M. Cubie (Public Citizen), January 19, 1976.
37. H. E. Lyon (U.S. Energy Research and Development Administration) letter to James M. Cubie (Public Citizen), January 19, 1976.
38. John G. Davis, letter.
39. J. G. Speth, statement, p. 3.
40. *Public Interest Report: Nuclear Terrorism"* (Los Angeles, Calif.: Environmental Alert Group, n.d.).
41. J. G. Speth, statement.
42. "Spectrum," *Environment,* October 1974, p. 21.
43. John G. Davis, letter.
44. "Bombs Hit Nuclear Site in France," *Washington Post,* May 4, 1975, p. A1.
45. "Spectrum," *Environment,* December 1975, p. 23.
46. "A Special Safeguards Study," S 6623.
47. J. Gustave Speth, statement, p. 6.
48. Donald Geesaman, "Statement in Support of a Nuclear Moratorium for the State of Minnesota," December 12, 1974.
49. "NRC Commissioners Stress They Have Open Mind on Plutonium," *Nucleonics Week,* May 29, 1975, p. 1.
50. H. Peter Metzger, *The Atomic Establishment* (New York: Simon and Schuster), 1972, p. 287, note 225, states that 6×10^{-9} grams of botulin toxin would be lethal for a 60 kg (132-pound) man. *Radiation Standards for Hot Particles* (Washington, D.C.: Natural Resources Defense Council, February 14, 1974), gives the following information: On p. 5, 0.2 microcuries of Pu-239 is 3 micrograms. On p. 33–34, the limiting activity of a hot particle is set at 0.07 picocuries (0.07×10^{-12} curies). On p. 43, the risk of cancer from 1 hot particle is estimated at 1 in 2000. The Pu-239 dose that would cause cancer is therefore $.07 \times 10^{-12}\ c \times \frac{3\ g}{.2\ c} \times 2000 = 2.1 \times 10^{-9}$ g, which is comparable to the lethal amount for botulin toxin.
51. Committee of Inquiry: The Plutonium Economy, *The Plutonium Economy: A Statement of Concern,* background report, submitted to the National Council of Churches of Christ of the U.S.A., New York, September 1975, p. 14.

52. U.S. Nuclear Regulatory Commission, press release 75-270 (Washington, D.C., November 12, 1975).
53. David Burnham, "Court Bars Plutonium's Commercial Use," *New York Times*, May 28, 1976, p. A1.
54. See, for example, J. Gustave Speth (Natural Resources Defense Council, Washington, D.C.), memorandum, November 13, 1975.
55. Convention of the National Council of Churches of Christ in the U.S.A., Atlanta, resolution adopted March 4, 1976.
56. *The Plutonium Economy: A Statement of Concern.*

XIII. NUCLEAR ECONOMICS

1. Charles Komanoff, *Responding to Con Edison: An Analysis of the 1974 Costs of Indian Point and Alternatives* (New York: Council on Economic Priorities, August 1975), p. ii.
2. Ibid., p. 7.
3. Ibid., p. 10.
4. Ibid., p. 4.
5. Ibid., p. 14.
6. Charles Komanoff (Council on Economic Priorities, New York), memorandum to Nancy Matthews (staff of Representative Richard L. Ottinger), February 28, 1975, p. 5.
7. Atomic Industrial Forum, press release (New York, April 2, 1975).
8. "Problems in the City," WAIT-AM Radio, Chicago, broadcast May 25, 1975; George Travers, telephone conversation with Public Interest Research Group (Washington, D.C.), September 8, 1975.
9. A. David Rossin (Station Nuclear Engineering Department, Commonwealth Edison, Chicago), letter to David Turner (U.S. Federal Energy Administration, Washington, D.C.), November 13, 1975.
10. Daniel F. Ford (Union of Concerned Scientists, Cambridge, Mass.), "Nuclear Power: Some Basic Economic Issues," testimony before House Committee on Interior and Insular Affairs, U.S. Congress, April 28, 1975, p. 14.
11. "Why Atomic Power Dims Today," *Business Week*, November 17, 1975, p. 98.
12. Irvin C. Bupp, et al., "The Economics of Nuclear Power," *Technology Review*, February 1975, p. 15.
13. Ibid., p. 18.
14. Ibid., p. 21.
15. Daniel F. Ford, testimony, pp. 15–16.
16. "Inoperative A-Power Plant Said to Cost TVA $18 million," *Washington Post*, August 19, 1976.
17. David Dinsmore Comey, *The Incident at Browns Ferry* (Washington, D.C.: Friends of the Earth).
18. Investor Responsibility Research Center, *The Nuclear Power Alternative* (Washington, D.C., January 1975), p. 14.
19. Ibid.
20. Ibid., p. 93.
21. "The Economics of Nuclear Power," p. 19.
22. Ibid., p. 21.
23. Daniel F. Ford, testimony, pp. 13–14.
24. David Dinsmore Comey, "Nuclear Plant Reliability: The 1973–1974 Record," (Chicago: Business and Professional People for the Public Interest, February 14, 1975), p. 3.
25. Ibid., p. 1.
26. Ibid., p. 4.
27. Ibid., p. 5.
28. Ibid., Table 8.

29. David Burnham, "Hope for Cheap Power from Atom Is Fading," *New York Times*, November 16, 1975, p. 1.
30. David Dinsmore Comey, "Capacity Factors Stay Constant in 1975," in *Not Man Apart* (Washington, D.C.: Friends of the Earth, March 1976).
31. Thomas Ehrich, "Atomic Lemons: Breadowns and Errors In Operation Plague Nuclear Power Plants," *The Wall Street Journal*, May 3, 1973.
32. David Dinsmore Comey, "Will Idle Capacity Kill Nuclear Power?" *Bulletin of the Atomic Scientists*, November 1974, p. 27.
33. "Nuclear Plant Reliability," p. 6.
34. Daniel F. Ford (Union of Concerned Scientists, Cambridge, Mass.), testimony before House Ways and Means Committee, U.S. Congress, March 7, 1975, p. 7.
35. David G. Snow, *The Uranium Stocks: Nuclear Industry Kaleidoscope Coming Together* (New York: Mitchell, Hutchins, January 30, 1976, p. 164.
36. Ralph Nader, "Nuclear Power: It Just Doesn't Pay," remarks at Critical Mass '75, Washington, D.C., November 1975, p. 5.
37. *The Uranium Stocks*, p. 1.
38. See, for example, *Congressional Record* 121: S 1159, January 29, 1975.
39. "Uranium Backfires on Westinghouse," *Business Week*, August 18, 1975, pp. 29–32.
40. *The Uranium Stocks*, p. 5.
41. Byron E. Calame, "Westinghouse Charges 29 Firms in Uranium Suit," *Wall Street Journal*, October 18, 1976, p. 2.
42. Reginald Stuart, "Utility to Buy Uranium at a Quadrupled Price," *New York Times*, December 19, 1975, p. 59.
43. U.S. General Accounting Office, "Selected Aspects of Nuclear Powerplant Reliability and Economics" (Washington, D.C., August 15, 1975), Appendix I, p. 5.
44. *Nucleonics Week*, March 13, 1975.
45. The premiums are calculated by:

$$\$40.5 \times 10^9 \text{ in damages} \times \frac{\$0.58 \times 10^3}{\$10^6 \text{ liability}} \text{ per yr} = \$23.5 \times 10^6,$$

which is \$23.5 million per year. See Herbert S. Denenberg, testimony before the Licensing Hearing for Three Mile Island Nuclear Power Plant, November 7, 1973, reprinted in Joint Committee on Atomic Energy, U.S. Congress, *Possible Modification or Extension of the Price-Anderson Insurance and Indemnity Act*, hearings, January and March, 1974, Phase II: Review, pp. 226–240.
46. Ibid.
47. The electricity from a 1000 Megawatt-electric nuclear plant, operating at a capacity factor of 70 percent, would be:

$$1000 \text{ MW} \times 365 \text{ da/yr} \times 24 \text{ hr/da} \times 0.70 \text{ capacity factor}$$
$$= 10^3 \times 3.65 \times 10^2 \times 2.4 \times 10^1 \times .7 = 6.14 \times 10^6 \text{ MWh/yr}$$

The limited liability distortion would then be:

$$\frac{\$23.5 \times 10^6}{6.14 \times 10^6 \text{ MWh}} = \$3.8/\text{MWh} \times \frac{1000 \text{ mill/\$}}{1000 \text{ kwh/MWh}} = 3.8 \text{ mill/kwh}$$

48. "Legalize Evasion," *The Power Line* (Washington, D.C.: Environmental Action Foundation, September 1975), p. 2.
49. Edward Kahn, et al., *Investment Planning in the Energy Sector* (Berkeley, Calif.: Energy and Environment Division, Lawrence Berkeley Laboratory, March 1, 1976), p. 89.
50. Ibid., p. 59.
51. "Portland GE States Nuclear Plant Qualifies in Test for Tax Break," *Wall Street Journal*, December 26, 1975, p. 15.
52. *The Nuclear Power Alternative*, p. 24.

53. From *Nuclear Safety* 16: 228–231 (March/April 1975) it can be calculated that the electrical generation of commercial power plants through the month of December 1974 was 359.097 million megawatt-hours, electric.

$$\frac{\$5 \times 10^9}{0.359 \times 10^9} = \$13.90/\text{MWh} = 13.90 \text{ mills/kwh}$$

54. U.S. Atomic Energy Commission, WASH 1174-74, *The Nuclear Industry 1974* (Washington, D.C., January 1975), p. 20.
55. Marvin Resnikoff, "Expensive Enrichment," *Environment*, July/August 1975, p. 31.
56. C. R. Moore (Nuclear Fuel Services, Rockville, Md.), letter to Ronald L. Heiks (Consumers Power Company, Jackson, Mich.), July 13, 1976.
57. Adapted from Garry DeLoss, "Jacksonville Grocer Leads Battle Against Floating Nukes," *Critical Mass* (Washington, D.C.), January 1976, p. 6.
58. Jeffrey A. Tannenbaum, "Big Plant to Recycle Nuclear Fuel Is Hit by Delays, Cost Rises," *Wall Street Journal*, February 17, 1976, pp. 1, 21.
59. "Utility Regulators Review Costs of Uranium Power," *New York Times*, February 16, 1976, p. 29.

XIV. ALTERNATIVES TO NUCLEAR

1. See, for example, Denis Hayes, *Energy: The Case for Conservation*, Worldwatch Paper 4 (Washington, D.C.: Worldwatch Institute, January 1976). Readers who want to learn more about energy efficiency and its enormous potential can choose from many other references:

Energy Efficiency in General

Energy Policy Project of the Ford Foundation, *A Time to Choose* (Cambridge, Mass.: Ballinger, 1974); Robert H. Williams, ed., *The Energy Conservation Papers* (Cambridge, Mass.: Ballinger, 1975); American Physical Society, *Efficient Use of Energy: A Physics Perspective*, (January 1975), reprinted in House Committee on Science and Technology, U.S. Congress, *ERDA Authorization—Part I, 1976 and Transition Period: Conservation*, hearings, February 1975, pp. 397–659; Lee Schipper, *Energy Conservation—Its Nature, Hidden Benefits, Hidden Barriers* (June 1975) and *Towards More Productive Energy Utilization* (October 1975) (Energy and Environment Division, Lawrence Berkeley Laboratory, University of California at Berkeley); Amory Lovings, "Energy Strategy: The Road Not Taken?" *Foreign Affairs*, October 1976.

Residential Energy Efficiency

Eric Hirst, ORNL/Con-2, *Residential Energy Conservation Strategies* (Oak Ridge, Tenn.: Oak Ridge National Laboratory, September 1976); B. B. Hamel and H. L. Brown (Drexel University) and W. Steigelmann (Franklin Institute Research Laboratories), *Pennsylvania Power & Light Co. Energy Efficient House: Simulation of System Performance* (June 1976); *Energy Conservation Ideas to Build on: Report No. 1, The Arkansas Story*, Owens-Corning Fiberglass Corporation (Toledo, Ohio, 1975).

Commercial and Industrial Energy Efficiency

American Institute of Architects, "Energy and the Built Environment: A Gap in Current Strategies" and "A Nation of Energy Efficient Buildings by 1990" (Washington, D.C., 1975); Ralph Nader and Garry DeLoss, statement before Senate Commerce Committee, February 25, 1976 (contains a review of the Ohio State experience). Dubin-Minden-Bloome Associates, "Energy Conservation Guidelines for Buildings" (Washington, D.C.: U.S. General Services Administration, January 1974); E. P. Gyftopoulos, L. J. Lazaridis, T. F. Widmer, *Potential Fuel Effectiveness in Industry* (Cambridge, Mass.: Ballinger, 1974).

2. Energy Policy Project of the Ford Foundation, *A Time to Choose* (Cambridge, Mass.: Ballinger, 1974), pp. 5, 20.
3. Committee on Mineral Resources and the Environment, National Research Council, National Academy of Sciences, *Mineral Resources and the Environment* (Washington, D.C., 1975), p. 267.
4. Bureau of Mines, Department of the Interior, "Annual U.S. Energy Use Drops Again," news release, April 5, 1976, p. 1.
5. Federal Energy Administration, *National Energy Outlook* (Washington, D.C., February 1976), Table V-2, p. 223.
6. "An Energy Aide Tells of Three Rebuffs by White House in Conservation Plea," *New York Times*, April 28, 1976, p. 29.
7. U.S. Federal Energy Administration, *Federal Energy Management Program: Third Quarter Report (Fiscal Year 1975)*, November, 1975.
8. U.S. Federal Energy Administration, WN-8866-2-FEA, *How Business in Los Angeles Cut Energy Use by 20 Percent* (Washington, D.C., January 1975), p. 5.
9. Ralph Nader and Garry DeLoss, statement before Senate Commerce Committee, February 25, 1976, pp. 12–16.
10. E. P. Gyftopoulos, L. J. Lazaridis, T. F. Widmer, *Potential Fuel Effectiveness in Industry* (Cambridge, Mass.: Ballinger, 1974), p. 1.
11. Lee Schipper, "Conservation and the Sun: Our Most Energetic Sources of Well-Being." This is a short summary of Schipper's more extensive study, *Energy Conservation—Its Nature, Hidden Benefits, Hidden Barriers* (Energy and Environment Division, Lawrence Berkeley Laboratory, University of California at Berkeley, June 1975).
12. American Institute of Architects, "Energy and the Built Environment: A Gap in Current Strategies" and "A Nation of Energy Efficient Buildings by 1990" (Washington, D.C., 1975).
13. See, for example, *Energy Conservation*, and Ron Lanoue, *Nuclear Plants: The More They Build, the More You Pay* (Washington, D.C.: Center for Study of Responsive Law, 1976), Chap. I.
14. John Holdren, Ph.D. (Energy and Resources Group, University of California at Berkeley), testimony before Subcommittee to Review the National Breeder Reactor Program, Joint Committee on Atomic Energy, June 10, 1975.
15. *Energy Conservation*, p. 31.
16. Ibid., pp. 33–36.
17. *A Time to Choose*, p. 136.
18. U.S. Department of the Interior, *Geological Estimates of Undiscovered Recoverable Oil and Gas Resources in the United States*, U.S. Geological Survey Circular 725 (Washington, D.C., p. 2).
19. Ibid., p. 1.
20. Ibid.
21. *Mineral Resources and the Environment*, p. 8.
22. Ibid., p. 81.
23. *Wall Street Journal*, May 15, 1975, p. 34; Sanford Ross, "Our Vast Hidden Oil Resources," *Fortune*, April 1974, p. 105.
24. *Mineral Resources and the Environment*, p. 268. See also discussion, pp. 266–274.
25. Ibid., p. 3.
26. Edward Cowan, "Reserves Down for Oil and Gas," *New York Times*, April 2, 1975, p. 47.
27. Hendrick S. Houthakker and Michael Kennedy, "The Outlook for Petroleum Prices," reprinted in *Congressional Record*, 120: S 2041 (February 21, 1974).
28. Thomas H. Maugh II, "Natural Gas: United States Has It If the Price Is Right," *Science* 191: 550 (February 13, 1976).
29. See, for example, Institute for Policy Studies, *The Elements* (Washington, D.C., February 1975), p. 4. The chart printed there reportedly was prepared by the Independent Petroleum Association of America in 1952.
30. *Geological Estimates*, pp. 46, 47.

31. Union of Concerned Scientists, news release and full statement (Cambridge, Mass., August 6, 1975).
32. Dick Prouty, " 'Convert' Stresses N-Power Hazards," *Denver Post*, February 4, 1975, p. 31.
33. Harry B. Ellis, "Solar Heating: An Expensive Bargain," *Christian Science Monitor*, April 23, 1975, p. 15.
34. David Brand, "Power Pioneers," *Wall Street Journal*, March 18, 1975.
35. For other information on solar heat's economics, see F. A. Tybout and G. O. G. Löf, "Solar House Heating," *Natural Resources Journal*, Vol. 10, No. 2 (April 1970), pp. 268–326; and material inserted into the record by Senator Gary W. Hart (D., Colo.), *Congressional Record* 121: S 2742, February 27, 1975.
36. J. Richard Williams, *Solar Energy: Technology and Applications* (Ann Arbor, Mich.: Ann Arbor Science Publishers, 1974), p. 44.
37. "Power Pioneers."
38. J. J. Mutch, *Residential Water Heating, Fuel Conservation, and Public Policy* (Santa Monica, Calif.: RAND Corporation, May 1974), p. 7.
39. John H. Ingersoll, "Low-Cost Comfort," *House Beautiful*, February 1975, p. 8.
40. Lance Carden, "Northeast Wind Power Not a Gale Force Yet," *Christian Science Monitor*, July 21, 1975, p. 8.
41. "Solar Energy III: The Do-It-Yourself Guide to Solar Living," KNME-TV, Albuquerque, transcript, p. 5.
42. "In Rep. Reuss's Dream Utilities Pay Windmillers," *New York Times*, October 6, 1975, p. 32.
43. "Tilting at Con Ed," *Environmental Action*, December 18, 1976, pp. 9–10.
44. David R. Inglis (professor of physics, University of Massachusetts), letter to the Honorable James A. Haley (chairman, House Committee on Interior and Insular Affairs, U.S. Congress), June 19, 1975, p. 2.
45. Ibid., p. 20.
46. *Subpanel IX: Solar and Other Energy Sources*, report to Dixy Lee Ray, chairman, U.S. Atomic Energy Commission, December 1973, quoted by Senator Mike Gravel (D., Alaska) in the *Congressional Record*, April 11, 1974.
47. Skip Laitner, "Other Than Nuclear: Photovoltaic Systems," *Critical Mass*, February 1976, p. 8.
48. Dr. Paul Rappaport (RCA Corporation), "Solar Photovoltaic Energy," statement at hearings, House Committee on Science and Astronautics, U.S. Congress, June 6, 1974, pp. 41–42.
49. Edward Schumacher, "Sun Power Inches Closer," *Washington Post*, June 5, 1976, pp. A1, A5.
50. U.S. Department of the Interior, *Conversion of Cellulosic Wastes to Oil*, Bureau of Mines Report RI-8013 (Washington, D.C., 1975).
51. Farno L. Green, "Energy Potential from Agricultural Field Residues" (Detroit: General Motors Corporation, June 1975), p. 8.
52. Melvin Calvin, "Photosynthesis as a Resource for Energy and Materials," *American Scientist* 64: 273 (May/June 1976).
53. "Energy Potential," p. 12.
54. Dr. Thomas B. Reed (Massachusetts Institute of Technology), "Bioconversion," statement at hearings, House Committee on Science and Astronautics, U.S. Congress, June 13, 1974, pp. 48–55; Dr. Thomas B. Reed, "When the Oil Runs Out," in *Capturing the Sun Through Bioconversion: Proceedings* (Washington, D.C.: Washington Center for Metropolitan Studies, March 10–12, 1976), pp. 366–388.
55. Dr. Howard Wilcox (director, Ocean Farm Project, Naval Undersea Center), "Ocean Farming," in *Capturing the Sun*, pp. 255–276.
56. Clarence Zener, "Solar Sea Power," *Bulletin of the Atomic Scientists*, January 1976, p. 17.
57. Ibid., p. 24.
58. Ralph Nader, "Putting the Sun on Center Stage," *Washington Star*, May 1, 1976.

59. A. B. Meinel and M. P. Meinel, *Energy Research and Development*, statement at hearings, House Committee on Science and Astronautics, U.S. Congress, May 1972, p. 584.
60. U.S. Federal Energy Administration, *1977 National Energy Outlook* (Washington, D.C., January 15, 1977), p. 26.
61. Donald E. Carr, "The Lost Art of Conservation," *Atlantic Monthly*, December 1975, p. 70.
62. Ibid.
63. Dow Chemical Company, Environmental Research Institute of Michigan; Townsend-Greenspan and Company; and Cravath, Swaine, and Moore, *Energy Industrial Study*, report to the National Science Foundation (June 1975).
64. James H. Kreiger, "Geothermal Energy Stirs Worldwide Action," *Chemical & Engineering News*, June 9, 1975, p. 23.
65. National Science Foundation (sponsor), NSF/RANN-73-003, *Geothermal Energy: A National Proposal for Geothermal Resource Research* (Washington, D.C., 1973), p. 1.
66. *Report of the Cornell Workshops on the Major Issues Of a National Energy Research and Development Program*, September–October 1973 (Ithaca, N.Y.: Cornell University, December 1973), p. 44.
67. Natural Resources Defense Council, *Bypassing the Breeder* (Washington, D.C., March 1975), p. 11.
68. *Geothermal Energy*, pp. iii, iv.
69. U.S. Atomic Energy Commission, WASH 1174-74, *The Nuclear Industry 1974* (Washington, D.C., January 1975), p. 15.
70. U.S. Department of the Interior, "Methane from Coalbeds to Be Marketed for Fuel, Mines Bureau Says," news release, Bureau of Mines (Washington, D.C., January 29, 1974).
71. U.S. Department of the Interior, Bureau of Mines Report RI-7968, *Degasification of the Mary Lee Coalbed Near Oak Grove, Ala., by Vertical Borehole in Adventure of Mining* (Washington, D.C., 1974); U.S. Department of the Interior, Bureau of Mines Report RI-7969, *Methane in the Pittsburgh Coalbed, Washington County, Pa.* (Washington, D.C., 1974).
72. Substantiation for this statement is provided in *A Time to Choose*, where the Historical Growth scenario for High Nuclear development predicts that nuclear will rise from its 1973 contribution of 1 quadrillion BTU (1Q) to 12Q by 1985 and 50Q by the year 2000. This is plotted on the graph below:

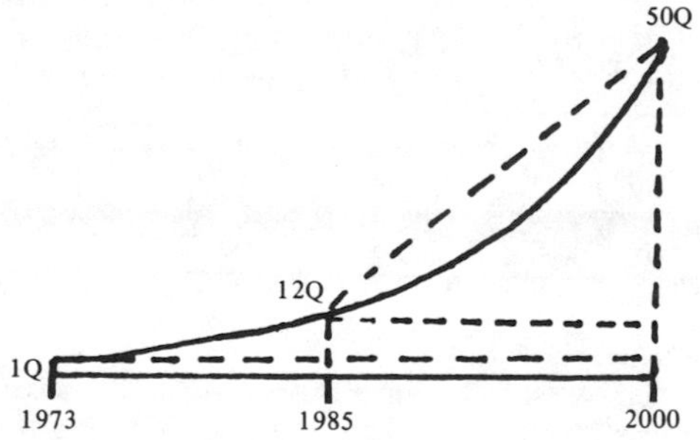

The total contribution from nuclear would be given by the area under the nuclear growth curve. This can be approximated by taking the area under the straight lines connecting the 1973, 1985, and 2000 points on the curve. This approximation will in fact overestimate the nuclear contribution.

The area under the curve is found to be:

$$
\begin{aligned}
&(1985-1973)1 + \tfrac{1}{2}(1985-1973)(12-1) \\
&\quad + (12)(2000-1985) + \tfrac{1}{2}(50-12)(2000-1985) \\
&= 12(1) + \tfrac{1}{2}(12)(11) + 12(15) + \tfrac{1}{2}(38)(15) \\
&= 12 + 66 + 180 + 285 \\
&= 543Q
\end{aligned}
$$

This figure for the nuclear contribution reflects units of heat energy from uranium fuel. The actual contribution in electricity would be one-third of 543Q, or *181Q*. The energy from the methane in coal seams would be derived from:

$$1 \text{ cubic foot} = 1031 \text{ BTU}$$
$$260 \times 10^{12}\text{cf} \times 1.031 \times 10^{3} \text{ BTU/cf} = 268 \times 10^{15} \text{ BTU} = 268Q$$

268Q is the energy available from burning the methane directly, which would be the most likely use for methane piped into commercial natural gas lines. It is interesting that if half this methane were used for direct heating, and half for the production of electricity, the end-use energy production would be about $134 + 134/3 = 134 + 44.6 = 178.6$Q, which is approximately equivalent to the optimistic nuclear projection for the remainder of the century.

73. *Cornell Workshops*, p. 8.
74. *A Time to Choose*, p. 183.
75. Center for Study of Responsive Law, *A Citizen's Manual on Nuclear Energy 1974* (Washington, D.C., 1974), p. 88.
76. U.S. Energy Research and Development Administration, "Fact Sheet: Fluidized Bed Combustion" (Washington, D.C.), n.d.
77. Allen L. Hammond, "Cleaning Up Coal: A New Entry in the Energy Sweepstakes," *Science* 189: 128 (July 11, 1975).

3. The Institutional Setting

XV. THE INDUSTRY PROFILED

1. Charles F. Zimmerman, "Competition in the Nuclear Industry" (Ph.D. dissertation, Department of Agricultural Economics, Cornell University, Ithaca, N.Y., 1975), p. 93.
2. U.S. Nuclear Regulatory Commission, *Facilities License Application Record* (Washington, D.C., June 30, 1976).
3. Ibid.
4. Adapted from "Competition in the Nuclear Industry," pp. 18–19.
5. U.S. Energy Research and Development Administration, ERDA-52, *LDC Nuclear Power Prospects, 1975–1990: Commercial, Economic & Security Implications* (Washington, D.C., 1975), Table C-3, p. C-7.
6. Ibid., Table C-13, p. C-64-a.
7. Daniel F. Ford (Union of Concerned Scientists, Cambridge, Mass.), testimony, *Oversight Hearings on Nuclear Issues*, House Subcommittee on Energy and the Environment, U.S. Congress, Part I, April–May 1975, p. 177.
8. U.S. Atomic Energy Commission, WASH-1174, *The Nuclear Industry 1974* (Washington, D.C., 1975), p. 1.
9. Ibid., p. 25.
10. Ibid., pp. 54, 55.
11. John Woodmansee, et al. *The World of a Giant Corporation* (Seattle: North Country Press, 1975), p. 10.
12. "GE's Legal Strategy," *New York Times*, December 25, 1975, p. 33.
13. "Investor's Newsletter" (General Electric Corporation, January 1976), pp. 1–2.
14. Tom Zeman, "Nuclear Power: Lying Doesn't Make It Safe," *Ramparts*, August 1974, p. 23.
15. Energy Policy Project of the Ford Foundation, *A Time to Choose* (Cambridge, Mass.: Ballinger, 1974), p. 234.
16. *Congressional Record* 121: S 1161, January 29, 1975.
17. *A Time to Choose*, p. 232.
18. Ibid.
19. Ibid., p. 234.
20. Ibid.

21. David Snow, *The Uranium Stocks: Nuclear Industry Kaleidoscope Coming Together* (New York: Mitchell, Hutchins, Inc., January 1976), p. 190.
22. *Congressional Record* 121: S 1159, January 29, 1975.
23. See also Ron Lanoue, *Nuclear Plants: The More They Build, the More You Pay* (Washington, D.C.: Center for Study of Responsive Law, 1976), Chap. I.
24. See, for example, Skip Laitner, "The Impact of Solar and Conservation Technologies upon Labor Demand," *Critical Mass* (Washington, D.C.), May 1976; and Skidmore, Owings & Merrill; Architects, Planners & Engineers, *Bonneville Power Administration Electric Energy Conservation Study* (Portland, Ore., July 1976), p. 11.

XVI. ATOMIC PROMOTION: THE FEDERAL PUSH

1. Joint Committee on Atomic Energy, U.S. Congress, *Radiation Exposure of Uranium Miners*, hearings, Part 1, May–August 1967, p. 426.
2. Joint Committee on Atomic Energy, U.S. Congress, *Environmental Effects of Producing Electric Power*, hearings, October and November, 1969, Part 1, pp. 128–129.
3. A more detailed review of the cozy relationship between the AEC and the JCAE is given by H. Peter Metzger, *The Atomic Establishment* (New York: Simon and Schuster, 1972). See also Ralph Nader, *Nuclear Reactor Safety*, testimony at hearings, Joint Committee on Atomic Energy, U.S. Congress, Part 2, January 1974, Vol. I, p. 481.
4. Ibid., p. 19.
5. Atomic Energy Act of 1954, Section 3.d, *Atomic Energy Legislation Through 93rd Congress, 2nd Session*, Joint Committee on Atomic Energy, U.S. Congress, July 1975, p. 5.
6. *The Atomic Establishment*, p. 86.
7. See *Congressional Record* 120: H 10359–10363 (October 10, 1974).
8. For a detailed description of the Energy Reorganization Act of 1974, refer to *United States Code* (St. Paul, Minn.: West Publishing, 1974), 93rd Congress—Second Session, Vol. 3, pp. 5470 ff.
9. "Serving Two Masters" (Washington, D.C.: Common Cause, October 1976), pp. 24, 25,27.
10. Center for Study of Responsive Law, *A Citizen's Manual on Nuclear Energy 1974* (Washington, D.C., 1974), p. 101.
11. Source for the FY 76 Request: ERDA, press release 75-109, June 30, 1975; source for the FY 77 Request: "Energy: Receiving the Lion's Share," *Science News* 109: 52 (January 24, 1976).
12. Dr. Robert C. Seamans, Jr. (administrator, U.S. Energy Research and Development Administration), statement at FY 1976 authorization hearings, Joint Committee on Atomic Energy, U.S. Congress, February 4, 1976, Appendix, Chart II.
13. Les Gapay, "Solar-Energy Planning Is Being Slighted by Ford, Former Chief of Program Says," *Wall Street Journal*, February 26, 1976, p. 6.
14. Public Interest Research Group, "ERDA's Involvement Against the California Initiative" (Washington, D.C., July 1976).
15. Public Interest Research Group, "The Fuel Cycle Bail-Out: ERDA's Plutonium Pork Barrel" (Washington, D.C., October 1976).
16. "General Atomic Paying DelMarva $125-Million to Get Out of Contract," *Nucleonics Week*, October 30, 1975, p. 4.
17. "Energy Agency Starts Talks with Industry on Enriching Uranium," *Wall Street Journal*, January 22, 1976, p. 21.
18. Energy Research and Development Administration, *Federal Register* 40: 56477 and 56478 (December 3, 1975).
19. U.S. Atomic Energy Commission, WASH-1327, Draft, *Generic Environmental Statement Mixed Oxide Fuel* (Washington, D.C., August 1974).
20. Ibid., Vol. 4, pp. V-49, V-50.
21. Russell W. Peterson (chairman, Council on Environmental Quality), letter to U.S. Nuclear Regulatory Commission, January 20, 1975.
22. Edson G. Case (acting director of licensing, U.S. Atomic Energy Commission), letter to

Anthony Z. Roisman (representing Natural Resources Defense Council, Washington, D.C.), December 23, 1974; Lee V. Gossick (acting executive director for operations, U.S. Nuclear Regulatory Commission), letter to Russell W. Peterson (chairman, Council on Environmental Quality), February 24, 1975.

23. U.S. Nuclear Regulatory Commission, *Federal Register* 40: 20142 (May 8, 1975).
24. Robert Gillette, "Recycling Plutonium: The NRC Proposes a Second Look," *Science* 188: 818 (May 23, 1975).
25. Ibid.
26. See, for example, A. W. Wofford (vice president, Long Island Lighting Company, Hicksville, N.Y.), letter to Nuclear Regulatory Commission, July 21, 1975.
27. Atomic Industrial Forum, "Comments on NRC May 8, 1975, Provisional Decision," in "Summary of Findings of AIF Executive Committee," July 24, 1975.
28. U.S. Nuclear Regulatory Commission, "Notice on Mixed Oxide Fuel" (Washington, D.C., November 11, 1975), pp. 2–4.
29. Hunton, Williams, Gay, & Gibson (Richmond, Va.), "Comments on NRC Provisional Decision," submitted for fifteen utilities, July 24, 1975, pp. 7, 16–20; compare with "NRC Notice on Mixed Oxide Fuel," pp. 5–7.
30. Natural Resources Defense Council, press release (Washington, D.C., December 19, 1975).
31. David Burnham, "Court Bars Plutonium's Commercial Use," *New York Times*, May 28, 1976, p. Al.
32. Natural Resources Defense Council, "Comments on Proposed Environmental Analysis of Safeguards and Deferrals of Licensing Actions," submitted by Anthony Z. Roisman, Esq. (Washington, D.C., July 24, 1975).
33 Jim Cubie, "Congress Weighs Breeder," *Critical Mass*, Washington, D.C., May 1975.
34 Jim Cubie, "Rising Cong. Concern Slows Breeder Program," *Critical Mass*, July 1975.
35. Jim Cubie, "Congress Ignores Professional, Citizen Input," *Critical Mass*, August 1975.
36. Jim Cubie, "Safe Energy Advocates Gain Votes in Congress," *Critical Mass*, June 1976.
37. Jim Cubie, "Ford Administration Pushes Nuclear Insurance Act," *Critical Mass*, September 1975.
38. "Joint Committee Rushes Price-Anderson Vote," *Critical Mass*, November 1975, Special Issue.
39. Jim Cubie, "Nuclear Industry Wins Price-Anderson Vote," *Critical Mass*, January 1976.
40. *A Citizen's Manual*, p. 99.
41. U.S. General Accounting Office, *Evaluation of the Administration's Proposal for Government Assistance to Private Uranium Enrichment Groups* (Washington, D.C., October 31, 1975).
42. Richard L. Lyons, "House Bars Private Output of A-Fuel," *Washington Post*, July 31, 1976, p. A6.
43. Richard L. Lyons, "Reversal on A-Fuel Is Voted," *Washington Post*, August 5, 1976, pp. A1, A4.
44. Spencer Rich, "Senate Kills Measure on Enriched Uranium," *Washington Post*, October 1, 1976, p. A2.
45. Jim Cubie, "House Takes Stand Against Nuclear Proliferation," *Critical Mass*, October 1976.
46. *Congressional Record* 122: S 16573 (September 24,1976).
47. *Congressional Record* 122: S 17338 (September 30, 1976).
48. John Walsh, "Congress: House Redistributes Jurisdiction over Energy," *Science* 195: 562 (February 11, 1977).
49. United States Senator Mike Gravel, statement before Senate Foreign Relations Committee, September 29, 1969.
50. Anna Mayo, "Defusing the Atomic Establishment," *Ms.*, October 1973, pp. 28–29.

XVII. THE INTERNATIONAL SPREAD OF ATOMIC POWER

1. See Paul Jacobs, "What You Don't Know May Hurt You," *Mother Jones*, February/March

1976, pp. 35–39; Clifford K. Beck (director, Office of Government Liaison–Regulation, U.S. Atomic Energy Commission), memorandum to files, December 27, 1972; W. Kenneth Davis (Bechtel Corporation), interoffice memorandum, September 27, 1973.

2. Thomas O'Toole, "Spread of Plutonium Worries A-Scientists," *Washington Post*, June 23, 1974, p. A1.
3. Natural Resources Defense Council, press release (Washington, D.C., March 2, 1976).
4. A more detailed description of Agreements for Cooperation is found in U.S. Energy Research and Development Administration, ERDA-1542, Final Environmental Statement, *U.S. Nuclear Power Export Activities* (Washington, D.C., April 1976), Vol. I, pp. 3-91–3-136.
5. Ibid., Vol. I, Table 3.16, p. 3-92.
6. Ibid., Vol. I, p. 3-117.
7. Ibid, Vol. I, p.3-122, 3-123.
8. Ibid., Vol. I, p. 3-152; and Title 10: Energy, *Code of Federal Regulations* (Washington, D.C.: U.S. Government Printing Office, January 1, 1976), Section 50.65.
9. ERDA-1542, Vol. I, p. 3-153.
10. Ibid., Vol. I, p. 3-150.
11. *Congressional Record* 120: H 7434 (July 31, 1974); and 120: H 10363 (October 10, 1974).
12. *Federal Register* 40: 45463, 45464 (October 2, 1975). See also U.S. Energy Research and Development Administration, ERDA- 80, *Findings Supporting Determination Related To International Nuclear Power Export Activities Pending Preparation of a Section 102(2)(c) NEPA Environmental Statement* (Washington, D.C., September 26, 1976).
13. See, for example, "Nuclear Power for Cairo," *New York Times*, June 17, 1974, p. 30; Dan Morgan, "Nixon Promises Egypt Aid on Nuclear Energy," *Washington Post*, June 15, 1974, p. 1.
14. ERDA-1542, Vol. I, p. 3-157.
15. Ibid., Vol. I, pp. 3-157, 3-162.
16. Senator Abraham A. Ribicoff, "NRC's Historic Opinion and Dissent on Nuclear Export Policy," *Congressional Record* 122: S 10042–S 10050 (June 21, 1976); Murrey Marder, "U.S. to Ship Uranium Fuel to India," *Washington Post*, July 3, 1976, pp. A1, A5.
17. William A. Anders, *The Export Reorganization Act-1975*, testimony at hearings, Senate Committee on Government Operations, U.S. Congress, April and May 1975, p. 94.
18. Congressional Research Service, "International Proliferation of Nuclear Technology," prepared for Subcommittee on Energy and the Environment, House Committee on Interior and Insular Affairs, U.S. Congress, April 15, 1976, p. 47.
19. Office of Congressman Udall, press release (Washington, D.C., June 21, 1976).
20. Albert Wohlstetter, et al., *Moving Toward Life in a Nuclear Armed Crowd?*, prepared by Pan Heuristics Division of Science Applications, Inc., Los Angeles, for U.S. Arms Control and Disarmament Agency, December 4, 1975, revised April 22, 1976, pp. 12, 13.
21. Ibid., p. 8.
22. Don Oberdorfer, "U.S. Aid to Private A-Fuel Plant Eyed," *Washington Post*, October 10, 1976, pp. A1, A16.
23. "Statement by the President on Nuclear Policy," The White House, October 28, 1976.
24. "The Holes in Atomic Inspection," *San Francisco Chronicle*, August 21, 1975. © Chronicle Publishing, 1976. Reprinted by permission.
25. Eldon V. C. Greenberg and Geoffrey A. Goodman (Center for Law and Social Policy), and S. Jacob Scherr (Natural Resources Defense Council, Washington, D.C.), "Comments on ERDA-1542," October 22, 1975, pp. 84–85.
26. David Burnham, "Study Asks for Caution in Sale of Atom Reactors," *New York Times*, August 20, 1975, p. 58.
27. U.S. General Accounting Office, "Assessment of U.S. And International Controls Over the Peaceful Uses of Nuclear Energy" (Washington, D.C., September 14, 1976), p. iii.
28. Ibid., p. 38.
29. Ibid., p. 39.

30. "Study Asks for Caution."
31. "Comments on ERDA-1542," p. 20.
32. Richard J. Barber Associates, Inc., ERDA-52, *LDC Nuclear Power Prospects, 1975–1990: Commercial, Economic, and Security Implications* (Washington, D.C., 1975), p. V-52.
33. Ibid., p. V-53.
34. James C. Phillips, "Safeguards, Recycling Broaden Nuclear Power Debate," *National Journal Reports*, March 22, 1975, p. 421.
35. ERDA-52, pp. V-53–V-54.
36. Ibid., pp. V-34–V-35.
37. Ibid., pp. V-35–V-36.
38. Ibid., p. V-54.
39. Ibid., p. V-58.
40. *Congressional Record* 122: S 9632–9637 (June 16, 1976).
41. "Comments on ERDA-1542," p. 52.
42. *Congressional Record* 122: S 9632–9637.
43. ERDA-52, p. C-10.
44. "GE Unit Nears Accord on the Sale to Canada of Heavy Water Plant," *Wall Street Journal*, January 13, 1975; telephone conversation between Public Interest Research Group and Canadian Embassy, Washington, D.C., November 1976.
45. Thomas O'Toole, "French Get S. African A-Plant Sale," *Washington Post*, May 30, 1976, p. A1; ERDA-52, p. C-17–C-19.
46. Ibid., Appendix C, passim.
47. Ibid., p. C-6.
48. Ibid., Appendix D.
49. H. J. Watters (special assistant to the director, Office of Program Analysis), memorandum to L. Manning Muntzing (director of regulation, U.S. Atomic Energy Commission), October 22, 1974, p. 3.
50. ERDA-52, p. V-85.
51. Amory B. Lovins and John H. Price, "Non-Nuclear Futures" (San Francisco: Friends of the Earth, p. 6).
52. ERDA-52, pp. II-47, II-48.
53. "Comments on ERDA-1542," pp. 63–64.
54. David E. Lilienthal, statement before Senate Committee on Government Operations, January 19, 1976.

4. What Can a Citizen Do?

XVIII. ACTION AT THE FEDERAL LEVEL: THE CONGRESS

1. Adapted from Ralph Nader, "A Youth Battles Nukes," *Washington Star*, April 26, 1975.
2. See, for example, Mary Russell, "ERDA Scraps Controversial Pamphlet," *Washington Post*, October 12, 1976, p. A10.

XIX. ACTION AT THE FEDERAL LEVEL: THE NUCLEAR REGULATORY COMMISSION

1. The description of the licensing process which follows comes from Atomic Energy Commission, EDM-530 (8–74), "Licensing of Nuclear Power Reactors" (Washington, D.C., 1974); and U.S. Nuclear Regulatory Commission, *Annual Report 1975* (Washington, D.C., 1976), pp. 14–15.
2. Stevin Ebbin and Raphael Kasper, *Citizen Groups and the Nuclear Power Controversy* (Cambridge, Mass.: MIT Press, 1974), p. 140.
3. *Izaak Walton v. AEC, U.S. Supreme Court Reports*, Vol. 423, p. 12, 1975. Abbreviated 423 US 12 (1975).

4. "Kleppe Balked in Effort to Block Nuclear Plant," *Washington Post*, July 23, 1976, p. A12.
5. David Burnham, "Court Bars Plutonium's Commercial Use," *New York Times*, May 28, 1976, p. A1.
6. Federal Court of Appeals, District of Columbia, No. 73-1776, *Aeschliman et al. v. U.S. Nuclear Regulatory Commission*, July 21, 1976; No. 73-1867, *Saginaw Valley Nuclear Study Group et al. v. U.S. Nuclear Regulatory Commission*, July 21, 1976; No. 74-1385, *Natural Resources Defense Council v. U.S. Nuclear Regulatory Commission*, July 21, 1976; No. 74-1586, *Natural Resources Defense Council et al. v. U.S. Nuclear Regulatory Commission*, July 21, 1976.
7. U.S. Nuclear Regulatory Commission, press release 76-252 (Washington, D.C., November 15, 1976).
8. U.S. Nuclear Regulatory Commission, WASH-1400 (NUREG-75/014), *Reactor Safety Study* (Washington, D.C., October 1975), Appendix VI, p. 13-34.
9. Terry Wolkerstorfer, "Report Questions Preparedness for Disaster," *Minneapolis Star*, June 10, 1970, pp. 1A, 5A.
10. "Atom Plant Holds Mock Test," *New York Times*, August 19, 1975.
11. Public Interest Research Group, "Indian Point Nuclear License Challenged, Emergency Plans Inadequate," press release (Washington, D.C., February 8, 1976).
12. Public Interest Research Group, "Maine Yankee Nuclear License Challenged; Emergency Plans Inadequate," press release (Washington, D.C., February 15, 1976).

XX. ACTION AT THE STATE AND COMMUNITY LEVEL

1. Skip Laitner, *Citizens' Guide to Nuclear Power* (Washington, D.C.: Critical Mass, 1975), p. 58.
2. Arthur W. Murphy and D. Bruce LaPierre, "Nuclear Moratorium Legislation in the States and the Supremacy Clause: A Case of Express Pre-Emption," Atomic Industrial Forum (Washington, D.C.), November 1975.
3. *Northern States Power Company v. The State of Minnesota, Federal Reporter*, 2nd Ser., Vol. 447, p. 1143, September 7, 1971. (This citation is abbreviated 447 F 2d 1143.) The opinion was subsequently upheld by the U.S. Supreme Court.
4. The story of Northern States Power is told in more detail in Richard Lewis, *The Nuclear Power Rebellion*, (New York: Viking 1972), pp. 122–135.
5. Ibid., pp. 122, 124.
6. *Northern States v. Minnesota*, 447 F 2d 1145.
7. Section 274.c of the Atomic Energy Act of 1954, as amended, in *Atomic Energy Legislation Through 93rd Congress, 2d Session*, Joint Committee on Atomic Energy, U.S. Congress, July 1975, p. 100.
8. *Northern States v. Minnesota*, 447 F 2d 1154.
9. *The Nuclear Power Rebellion*, p. 133.
10. Ibid., pp. 133–135.
11. *Congressional Record* 119: H 5284–H 5293 (June 5, 1973) and 118: S 8053–S 8061 (May 17, 1972).
12. Section 274.k, Atomic Energy Act of 1954, as amended in *Atomic Energy Legislation*, p. 103.
13. See, for example, "Vt. Yankee Exceeded Release Guidelines on Iodine Emissions," *Brattleboro Reformer* July 17, 1974.
14. Institute for Policy Studies, *The Elements* (Washington, D.C., November 1975), p. 9.
15. "Vermont Passes Curb on A-Power," *New York Times* April 1, 1975, p. 7.
16. *Times-Argus* (Barre, Vt.), March 5, 1975.
17. Taken from "Legislative Counsel's Digest of Assembly Bills" #2822, 2820, and 2821, respectively, Assembly of the State of California, 1976.
18. Center for Governmental Responsibility, *Energy: The Power of the States* (Gainesville: Holland Law Center, University of Florida, January 1975), pp. 199–201.
19. Environmental Action Foundation, *The Power Line* (Washington, D.C., October 1975), p. 4.
20. *Energy: Power of the States*, p. 201.

21. Center for Science in the Public Interest, *People & Energy* (Washington, D.C., August 1976), p. 10.
22. "Proposal to Restrict Use of Nuclear Power in California Is Defeated by 2-to-1 Margin," *Wall Street Journal*, June 10, 1976.
23. Larry Pryor, "Antinuclear Issues Defeated in 6 States," *Los Angeles Times*, November 4, 1976, p. 34.
24. "Missouri Shows 'Em," in Environmental Action Foundation, *The Power Line* (Washington, D.C., November 1976), p. 2.
25. Les Brown, "CBS Shuns a Request to Delay Bomb Show," *New York Times*, October 22, 1976, p. 28.
26. Luther J. Carter, "Nuclear Initiatives: Two Sides Disagree on Meaning of Defeat," *Science* 194: 812 (November 19, 1976).
27. See Atomic Energy Commission, WASH 1174-74, *The Nuclear Industry 1974* (Washington, D.C., January 1975), p. 45.
28. M. F. O'Connell and W. G. Holcomb, "A Summary of Low-Level Radioactive Wastes Buried at Commercial Sites Between 1962–1973, with Projections to the Year 2000," *Radiation Data and Reports* (Washington, D.C.: U.S. Environmental Protection Agency, December 1974), Vol. 15, No. 12, p. 759.
29. Roger Strelow (assistant administrator for air and waste management, U.S. Environmental Protection Agency), statement before Joint Committee on Atomic Energy, U.S. Congress, November 19, 1975, pp. 5–6.
30. "Suprising Findings About Plutonium Dangers to Man Reported," International Atomic Energy Agency meeting, San Francisco, November 17–21, 1975, summary by John Cobb, M.D. (University of Colorado Medical School, Denver), p. 3.
31. John J. Fialka, "Pilfered Radioactive Tools Turn Town Hot," *The Washington Star*, March 16, 1976, p. 1.
32. Ibid.
33. International Bridge, Tunnel and Turnpike Association, "Compendium of Regulations: Shipments of Radioactive Materials Over Toll Roads, Bridges and Tunnels" (Washington, D.C., February 1974).

XXI. CHALLENGING ELECTRIC UTILITIES

1. Federal Energy Regulation Study Team, chaired by William O. Doub (Atomic Energy Commission), *Federal Energy Regulation: An Organization Study* (Washington, D.C.: U.S. Government Printing Office, April 1974), p. 19.
2. MITRE Corporation, MTR 75-18, *Legislation and Regulation of Utility Corridors and Power Plant Siting* (McLean, Va., February 1975), p. 34; John H. Williams (Administrator, Wisconsin Public Service Commission), testimony before Subcommittee on Energy and Environment, House Committee on Interior and Insular Affairs, U.S. Congress, May 24, 1978, Attachment.
3. "28th Annual Electrical Industry Forecast," *Electrical World*, September 15, 1977, p. 54.
4. James Rowen, "Wisconsin Sets the Atom Back," *The Nation*, December 18, 1976, p. 659.
5. U.S. Nuclear Regulatory Commission, "Wisconsin Electric Power Company, et al.," *Federal Register*, 42: 54892, October 11, 1977.
6. Skip Laitner, *Citizens' Guide to Nuclear Power* (Washington, D.C.: Critical Mass, 1975), p. 56.
7. "Iowa Commission Focuses Concern on Nuclear Plant," *Critical Mass*, October 1975, p. 6.
8. Maurice Van Nostrand (Chairman, Iowa Commerce Commission), testimony in *Nuclear Power Costs*, Hearings, House Committee on Government Operations, U.S. Congress, September 1977, pp. 479 ff, 522–524.
9. Environmental Action Foundation, "Conn. PUC Lowers Rates," *The Power Line* (Washington, D.C., January 1977), p. 1.
10. Connecticut Citizen Action Group, "CCAG Wins Long-term Victories in Rate Ruling," *CCAG Reports* (Hartford: November 1977), p. 2.

11. Environmental Action Foundation, *How to Challenge Your Local Electric Utility* (Washington, D.C., 1974), pp. 77, 78.
12. "Antinuclear Ads Aired on Radio," *Public Utilities Fortnightly*, November 24, 1977, p. 48.
13. Federal Communications Commission, Memorandum Opinion and Order, Docket 76–453, May 18, 1976, pp. 26, 39.
14. Public Media Center, "Public Interest/Environmental Group Petitioners Hail Fairness Doctrine Victory on Nuclear Power Advertising," press release (Washington, D.C., May 18, 1976).
15. Georgette Jasen, "In the Fight Over Nuclear Energy's Role, Friends and Foes Are Deeply Committed," *Wall Street Journal*, July 21, 1977, p. 32; and Richard Morgan, "Boston Ed Gets Rapped," *The Power Line*, Environmental Action Foundation (Washington, D.C., April 1978), p. 7.
16. "New York State Utilities Lose a 'Captive Audience,' " *Wall Street Journal*, March 3, 1977, p. 7.
17. *How to Challenge Your Local Electric Utility*, p. 79.
18. "Minding the Corporate Conscience, 1978," Council on Economic Priorities (New York: April 20, 1978).
19. Ibid.
20. *How to Challenge Your Local Electric Utility*, pp. 80, 81.
21. "An Omaha Public Power District Director Has Resigned," *Nucleonics Week*, February 10, 1977, p. 7.
22. Edward Walsh, "President in Tennessee is Mum on Clinch River," *Washington Post*, May 23, 1978, p. A3.

UPDATE 1978

1. *Carolina Environmental Study Group v. Atomic Energy Commission*, 431 F. Supp. 230, W.D.N.C (1977).
2. Memorandum of Points and Authorities in Support of Defendant Duke Power Company's Motion for Stay of Judgment, *Carolina Environmental Study Group v. Atomic Energy Commission*, April 8, 1977.
3. *The National Energy Plan*, The White House, Washington, D.C. April 1977, p. 70.
4. "Carter Proposes to Ban the Use of Plutonium," *Wall Street Journal*, April 8, 1977, p. 2.
5. Manson Benedict and Norman C. Rasmussen (Massachusetts Institute of Technology), letter to Stephen Hanauer (Atomic Energy Commission), March 17, 1972. Source Document C in Daniel F. Ford, *A History of Federal Nuclear Safety Assessments*, Union of Concerned Scientists (Cambridge, Mass., April 1977).
6. Ed Gilbert to S. Levine and others (Atomic Energy Commission), "Q–A/Q–C Portions of Design Adequacy Task: Options and Recommendations," October 23, 1973, p. 2. Source Document I in *History*.
7. President Richard M. Nixon, remarks at the Atomic Energy Commission's Hanford works, Hanford, Washington, September 26, 1971.
8. "House Panel Defies Carter Again," *Energy Daily*, April 14, 1978, p. 2.
9. "Vast New Powers for Schlesinger?," *The Elements*, Public Resource Center (Washington, D.C., May 1978), pp. 8, 9.
10. John J. Fialka, "Did Atom Agencies Squelch Report on Cancer Risk?," *Washington Star*, January 19, 1978, p. A1.
11. Thomas F. Mancuso, M.D., *Study of the Lifetime Health and Mortality Experience of Employees of ERDA Contractors*, Final Report, No. 13 (Graduate School of Public Health, University of Pittsburgh, September 30, 1977), p. 3.
12. Walter Pincus, "House Panel Told That Exposure Limit for Radiation is 10 Times Too High," *Washington Post*, February 9, 1978, p. A2.
13. U.S. Department of Energy, press release, October 18, 1977.

14. J. D. Bredehoeft, et al., *Geologic Disposal of High-Level Radioactive Wastes—Earth-Science Perspectives,* U.S. Geological Survey, Circular 779 (Washington, D.C., 1978), p. 12.
15. Ibid., pp. 12, 13.
16. *Nuclear Power Costs,* House Committee on Government Operations, U.S. Congress, April 26, 1978, p. 74.
17. Ibid., p. 75.
18. "28th Annual Electrical Industry Forecast," *Electrical World,* September 15, 1977, p. 54.
19. "Bleak Outlook," *Electrical World,* October 15, 1977, p. 32.
20. Tim Metz, "Panel Finds Cartel Had a Direct Impact on Uranium Purchases by U.S. Utilities," *Wall Street Journal,* August 15, 1977, p. 3.
21. "Gulf Oil Corp. Won't Contest Trust Charges," *Wall Street Journal,* June 5, 1978, p. 2.
22. Report of the Nuclear Energy Policy Study Group, *Nuclear Power Issues and Choices* (Cambridge, Mass.: Ballinger, 1977), pp. 56, 62, 63.
23. Vernon E. Jordan, Jr., President, National Urban League, "Energy Policy and Black People," Address, Minneapolis, Minnesota, January 20, 1978, p. 10.
24. Congressional Black Caucus Statement on Natural Gas Deregulation and Conservation (Washington, D.C., January 23, 1978).
25. *Preliminary Analysis of an Option for the Federal Photovoltaic Utilization Program,* U.S. Federal Energy Administration (Washington, D.C., July 1977), pp. 1, 2, 4, 6, 7, 8, 14, 16.
26. *Distributed Energy Systems in California's Future: A Preliminary Report,* Distributed Energy Systems Study Group, University of California, September 1977.
27. *Solar Energy: Progress and Promise,* Council on Environmental Quality (Washington, D.C., April 1978).
28. "World News Beat," *Electrical World,* January 15, 1977, p. 19.
29. "Australia Endorses Mining and Exports of Uranium, Evokes Soviet Denunciation," *Wall Street Journal,* August 26, 1977, p. 28.
30. "Saint-Gobain Selling Nuclear Unit to France," *Wall Street Journal,* March 18, 1977, p. 5.
31. "France Will Study Financing Exports of Nuclear Plants," *Wall Street Journal,* May 18, 1977, p. 16.
32. "Nuclear Project of San Diego G&E Scrapped by Board," *Wall Street Journal,* May 4, 1978, p. 2.

Index